XXIII International Conference on Spectral Line Shapes (ICSLS 2016)

Journal of Physics: Conference Series Volume 810

Torun, Poland
19 – 24 June 2016

ISBN: 978-1-5108-3857-4
ISSN: 1742-6588

Printed from e-media with permission by:

Curran Associates, Inc.
57 Morehouse Lane
Red Hook, NY 12571

Some format issues inherent in the e-media version may also appear in this print version.

Printed by Curran Associates, Inc. (2017)

For permission requests, please contact the Institute of Physics
at the address below.

Institute of Physics
Dirac House, Temple Back
Bristol BS1 6BE UK

Phone: 44 1 17 929 7481
Fax: 44 1 17 920 0979

techtracking@iop.org

Additional copies of this publication are available from:

Curran Associates, Inc.
57 Morehouse Lane
Red Hook, NY 12571 USA
Phone: 845-758-0400
Fax: 845-758-2633
Email: curran@proceedings.com
Web: www.proceedings.com

Table of contents

Volume 810

XXIII International Conference on Spectral Line Shapes
19–24 June 2016, Torun, Poland

Accepted papers received: 8 February 2017
Published online: 30 March 2017

Preface

Papers

Invited

Contributed

Posters

23rd ICSLS
19 – 24 June 2016, Toruń, Poland

NICOLAUS COPERNICUS UNIVERSITY IN TORUŃ
FACULTY OF PHYSICS, ASTRONOMY AND INFORMATICS

June 19-24, 2016

International Conference on Spectral Line Shapes

SCOPE

- Line shapes of atomic and molecular transitions in neutral gaseous mixtures
- High and low temperature plasmas
- Clusters and helium droplets
- Spectroscopy of stellar atmospheres and interstellar media
- Spectroscopy of planetary atmospheres and exoplanets
- Cold atoms and molecules
- Collision-induced spectra
- Processes in laser fields
- Innovative techniques of line shape applications for diagnostic purposes
- Fundamental studies with narrow optical resonances

INVITED SPEAKERS

Annette Calisti (Aix-Marseille Université, Marsille, France)
Alain Campargue (LIPhy, Universite Grenoble Alpes, Grenoble, France)
Elisabeth Dalimier (Pierre and Marie Curie University - Paris 6, Paris, France)
Thibault Delahaye (Université Paris Est Créteil, Paris, France)
Aleksandra Foltynowicz (Umea Universitet, Umea, Sweden)
Iouli E. Gordon (Harvard-Smithsonian Center for Astrophysics, Cambridge, USA)
Robab Hashemi (University of Lethbridge, Alberta, Canada)
Takeshi Higashiguchi (Utsunomiya University, Utsunomiya, Japan)
Paul S. Julienne (University of Maryland, College Park, Maryland, USA)
Daniel Lisak (Nicolaus Copernicus University, Toruń, Poland)
Valery S. Lisitsa (Moscow Institute of Physics and Technology, Moscow, Russia)
Andrew Ludlow (National Institute of Standards and Technology, Boulder, CO, USA)
Luigi Moretti (Seconda Universita Degli Studi di Napoli, Napoli, Italy)
Eugene Oks (Auburn University, Auburn, Alabama, United States of America)
Christian G. Parigger (University of Tennessee Space Institute, Tullahoma, USA)
Gillian Peach (University College London, London, United Kingdom)
Joël Rosato (Aix-Marseille Université, Marseille, France)
Stefan A. Schäffer (University of Copenhagen, Copenhagen, Denmark)
Uwe Sterr (Physikalisch-Technische Bundesanstalt, Braunschweig, Germany)
Satoshi Tojo (Chuo University, Tokyo, Japan)
Wim Ubachs (VU University, Amsterdam, Netherlands)
Ad van der Avoird (Radboud Universiteit, Nijmegen, Netherlands)
Tigran Vartanyan (ITMO University, Saint Petersburg, Russia)
Sergei Yurchenko (University College London, London, United Kingdom)

Organizers

Aleksander Jabłoński Foundation

Sponsors

Toruń · COMEF · Lasotronix · Ministry of Science and Higher Education · moglabs · TOPTICA Photonics · MenloSystems

Contact

HTTP://ICSLS23.FIZYKA.UMK.PL
E-MAIL: ICSLS23@FIZYKA.UMK.PL

IOP Conf. Series: Journal of Physics: Conf. Series **810** (2017) 011001 doi:10.1088/1742-6596/810/1/011001

23rd ICSLS

19 - 24 June 2016, Toruń, Poland

The 23rd International Conference on Spectral Line Shapes was organized by
Institute of Physics, Faculty of Physics, Astronomy and Informatics,
Nicolaus Copernicus University in cooperation with Aleksander Jabłoński Foundation.

INTERNATIONAL PROGRAMME COMMITTEE

Dionisio Bermejo (Spain)
Roman Ciuryło (Poland)
Elisabeth Dalimier (France)
Alexander Devdariani (Russia)
Robert Gamache (USA)
Motoshi Goto (Japan)
Magnus Gustafsson (Sweden)
Carlos Iglesias (USA)
John Kielkopf (USA)
John C. Lewis (Canada)
Valery Lisitsa (Russia)
Eugene Oks (USA)
Christian Parigger (USA)
Gillian Peach (UK)
Luka Popovic (Serbia)
Adriana Predoi-Cross (Canada)
Roland Stamm (France)
Ha Tran (France)

LOCAL ORGANIZING COMMITTEE

Roman Ciuryło (Chair)
Daniel Lisak (Deputy Chair)
Piotr Masłowski (Deputy Chair)
Józef Szudy
Ewa Kaszewska

Piotr Ablewski
Mateusz Borkowski
Katarzyna Bielska
Hubert Cybulski
Agata Cygan
Monika Czajkowska
Jolanta Domysławska
Grzegorz Kowzan
Ryszard S. Trawiński
Piotr Wcisło
Szymon Wójtewicz
Mikołaj Zaborowski

Contact:

Institute of Physics, Faculty of Physics, Astronomy and Informatics, Nicolaus Copernicus University,
Grudziadzka 5/7, Toruń, Poland, e-mail: icsls23@fizyka.umk.pl, webpage: icsls23.fizyka.umk.pl
Aleksander Jabłoński Foundation, Grudziadzka 5/7, Toruń, Poland,
e-mail: kontakt@faj.org.pl, webpage: faj.org.pl

NICOLAUS COPERNICUS UNIVERSITY IN TORUŃ
Faculty of Physics, Astronomy and Informatics

ALEKSANDER JABŁOŃSKI FOUNDATION

IOP Conf. Series: Journal of Physics: Conf. Series **810** (2017) 011001 doi:10.1088/1742-6596/810/1/011001

The ICSLS conference was co-financed by the Ministry of Science and Higher Education Republic of Poland and by the Toruń City's funds, and supported by companies: COMEF Sp. z o.o. Sp. k., Lasotronix, Menlo Systems GmbH, MOGLabs Europe and TOPTICA Photonics AG.

Ministry of Science
and Higher Education
Republic of Poland

COMEF

Lasotronix

MenloSystems

moglabs

TOPTICA
PHOTONICS

Miasto Toruń

The patron of the conference was the President of the Municipal Office of Toruń.

Publishing conference papers was funding from a conference budget
with support of the Kujawsko-Pomorskie Region's funds.

KUJAWSKO-POMORSKIE
REGION

23rd International Conference on Spectral Line Shapes (ICSLS)

In common use the "Europhysics Study Conference on Spectral Line Broadening and Related Topics" held in Meudon in 1973 is referred to as the first in the current sequence of ICSLS meetings. It should be noted, however, that "The First International Conference on Spectral Lines" took place in 1972 at the University of Tennessee at Knoxville, TN is remembered as a meeting fully devoted to line shape problems, and as such should be regarded, in addition to the Meudon conference, as one of the roots of the line-shape community. The ICSLS be-annual series is held alternating European and North-American conference venues. The last three European venues were Paris, France (2004), Valladolid, Spain (2008), and St. Petersburg, Russia (2012). The last three North American venues were Auburn, Alabama USA (2006), St. John's, Canada (2010) and Tullahoma, Tennessee, USA (2014).

The 23rd ICSLS conference was organized by the Institute of Physics Nicolaus Copernicus University in cooperation with Aleksander Jabłoński Foundation. Nicolaus Copernicus University was previously an organizer of the 9th edition of ICSLS in 1988 chaired by Józef Szudy.

***Figure 1**. The conference photo form 9th ICSLS in Toruń, 1988.*

The conference covered a broad range of topics related to spectral line shape studies, including: line shapes of atomic and molecular transitions in neutral gaseous mixtures, high and low temperature plasmas, clusters and helium droplets, spectroscopy of stellar atmospheres and interstellar media, spectroscopy of planetary atmospheres and exoplanets, cold atoms and molecules, collision-induced spectra, processes in laser fields, innovative techniques of line shape applications for diagnostic purposes and fundamental studies with narrow optical resonances. The conference program included invited lectures, contributed talks, poster presentations, and round table discussions.

The 23rd ICSLS was held in Toruń, Poland, from Sunday, June 19 to Friday, June 24, 2016 in newly built The Culture and Congress Centre - Jordanki. The Conference was attended by 99 participants representing 60 research and academic institutions as well as high-tech companies from 19 different countries. The schedule included 116 contributions: 26 invited talks, 15 contributed oral and 75 poster presentations as well as 5 exhibitions. The abstracts are available at the conference website: http://icsls23.fizyka.umk.pl. After the conference there were submitted 64 papers which are published by IOP in the Journal of Physics: Conference Series.

The opening of the conference was attended by Professor Ireneusz Grabowski, Director of Institute of Physics, NCU and by Zbigniew Fiderewicz, Vice-president of the City of Toruń. It was preceded by informal welcome in Collegium Maximu day before. Various social activities covered a tour in the old town of Toruń including birthplace of Nicolaus Copernicus, evening in ruins of Teutonic Knights castle including Chivalrous Competition Show, and Conference Diner in the City Townhall.

During the poster sessions young scientists competed for the Best Poster Presentation Award of the 23rd International Conference on Spectral Line Shapes. It was awarded to Alexandra Johansson from Umeå University, Sweden for poster presentation "Optical frequency comb spectroscopy of H_2O and OH in a flame".

We would like to thank the many people who helped in organizing the conference, especially the members of the International Program Committee and the Local Organizing Committee. Special thanks are due to the Institute of Physics, Nicolaus Copernicus University, and the Aleksander Jabłoński Foundation, as well as sponsors that have supported the conference, in particular the Ministry of Science and Higher Education Republic of Poland, the City of Toruń , and companies: COMEF Sp. z o.o. Sp. k., Lasotronix, Menlo Systems GmbH, MOGLabs Europe and TOPTICA Photonics AG. Conference Proceedings were published with the support of the Kujawsko-Pomorskie Region. The patron of the conference was the President of the City of Toruń.

The International Program Committee met to discuss future of the ICSLS and role of this conference for the spectral line shape community. The next 24th ICSLS was scheduled in Dublin, Ireland in the second half of June, 2018. John T. Costello from the Dublin City University accepted the huge task of organizing the next conference. Our best wishes for a successful conference go to him and his colleagues.

Roman Ciuryło,
Daniel Lisak,
Piotr Masłowski

Minutes of the meeting of the International Program Committee of the 23rd ICSLS

The meeting of the International Program Committee member took place in the Institute of Physics, Nicolaus Copernicus University in Toruń, Poland, on Monday, June 20, 2016. Members attending the meeting were: Elisabeth Dalimier (France), Alexander Devdariani (Russia), Robert Gamache (USA), Motoshi Goto (Japan), Magnus Gustafsson (Sweden), Valery Lisitsa (Russia), Eugene Oks (USA), Christian Parigger (USA), Gillian Peach (UK), Luka Popovic (Serbia) was represented by Zoran Simic, Roland Stamm (France), and Roman Ciuryło (Poland).

The meeting started with short introduction and presentation of statistics and budget of the current edition of the Conference. The schedule included 116 contributions: 26 invited talks, 15 contributed oral and 75 poster presentations as well as 5 exhibitions. The 23rd ICSLS was attendant by 99 participants representing 60 research and academic institutions as well as high-tech companies from 19 different countries.

John Costello presented plans for organization of the next 24th ICSLS in Dublin, Ireland in the second half of June, 2018. After some discussion and voting Committee members gratefully accepted his offer and confirmed that the 24th International Conference on Spectral Line Shapes will be organized by John T. Costello from the Dublin City University. John T. Costello was invited to join the Committee and he accepted.

For the 25th ICSLS in 2020 two proposals were received before the International Program Committee meeting. Authors of these proposals were asked to give short presentation during the meeting. Firstly, Magnus Gustafsson from Luleå University of Technology offered to organize the 25th ICSLS in Luleå, Sweden. Secondly, Luigi Moretti representing Livio Gianfrani from the Second University of Naples proposed Caserta, Italy as an alternative option. Both presentations were acknowledged by Committee members. After some discussion and preliminary voting the Committee decided that further action will be deferred to the Committee meeting in Dublin in 2018.

Next the Scientific Committee discussed progress as well as changes in the scientific scope of physical problems related to spectral line shapes. It was indicated that the future conferences of this series should put particular attention to fields such as: astrophysics, exoplanets, optical clocks development, ultra cold physics, nuclear fusion, basic plasma physics and its spectroscopic aspects. Moreover, representation of these specializations should be strengthened in the future committee. Committee members looked very optimistically in the future development of the scientific interest in areas which were changed from the origin of ICSLS. After discussion the meeting was finished.

23rd ICSLSh
19 - 24 June 2016, Toruń, Poland

Participants of the 23rd International Conference on Spectral Line Shapes

IOP Conf. Series: Journal of Physics: Conf. Series **810** (2017) 011002 doi:10.1088/1742-6596/810/1/011002

Peer review statement

All papers published in this volume of *Journal of Physics: Conference Series* have been peer reviewed through processes administered by the proceedings Editors. Reviews were conducted by expert referees to the professional and scientific standards expected of a proceedings journal published by IOP Publishing.

IOP Conf. Series: Journal of Physics: Conf. Series **810** (2017) 012001 doi:10.1088/1742-6596/810/1/012001

Reminiscences and Reflections on the History of International Conferences on Spectral Line Shapes

J Szudy
Institute of Physics, Faculty of Physics, Astronomy and Informatics, Nicolaus Copernicus University, Grudziądzka 5, 87-100 Toruń, Poland

Abstract. A brief account of the history of International Conferences on Spectral Line Shapes (ICSLS) is given. Although in common use the "Europhysics Study Conference on Spectral Line Broadening and Related Topics" held in Meudon in 1973 is referred to as the first in the current sequence of ICSLS meetings, it is noted that five conferences dealing with line shape topics were organized before 1973 both in the USA and in Europe. Some details are given about their format and program. In particular, "The First International Conference on Spectral Lines" held in 1972 at the University of Tennessee at Knoxville is remembered as a meeting fully devoted to line shape problems, and as such should be regarded, in addition to the Meudon conference, as one of the roots of the line-shape community. Some of the highlights of particular ICSLS conferences as well as characteristics of their proceedings are briefly reviewed.

1. Introduction

I feel very honoured at having been asked to review the history of International Conferences on Spectral Line Shapes (ICSLS), but in truth I am aware that I have been asked because I am now one of the oldest remaining members of the line shape community who still has a recollection of the very beginning of this conference series. In particular, I remember two conferences dealing with line shape topics that I attended in 1968 and 1972, but which seem to be forgotten nowadays as they are not included in the current sequence of line-shape conferences. One of the invited speakers at the 1968 Conference, held in Warsaw, was the late Professor Aleksander Jabłoński, a pioneer of the field of pressure broadening. It was he who in 1946 established the Physics Department of the Nicolaus Copernicus University. So, the year 2016 marks the 70th anniversary of the Jabłoński school of Physics in Toruń. This allows me to dedicate this review to the memory of Aleksander Jabłoński who as early as in 1930's initiated the line-shape research in Poland.

2. About the First ICSLS

In his talk delivered at the 17th ICSLS in Paris in 2004, Dietrich Voslamber expressed a view that "the first in the current sequence of line-shape conferences was held in 1973 at Henri Van Regemorter's Laboratory at the Observatory of Paris-Meudon" [1]. The same view was formulated as early as 1980 by Burkhard Wende in his Preface to the Proceedings of the 5th ICSLS [2] and later in all materials related to the next line-shape conferences including the present one, the 23rd, held in 2016 in Toruń. It should be noted, however, that the 1973 Conference was named "Europhysics Study Conference on Spectral Line Broadening and Related Topics" and not Conference on Spectral Line Shapes. It was held in Meudon from the 2nd to the 6th of July 1973, and D. Voslamber was the chairman of the Organizing Committee.

According to Voslamber, the name "International Conference on Spectral Line Shapes" (ICSLS), which is presently still in use, was given after "The Second International Conference on Spectral Lines" which was held from the 26th through the 30th August 1974 at the campus of the University of Oregon in Eugene, Oregon and was chaired by Shang Yi Ch'en. To my knowledge, however, the Eugene Conference was not related directly to the Meudon meeting but rather to "The First

International Conference on Spectral Lines" which took place from August 28 to September 1, 1972 at the University of Tennessee in Knoxville, Tennessee. Its chairman was Harry C. Jacobson. I was privileged to take part at that conference as the postdoc working with William E. Baylis at the University of Windsor in Canada and I remember up to the present day the impression exerted upon me by the lectures by the then most eminent experts in the field of pressure broadening of spectral lines, to mention only Theodore Holstein (Los Angeles), Boris Vodar (High Pressure Laboratory in Bellevue, France), Alan Gallagher (JILA, Boulder), William R. Bennett (Yale University), Arthur L. Schawlow (Stanford), Abraham Ben-Reuven (Weizmann Institute in Rehovot), Charles H. Hooper (University of Florida), Edwin L. Lewis (Newcastle upon Tyne, UK), Walter Lochte-Holtgreven (Kiel), Shang Yi Ch'en (Eugene), Paul R. Berman (New York), Marie-Claude Castex (Bellevue), Antoine Royer (Montreal), François Rostas and Evelyne Roueff (Meudon). I also remember two Roundtable Discussions during the Knoxville Conference; the first one was devoted to Satellite Bands, and the second one dealt with Stark Broadening. W.E. Baylis (Windsor) was the leader of the panel on satellite bands, in which R. Exton (NASA), R. Futrelle (JILA), B. Vodar and R. Granier (Bellevue), K.M. Sando (Un. of Iowa), H. C. Jacobson, A. Royer, and M. Takeo (Portland) took part. In the panel on Stark Broadening headed by C.H. Hooper the following researches participated: W. Cooper (Berkeley), C. Deutsch (Orsay), P. Kepple (Washington, DC), W. Wiese (NBS, Gaithersburg), H. Van Regemorter (Meudon), D. Voslamber (Euratom – CEA, Fontenay-aux-Roses), and W. Jones (Un. of Maryland) . It should be noted that although the name "line shapes" was not adopted officially to the Knoxville Conference, it surely deserved such a name since almost 100% of invited and contributed papers dealt with topics on line broadening. All presentations were oral since posters were not in common use at that time. The name "International Conference on Spectral Line Shapes" was officially used for the first time four years later when the 3rd ICSLS was organized by David Burgess during September 13 – 17, 1976 at the Imperial College in London.

3. A Bit of Prehistory

The story about the beginnings of line-shape conferences would not be complete unless we mention meetings organized in the 1950's in the USA and in the 1960's in Europe. In particular, the Conference on the Broadening of Spectral Lines was held at the University of Pittsburgh from September 15 to 17, 1955. This conference is still remembered because of the famous paper by P.W. Anderson and J.D. Talman, in which they presented an attempt to span a bridge between the impact and quasistatic theory. Usually it is cited as "unpublished" paper with reference to Bell Teleph. Syst. Tech. Publ. No 3117. Fortunately, the main results of the Anderson-Talman approach were thoroughly discussed in the review article by S.Y. Ch'en and M. Takeo [3] and in the monograph by G. Traving [4].

In 1965 H. G. Kuhn, G. Series and G.K. Woodgate organized the Conference on Atomic Spectra and Radiative Processes which was held from the12th to the 14th April at the Clarendon Laboratory in Oxford. It was a conference by invitation, international in character and wide in scope. Besides the session on line broadening the subject of the Conference included sessions on isotope shifts and hyperfine structure of spectral lines as well as optical pumping. Hans R. Griem delivered an introductory paper "On the theory of pressure broadening" which was followed by talks by S.Y. Ch'en on pressure effects of foreign gases on the absorption lines of caesium, and by W. Lochte-Holtgreven on the self-broadening of helium lines as well as by J.M. Vaughan on resonance broadening at low pressures. The proceedings of this conference have not been published, but the list of all contributions is collated in a report by Kuhn, Series and Woodgate [5].

After the Oxford Conference two meetings took place in the USA at the Proctor Academy, Andover, New Hampshire in the framework of the Gordon Research Conferences on Spectral Line Shapes, the first one during August 29 – September 2, 1966, and the second one July 8 – 12, 1968. In the same year Tadeusz Skaliński from the Institute of Physics of the Polish Academy of Sciences organized the International Conference on Optical Pumping and Atomic Line Shape which was held from the 25th to the 28th June 1968 at the University of Warsaw. Roughly half of invited and contributed papers dealt with pressure-broadened spectral line shapes. The invited papers were presented by such experts of the field as A. Kastler, B. Cagnac, and A. Omont (all from Paris), S.Y. Ch'en (Eugene), A. Corney, H.G. Kuhn, E.L. Lewis, and D. Stacey from Oxford, J. Fiutak and E.

Czuchaj (Gdańsk), H.R. Griem (Un. of Maryland), W. Happer (Columbia Un. New York), A. Jabłoński (Toruń), G. zu Putlitz (Heidelberg), and R.M. Herman (Penn State Un.).

4. From the 4th to the 19th ICSLS

The 4th ICSLS was held from July 31 to August 4, 1978 at the University of Windsor, Windsor, Ontario in Canada. The chairman of the Program Committee was Wofgang Wiese (NBS, Gaithersburg) and William E. Baylis (Windsor) was the organizing chairman. The remaining members of the Committee were: D.D. Burgess (Imperial College, UK) acting as the past chairman, D. Voslamber (EURATOM-CEA, Fontenay-aux-Roses) as European Treasurer, G.L. Hammond (White Oak, Maryland) as North American Treasurer, B. Wende (Berlin) as European Coordinator, A. Ben-Reuven (Tel-Aviv) and R.J. Exton (NASA, Hampton, Virginia). As the participant of that conference I remember the amazing inaugural lecture delivered by Derek R. Stacey (Oxford) under the title "Resonance broadening – the truth at last" as well as the invited talk by John (Jinx) Cooper (JILA, Boulder) on "Collisional redistribution of radiation" which was the hot topic in those days.

At the business meeting in Windsor it was confirmed that the next, the 5th, ICSLS would be organized in 1980 in Berlin (West) by Burkhard Wende from the Physikalisch-Technische Bundesanstalt (PTB). This Conference was held in the Reichstag building from the 7th to the 11th of July 1980, but originally it was scheduled to be held in the saucer-shaped Building of the Kongresshalle in the eastern part of the Tiergarten. However, on May 21, 1980 part of the roof of the Kongresshalle collapsed and therefore the Conference was moved to the Reichstag. I attended that Conference and have a recollection of a pleasant atmosphere both during the sessions and the Sightseeing Tour as well as at the Conference dinner and at a warm reception we got from the Senat von Berlin. The Program Committee was almost the same as in the 4th ICSLS except that G.L. Hammond was replaced by C.H. Hooper and H. van Regemorter became the new member. The Panel Session which closed the Conference on July 11 was devoted to summarizing the results of the conference as well as future needs. H.R. Griem was the Chairman of the Panel, the members of which were: W.E. Baylis and D.N. Stacey (Neutrals), D.D. Burgess and J. Seidel (Plasmas), and D. Voslamber (Collisional Redistribution). For the first time in this series of conferences the proceedings of all invited and contributed papers presented in Berlin have been published as a volume (1200 pages) edited by Walter de Gruyter & Co. [2]. At the business meeting it was confirmed that the 6th ICSLS would be organized by Keith Burnett in 1982 at JILA in Boulder, Colorado. At the same meeting I offered, on behalf of the Nicolaus Copernicus University, to organize the 7th ICSLS in Toruń in 1984 and my offer was accepted.

Unfortunately, because of the martial law imposed in Poland at the end of 1981, I was unable to attend the 6th ICSLS, which was held in Boulder during July 12-16, 1982. The proceedings of this Conference were published by Walter de Gruyter as *Spectral Line Shapes, Volume 2* [6]. Since the political situation in Poland was still unclear it was confirmed at the business meeting in Boulder that the next meeting, the 7th ICSLS, would be held not in Toruń as previously planned, but in France. François Rostas (Meudon) took on the task of organizing the conference which eventually was held at the CNRS Conference Center of Aussois in the French Alps from the 11th to the 15th of June 1984. It was attended by 175 participants from 13 countries. A new format was tried for the first time in this series of meetings; it consisted of invited lectures in the morning, free time in the early afternoon, posters before dinner and round-table discussions after dinner. Proceedings were published as *Spectral Line Shapes - Volume 3* [7]. This time I was present at the conference as well as at the business meeting so that I could personally renew the offer to organize the 9th ICSLS in 1988 in Toruń. At the business meeting Reginald J. Exton (NASA, Langley Research Center) accepted the organization of the 8th ICSLS which was held from the 9th to the 13th of July 1986 at the campus of the College of William and Mary in Williamsburg, Virginia. Proceedings were published as *Spectral Line Shapes - Volume 4* by A. Deepack Publishing, Hampton, Virginia [8].

The 9th ICSLS was held at Nicolaus Copernicus University in Toruń during July 25-29, 1988 and was attended by 241 participants from 22 countries and this seems to be the largest attendance in the whole history of this series of line-shape conferences. Following the suggestion by Lothar Frommhold, the Toruń meeting – for the first time in the history of line shape conferences – brought together the communities of pressure broadening and collision-induced phenomena. A special round table discussion "On laser-assisted collisions and collision-induced spectra" was held. Co-chairmen of this joint discussion were Paul S. Julienne (NBS, Gaithersburg) and L. Frommhold (University of Texas, Austin). The full text of their NOTES and COMMENTS on this round table discussion as well as all invited and some contributed papers are contained in the Proceedings which were published as *Spectral Line Shapes - Volume 5* by Ossolineum Publishing House [9].

The 10th ICSLS was organized by L. Frommhold and John Keto at the campus of the University of Texas in Austin, Texas from June 25 through 29, 1990. I took part in this Conference so that I can agree with the statement of the co-chairmen, Lothar Frommhold and John Keto, who in the Preface to the Proceedings have written: "*Unfortunately, the American Southwest was in the midst of a heat wave making it very hot for colleagues from northern climates. Overall, even with this special „warm welcome", the meeting had its usual enthusiastic sessions and friendly, close discussions*". For the first time the Proceedings were published in the framework of American Institute of Physics (AIP) Conference Proceedings entitled *Spectral Line Shapes – Volume 6* [10].

It was confirmed at the business meeting in Austin that the 11th ICSLS would be hosted by Nikola Konjević, Belgrade, and Goran Pichler, Zagreb in Yugoslavia in the summer of 1992**.** It is to be regretted that due to the outbreak of the Yugoslav Wars, the International Program Committee had to move the Conference to France. Roland Stamm and Bernard Talin accepted the huge task of organizing the 11th ICSLS which eventually was held in a seaside resort Carry le Rouet, near Marseille from the 8th to the 12th of June 1992. Proceedings under the title *Spectral Line Shapes – Volume 7* were published by Nova Science Publishers, Commack, NY, USA [11]. Two years later, the 12th ICSLS was organized by A. David May in the Physics Department of University of Toronto in Canada during June 13 – 17, 1994. Proceedings entitled *Spectral Line Shapes - Volume 8* were again published in the framework of AIP Conference Proceedings [12].

The 13th ICSLS was organized by Marco Zoppi and Lorenzo Ulivi from Consiglio Nazionale Delle Recherche in Firenze (Italy) from June 16 through 21, 1996. It was held in the auditorium of Banca Toscana and attended by 160 participants from 21 countries. Besides the traditional sessions devoted to Stark broadening in plasmas, broadening by neutrals, optical collisions in strong laser fields and astrophysical applications a special session was dedicated to a symposium in honor of George Birnbaum. It was inaugurated by Lothar Frommhold who gave a talk entitled "George Birnbaum and the collision-induced Raman and infrared spectroscopies" and which was followed by two invited talks and 12 oral contributions in the field of collision induced spectroscopy. I still have in mind the impressive presentation by L. Frommhold and Wilfried Meyer (Kaiserslautern) on collision-induced emission and sonoluminescence. Proceedings under the title *Spectral Line Shapes - Volume 9* were published as AIP Conference Proceedings [13].

Roger M. Herman from Pennsylvania State University was the host of the 14th ICSLS which was held in State College, Pennsylvania from June 22 to 26, 1998. It was attended by 97 participants from 15 countries. In this Conference special attempt was made to incorporate novel techniques for studying line shapes, as well as a full discussion of line shape phenomena in the ultracold regime. The program of the Conference included a symposium in honor of the 100th anniversary of the birth of Aleksander Jabłoński which consisted of three invited and one oral contribution. *Spectral Line Shapes - Volume 10* was the title of the Proceedings published by AIP [14].

The 15th ICSL was held in Berlin from 10 to 14 July 2000 and was chaired by Joachim Seidel from PTB. All sessions took place in PTB's Hermann von Helmholtz building, named after the first president of the Physikalisch-Technische Reichsanstalt. It was attended by 121 participants from 24 countries. As in the previous four conferences the Proceedings *Spectral Line Shapes - Volume 11* were published by AIP [15].

In 2002 (during June 3-7) the 16th ICSLS was organized by Christina A. Back from Lawrence Livermore National Laboratory and University of California, Davis and was held in Berkeley, CA at the Clark Kerr Campus of the University of California. Participants represented 14 countries and delivered 68 papers, 61 of which are included in *Spectral Line Shapes - Volume 12* published by AIP [16]. Unfortunately, due to the September 11 attacks some people were unable to get visas in time to come for the conference. The organizers tried to rebalance the program, to recognize new efforts in astrophysics and ultra cold physics.

Elisabeth Dalimier from Université Pierre et Marie Curie organized the 17th ICSLS, which was held at the Henri Poincaré Institute in Paris from 21 to 25 June 2004. It was attended by 108 researches from 21 countries. An important event during the Conference was the symposium in memory of Henri van Regemorter. The symposium began with the talk by D. Voslamber, the organizer of the 1973 Meudon Conference, usually referred to as the 1st ICSLS. He appreciated Henri van Regemorter's kind assistance and steady encouragement during the whole preparatory phase. The next talk at the Symposium entitled "A Tribute to Henri Van Regemorter: Line Shapes, Collisions and Radiation" was given by S. Sahal-Brèchot. Then F. Masnou-Seeuws presented a paper dedicated to Henri van Regemorter on "Making Ultracold Molecules with Chirped Pulses". The invited and contributed papers presented at the Conference, including those given at Henri van Regemorter's Symposium were published by "Editions Frontières" [17]. As a participant I still keep in my memory the social events that contributed to the splendid atmosphere during the Paris Conference: the cocktail at the Observatory on the eve of the opening ceremony, the River Seine Cruise, and the Concert and the Banquet Dinner at the Hôtel de Varengeville.

The 18th ICSLS was held during June 4 – 9, 2006 at the Auburn University in Auburn, Alabama, USA and was attended by 51 participants from 8 countries. The co-chairmen were Eugene Oks and Michael Pindzola from Physics Department of the Auburn University. Kay Niemax from Dortmund gave a talk under the title "Is there anything left to do in the 100 years old field of pressure broadening of atomic lines ?", in which he presented his personal view on the status of the research field founded by Lorentz in 1906. He indicated several experimental observations still lacking understanding and proper theoretical description. All papers are collected in the proceedings published by AIP [18].

Marco Gigosos and Manuel Gonzales from the Universidad de Valladolid in Spain organized the 19th ICSLS, which was held from 15 to 20 June 2008 at the Palacio de Congress Conde Ansúres in Valladolid. It was attended by 102 participants from 23 countries.

Since this conference took place exactly 50 years after the appearance of fundamental papers by Baranger and by Kolb and Griem a special panel discussion entitled "On the first 50 years of quantum theories of Stark broadening" was included in its program. The members of the panel were: J. Dufty (Gainesville), N. Konjević (Belgrade),V. Lisitsa (Moscow), and R. Stamm (Marseille). As previously the proceedings were published by AIP [19].

5. Summary

Unfortunately, I was unable to attend the 20th ICSLS in 2010 (St. John's, Newfoundland, Canada) [20], the 21st in 2012 (St Petersburg, Russia) [21], and the 22nd in 2014 (Tullahoma, Tennessee, USA) [22]. However, from information I received second hand, I learned that these conferences were very successful. The main conclusion that can be drawn from the Proceedings of these Conferences is that despite of more than 100 years of its existence, the field of pressure broadening and shifting of spectral lines is still alive. Obviously, the field has changed considerably since 1956 when the first conference dealing with pressure effects on spectral lines took place. On the experimental side, the advent of new techniques such as cavity ring-down spectroscopy or optical-frequency comb spectroscopy, has yielded in recent years experimental data with such high accuracy that in many cases they represent a challenge to theory. On the theoretical side, much progress has been achieved in interatomic and intermolecular interaction calculations.

In the last three decades, there has been an explosion of new topics both in the field of Stark broadening in plasmas and the broadening by neutrals as well as in the area of their applications in astrophysics. Of particular importance are the topics dealing with line shapes associated with ultracold collisions between atoms confined in traps, especially those connected with problems of photoassociation of cold atoms and Bose-Einstein condensates. To my knowledge, the field of ultracold collisions appeared for the first time in the program of our conferences at the 9th ICSLS in 1988 when Paul S. Julienne and Alan Gallagher discussed the role of light interacting with very cold colliding atoms and showed that processes associated with resonance line broadening can play an important role in optical trap dynamics. Since then the research field of cold atoms has been constantly expanding and I am indeed glad to notice that the relevant topics are included in the program of the present conference. I also believe that as topics of major interest they should constitute issues for special sessions at future meetings in the ICSLS series.

References

[1] Voslamber D 2004 *The First ICSLS*, in: *17th International Conference on Spectral Line Shapes*, ed Dalimier E (Paris: Editions Frontier Group)

[2] Wende B 1981 *Preface*, in: *Spectral Line Shapes, Proceedings, Fifth International Conference, Berlin (West), July 7-11. 1980*, ed Wende B (Walter de Gruyter, Berlin-New York 1981)
[3] Che'n S Y and Takeo M 1957 *Revs. Mod. Phys.* **29** 20
[4] Traving G 1960 *Über die Theorie der Druckverbreiterung von Spektrallinien* (Verlag G. Braun Karlsruhe 1960)
[5] Kuhn H G, Series G W, and Woodgate G K 1965 *Physics Bulletin (IOP and Physical Society)* Vol. 16 Number 8 (August 1965)
[6] *Spectral Line Shapes – Volume 2 , Proceedings, Sixth International Conference, Boulder, Colorado, USA, July 12 – 16, 1982,* ed Burnett K (Walter de Gruyter, Berlin-New York 1983).
[7] *Spectral Line Shapes – Volume 3 , Proceedings, Seventh International Conference, Aussois, France, July 11 – 15, 1984,* ed Rostas F (Walter de Gruyter, Berlin-New York 1985).
[8] *Spectral Line Shapes - Volume 4, Proceedings, Eighth International Conference, Williamsburg, Virginia, June 9-13, 1986,* ed Exton R J (A. Deepack Publishing, Hampton, Virginia 1987).
[9] *Spectral Line Shapes - Volume 5, Proceedings, Ninth International Conference, Toruń, Poland, 25-29 July 1988,* ed Szudy J (Ossolineum Publishing House, Wrocław 1989)
[10] *Spectral Line Shapes – Volume 6*, AIP Conference Proceedings 216, eds Frommhold L and Keto J W (American Institute of Physics, New York 1990)
[11] *Spectral Line Shapes – Volume 7, Proceedings, Eleventh International Conference,* eds Stamm R and Talin B (Nova Science Publishers, Commack, NY 1993)
[12] *Spectral Line Shapes – Volume 8*, AIP Conference Proceedings 328, eds May A D, Drummond J R, and Oks E (American Institute of Physics, New York 1995)
[13] *Spectral Line Shapes – Volume 9*, AIP Conference Proceedings 386, eds Zoppi M, and Ulivi L (American Institute of Physics, Woodbury, New York 1997)
[14] *Spectral Line Shapes – Volume 10*, AIP Conference Proceedings 467, ed Herman R M (American Institute of Physics, Woodbury, New York 1999)
[15] *Spectral Line Shapes – Volume 11*, AIP Conference Proceedings 559, ed Seidel J (American Institute of Physics, Melville, New York 2001)
[16] *Spectral Line Shapes – Volume 12*, AIP Conference Proceedings 645, ed Back C A (American Institute of Physics, Melville, New York 2002)
[17] *17th International Conference on Spectral Line Shapes, Paris. June 21-25, 2004,* ed Dalimier E (Editions Frontier Group, Paris 2004)
[18] *Spectral Line Shapes – 18 th International Conference, Auburn, Alabama 4-9 June 2006,* AIP Conference Proceedings 874, eds Oks E, and Pindzola M (American Institute of Physics, Melville, New York 2006)
[19] *Spectral Line Shapes – Volume 15,* AIP Conference Proceedings 1058, eds Gigosos M A and González M Á (American Institute of Physics, Melville, New York 2008)
[20] *20th International Conference on Spectral Line Shapes, St. John's, Newfoundland, Canada, 6-11 June 2010,* AIP Conference Proceedings 1290, eds Lewis J K C, and Predoi-Cross A (American Institute of Physics, Melville, New York 2010)
[21] *XXI International Conference on Spectral Line Shapes (ICSLS 2012), 3-9 June 2012, St Petersburg, Russia,* Journal of Physics: Conference Series, Volume 397 2012, ed Devdariani A (IOP Publishing Bristol)
[22] *XXII International Conference on Spectral Line Shapes 2014, 1-6 June 2014, Tullahoma, Tennessee, USA,* Journal of Physics: Conference Series, Volume 548 2014, ed Parigger C (IOP Publishing Bristol)

Towards passive and active laser stabilization using cavity-enhanced atomic interaction

S A Schäffer, B T R Christensen, S M Rathmann, M H Appel, M R Henriksen and J W Thomsen

The Niels Bohr Institute, University of Copenhagen, Blegdamsvej 17, DK-2100, Denmark

E-mail: schaffer@nbi.dk

Abstract. Ultra stable frequency references such as the ones used in optical atomic clocks and for quantum metrology may be obtained by stabilizing a laser to an optical cavity that is stable over time. State-of-the-art frequency references are constructed in this way, but their stabilities are currently limited by thermally induced length fluctuations in the reference cavity. Several alternative approaches using the potential for frequency discriminating of highly forbidden narrow atomic transitions have been proposed in, e.g., [1] and [2]. In this proceeding we will present some of the ongoing experimental efforts derived from these proposals, to use cavity-enhanced interaction with atomic ^{88}Sr samples as a frequency reference for laser stabilization. Such systems can be realized using both passive and active approaches where either the atomic phase response is used as an error signal, or the narrow atomic transition itself is used as a source for a spectrally pure laser. Both approaches shows the promise of being able to compete with the current state of the art in stable lasers and have similar limitations on their ultimately achievable linewidths [1, 2].

1. Introduction

Quantum metrology and ultra-stable optical atomic clocks rely on the frequency stability of reference lasers [3, 4, 5]. These lasers have been demonstrated with linewidths down to tens of mHz [6], relying heavily on stabilization to ultra-stable reference cavities [7] whose fractional frequency stability is currently limited by the Brownian motion in the mirror substrates [8]. The possibility of using cavity-enhanced non-linear spectroscopy on narrow transition lines has been studied extensively over the years. In the context of laser stabilization recent experimental studies in [9, 10] have demonstrated the potential for a stability comparable to the state of the art. It has also been proposed to use the direct emission of radiation from atoms with narrow transition linewidths [2, 11]. Such radiation emitted directly from a narrow transition requires unrealistically large atomic samples due to the small decay rate, which renders it impractical for reference purposes. However, when operated as a laser in the bad cavity regime, not only can the emission rate be significantly increased, the laser linewidth can also experience a further spectral narrowing compared to the natural linewidth [2, 12, 13]. These effects arise if one considers the case of superradiant or superfluorescent emission of light. In this case the photon emission flux can be considerably increased by collective atomic decay, while simultaneously achieving a spectral narrowing of the emitted light inversely proportional to the single atom cooperativity [12, 13]. Pioneering advances have already been made in connection with proof-of-principle quasi-continuous superfluorescent systems [12, 14] using atoms loaded into an optical lattice at very low temperatures, as well as high temperature gas active laser systems [13, 15].

In this proceeding we will present recent experimental advances in both the passive and active approaches using ^{88}Sr. The passive approach follows [10] but uses an optimized system with an increased cavity finesse. For improved noise characteristics of the measured signals we employ the heterodyne detection technique, NICE-OHMS [16]. Here the atom-cavity system acts as a passive frequency reference in the bad cavity regime, and we show recent results concerning the maximally obtainable phase-response.

Additionally, we report the observation of superfluorescent-like behaviour of an ensemble of thermal strontium atoms freely moving at temperatures of about 5 mK. The atoms have a strong collective coupling to a single cavity mode, significantly enhancing their collective cooperativity, and allowing them to emit a burst of photons into the cavity mode. In this approach the atoms act as the active part of the laser allowing enhanced emission intensity on a narrow atomic transition. This system may also be important for entanglement studies of atomic ensembles [17, 18] as it can move easily between interesting regimes.

Before delving into the specifics of the passive and active approach respectively we will characterize the experimental system under investigation, which is common to the two cases.

2. Cavity enhancement

Since both the passive and the active systems we are investigating here can operate on essentially the same cavity-enhanced system, this section is dedicated to the description of such a system and the physical parameters of it. In figure 1 a sketch of the atom-cavity system can be seen. It consist of an ensemble of ^{88}Sr atoms that are laser-cooled and trapped in a Magneto-Optical Trap (MOT) inside an optical cavity. The atomic transition of interest to us is the narrow $(5s^2)^1S_0 - (5s5p)^3P_1$ dipole transition at 689 nm. This transition is dipole-forbidden and has a natural linewidth of $\Gamma/2\pi = 7.5$ kHz making the requirements on the probing laser relatively relaxed, while simultaneously providing very promising results for the final laser stability. The cavity decay rate is given by $\kappa/2\pi = 539$ kHz and the free spectral range is $\Omega/2\pi = 781$ MHz resulting in a cavity finesse of $\mathcal{F} = 1450$. The coupling between the atom and the cavity field is characterized by the coupling parameter $g/2\pi = 815$ Hz. Here $g = d/\hbar\sqrt{\hbar\omega/2\epsilon_0 V_{\text{eff}}}$, where d is the transition dipole moment, ω is the transition frequency, ϵ_0 is the vacuum permittivity, and V_{eff} is the effective mode-volume of the cavity field.

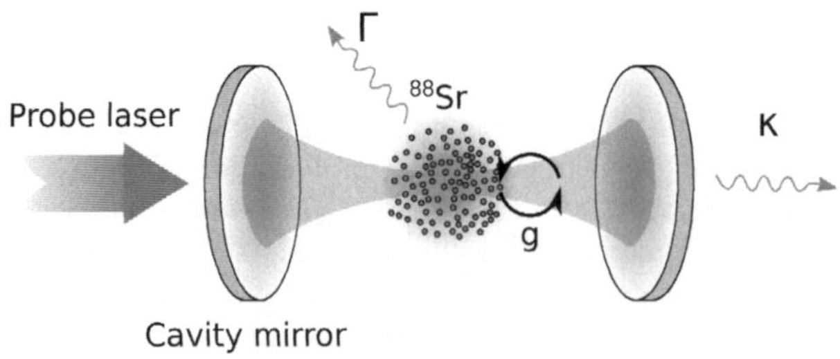

Figure 1. ^{88}Sr atoms are cooled and trapped inside a large-waist optical cavity. The atoms couple to the intra-cavity field with the coupling strength g. Cavity losses are given by the rate κ and the natural decay rate of the $(5s^2)^1S_0 - (5s5p)^3P_1$ atomic transition is given by Γ. During experiments the cooling light is switched off while cavity is kept in resonance with the probe light.

We wish to operate the system in the bad cavity regime, where the cavity linewidth κ is much larger than the natural linewidth of the atoms Γ. This ensures that the transition linewidth becomes decisive in terms of frequency discriminating properties and suppresses any line-pulling effects that may arise due to fluctuations in the cavity mirror positions. This bad-cavity limit is thus instrumental in obtaining the narrow laser linewidths we are targeting. The dynamics of the atom-cavity coupling in the system can be quantified by the single atom cooperativity $C_0 = 4g^2/\kappa\Gamma$ representing the cavity-field mediated interaction of an atom with itself, or the coherence buildup in the system relative to the coherence decay. This quantity can also be expressed by geometrical considerations as $C_0 = \frac{6}{\pi^3}\mathcal{F}\lambda^2/w_0^2$ where λ is the cavity field wavelength on resonance with the atomic transition, and w_0 is the waist radius of the cavity mode intensity. Since we are considering large atomic samples the system dynamics may be represented by the collective cooperativity $C_N = C_0N$ where the total number of atoms in the cavity mode is typically $N = 2\cdot10^7$. In the following experiments we investigate a regime where we have strong collective coupling $C_N\Gamma \gg \Gamma_{\text{decoherence}}$, but weak single-atom coupling $C_0\Gamma \ll \Gamma_{\text{decoherence}}$. By fulfilling these requirements we can simultaneously achieve large collective effects, and good suppression of cavity noise effects. Ideally then, we would want a high number of atoms N and a low cavity finesse $\mathcal{F}$.

Here the system behaviour is recorded by observing the field leaking out of the mirrors, with either a seeding field incident on the cavity, or by using a probe laser to record the absorption and phase response. The cavity length is controlled in order to ensure resonance conditions with the seed or probing laser at all times.

3. Passive approach

In a first approach to laser stabilization we will discuss passive stabilization which centers on the idea of having an external laser stabilized to a narrow frequency discriminator. This frequency discriminator should be able to provide a signal with high signal-to-noise ratio while simultaneously being very sensitive to any frequency-deviations. Here we follow [10] in using direct spectroscopy on the $(5s^2)^1S_0$ – $(5s5p)^3P_1$ transition in ^{88}Sr. The cavity enhances the effective interaction length with the atomic sample, by order of the finesse $\mathcal{F}$, while simultaneously increasing the saturation parameter by about a factor of 500. This places us deep in the saturated regime whose behaviour close to resonance was described in [10]. By sending in a probing laser and using the NICE-OHMS technique [16] we can record the phase-response of the system with good signal-to-noise ratio, see figure 2. We modulate the probing light in order to induce sidebands on the carrier frequency. These sidebands are used as reference in the heterodyne beating as they are far detuned from atomic resonance and do not interact with the atoms. They are separated by one free spectral range (FSR) of the cavity ensuring that they are transmitted through the cavity together with the carrier signal. A photodiode (PD) records the beat-signal, which is then demodulated using a mixer, thus producing the phase measurement. The antisymmetric phase-response acts as an error signal close to resonance where it has a linear dependence on the frequency detuning of the probing laser. In our case the probing laser is prestabilized to a linewidth of $\Gamma_{\text{probe}}/2\pi = 800$ Hz on the 100 µs timescale of a single measurements. By feeding the error signal obtained back to the laser it can thus be stabilized to the atom-cavity system. By ensuring that we are deep in the bad-cavity regime, the system resonance is primarily dictated by the atomic resonance, and all fluctuations on the cavity resonances are strongly suppressed.

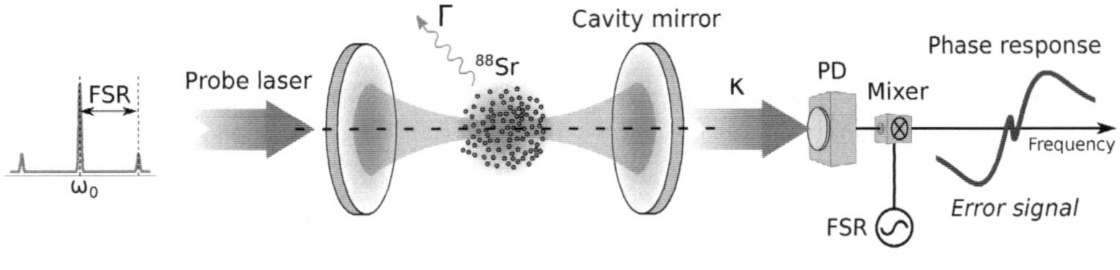

Figure 2. Sketch of a passive system where atoms inside the optical cavity are probed with an external laser having sidebands at ±FSR. The frequency sidebands are used for the NICE-OHMS technique, and the beat-signal recorded on a photodiode (PD) is then demodulated. The phase-response is recorded, producing an error signal that can be used to act back on the probing laser, correcting the probe laser frequency.

IOP Conf. Series: Journal of Physics: Conf. Series **810** (2017) 012002 doi:10.1088/1742-6596/810/1/012002

The atoms are trapped using a MOT with an anti-Helmholtz coil configuration and six beams of $I = 1.5$ mW/cm^2 blue light, red-detuned by about 40 MHz from the strong $(5s^2)^1S_0 - (5s5p)^3P_1$ transition in strontium-88. We then map out the phase response of the system by first turning on the trap for up to 800 ms in order to accumulate atoms, which will then have a temperature of typically $T = 5$ mK. The atoms are then released by shutting off the trapping light, and a single measurement at a given probe detuning is recorded over 100 μs. Reiterating this procedure while varying the probe laser detuning then produces traces such as the ones shown in figure 3. A theoretical model from [10] using our experimental parameters is also plotted. The signal-to-noise ratio is currently limited by the shot-to-shot variations in the atom number N, as well as the stability of the cavity field intensity.

For low atom numbers we observe structures similar to what was shown in [10] where a higher atom number and a cavity with much lower finesse $\mathcal{F}_{\text{low}} = 85$ was used. As the number of atoms is increased to a similar level as used in [10], the phase shift induced by the atom-cavity system is effectively much larger due to the higher finesse, $\mathcal{F} = 1450$, of the cavity in the current system. This causes the total phase to increase beyond $\pi/2$ which is seen in the measurements as a mirroring around some maximal value. In figure 3 this can be seen for $N = 1 \cdot 10^7$ (shown in pink) at about ±500 kHz, where two new peaks seem to have appeared. These features do not affect the slope at the center of the resonance feature which is decisive in order to provide a good error signal for laser feedback. This inversion for an absolute phase shift of π does, however, put a maximum bond on the amplitude of the error signal. As we require a significant output signal in order for a lock to not be limited by the photon shot noise on the detectors, simply increasing the finesse of these systems will thus not be an adequate solution.

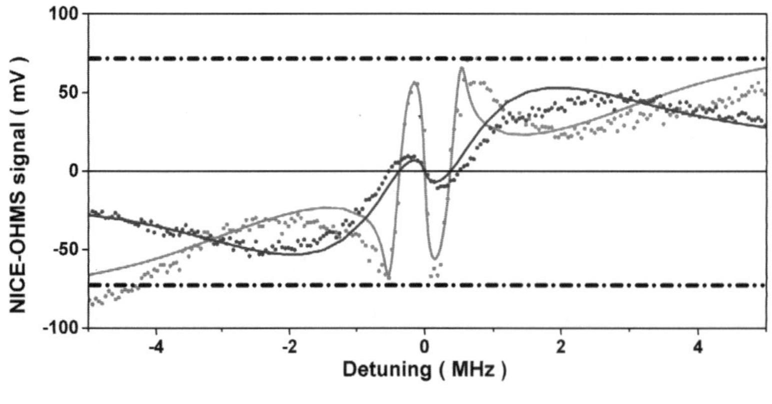

Figure 3. NICE-OHMS response of the passive system representing the phase for $N = 2 \cdot 10^6$ (blue) and $N = 1 \cdot 10^7$ (pink). Data is represented by dots whereas the full lines represents the theoretical model. For $N = 1 \cdot 10^7$ the phase shift becomes greater than $\pm\pi/2$ (corresponding to the black dot-dashed lines) at about ±500 kHz resulting in a mirroring of the signal at about ±70 mV. Both signals were recorded for $T = 7$ mK.

The shot noise limited linewidth of a laser perfectly locked to such a signal can be found following [1] and has been projected to be below 10 mHz [19] which is comparable with the state of the art in passive laser stabilization on empty optical cavities [6].

4. Active approach

Alternatively to a passively stabilized laser based on, e.g., a solid-state laser diode, the active approach seeks to use the narrow atomic transition directly as a lasing transition [2, 12, 14]. As an alternative to having effectively stationary atoms in an optical lattice at μK temperatures, we consider here a simpler system with freely moving atoms at three orders of magnitude higher temperatures. The ultimate goal is to realize a so-called optical maser where collective effects can cause an ensemble of optical dipoles to synchronize and emit light of high spectral purity [20, 11] into the cavity mode. In a first generation of experiments pursuing these phenomena in our setup the atoms are pumped by an external source, say by applying a pi-pulse, and can coherently emit light into the cavity mode. This can happen spontaneously in a process known as superfluorescence, or by stimulation from a weak seed field inside the cavity mode. Although the introduction of a weak driving field is of less importance for the optical maser it has interesting applications for quantum optics schemes targeting entanglement of atomic ensembles. It turns out superradience or superfluorescence is not sufficient to guarantee entanglement, but the introduction of a weak seed field may generate entanglement of the atomic ensemble [17, 18].

The condition for collective effects to be important is that the collectively enhanced emission rate is much larger than any decoherences of the system $C_N\Gamma \gg \Gamma_{\text{decoherence}}$. In the system investigated here the atoms are at a temperature of $T = 5$ mK resulting in the Doppler broadening of $\Gamma_{\text{Doppler}}/2\pi = 1$ MHz as the most important decoherence in the system. With the parameters given above we obtain a collective cooperativity of $C_N = 1 \cdot 10^4$ thus fulfilling the condition for strong collective coupling in the system, $C_N\Gamma/2\pi = 75$ MHz. As a preliminary approach we have resonant seed-light in the cavity mode ensuring that the cavity and atomic resonances overlap. We inject a resonant pi-pulse of light at an angle of 45° with respect to the cavity axis, see figure 4. The pulse causes ground-state atoms to be excited, and collectively emit a burst of light into the cavity mode stimulated by the seed light with a delay τ_D.

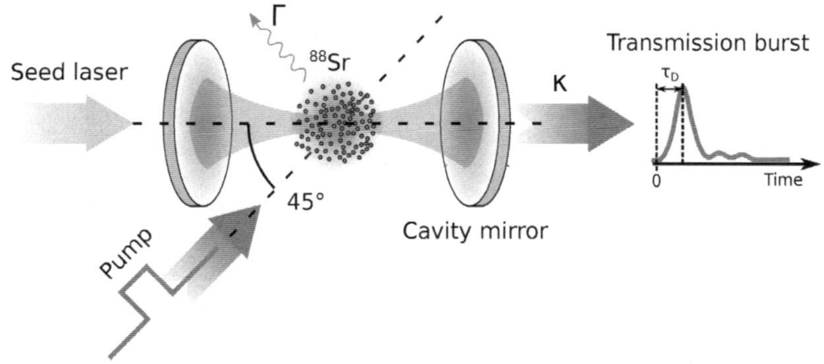

Figure 4. Sketch of an active system where ground state atoms are initially pumped into an excited state, and can subsequently be collectively stimulated to emit a burst of photons into the cavity mode after a time τ_D. The burst is observed on a photodetector through the field leaking out of the cavity.

The superfluorescent-like behaviour of the system with a seed-field in the cavity mode can be seen in figure 5 for a pump pulse of $\tau_\pi = 280$ ns. The data was averaged over 128 data series, and retains its characteristic shape. The time τ_π corresponds to a 3π-pulse, and was chosen due to technical limitations. The delay of about $\tau_D = 1.5$ μs is not limited by cavity leakage rate but attests the collective process that causes this emission of light. The decay rate of the burst can be seen to initially be far greater than the natural decay rate of the atomic transition, $1/\Gamma = 22$ μs, and after the initial burst small oscillations in the field intensity, partially washed out by the averaging, can be observed. Both features are characteristic for collective effects taking place, and reveals some level of coherence. Notice that the oscillations following the primary burst sit on top of two slopes. The first one is clearly visible from 2 to 4 μs and is expected to be caused by the averaging process where shot-to-shot variations in the atom number and seed power wash out some of the oscillatory behaviour. The second one is that caused by some atoms spontaneously decaying with the natural decay rate, a characteristic time of $\tau = 1/\Gamma = 22$ μs, and thus not participating

IOP Conf. Series: Journal of Physics: Conf. Series **810** (2017) 012002 doi:10.1088/1742-6596/810/1/012002

in the cooperative emission of light. This is attested by the fact that the final power in figure 5 is a few nW higher at 10 μs than the initial seed at $-1\ \mu$s.

We have developed a theoretical description of the system allowing us to simulate the expected behaviour of the cavity transmission. This model relies on a Jaynes-Cummings approach with a classical pump-field and incorporates a constant thermal velocity distribution of the atoms. A simulation of the cavity transmitted field for typical parameters can be seen in figure 6, where a non-zero seed field is present in the cavity initially. The initial burst is seen to have similar behaviour and delay τ_D to the measured burst. It is followed by a series of oscillations that are much more pronounced than what we see experimentally. We expect that the great difference in these oscillations is caused by the spatial profile of our pump field which is narrower compared the spatial extend of the atomic cloud. This results in different Rabi frequencies for atoms at different positions and the effective population inversion obtained by the pump pulse is thus smaller in the experiment than what the simulation assumes.

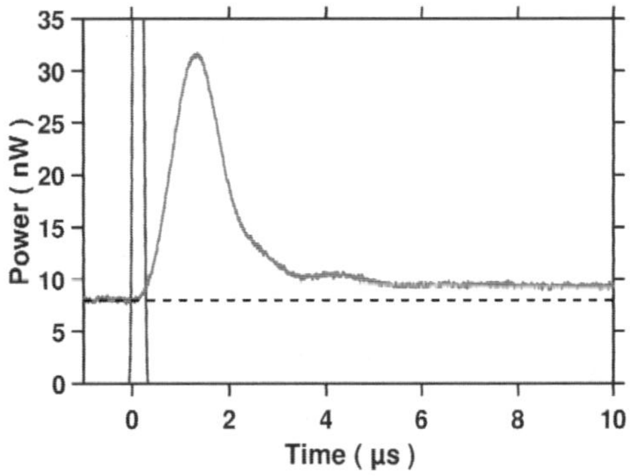

Figure 5. Recorded data of a light leaking out of the cavity (red) after a burst of light has been emitted by the atoms into the cavity mode. The initial pump-pulse (blue) is also shown. The initial decay rate of the flash is much faster than the natural decay rate Γ of the atomic transition. Though the signal is averaged over 128 data series, ringing effects are still visible. The non-zero start value is due to the seed field (dashed line) present in the cavity.

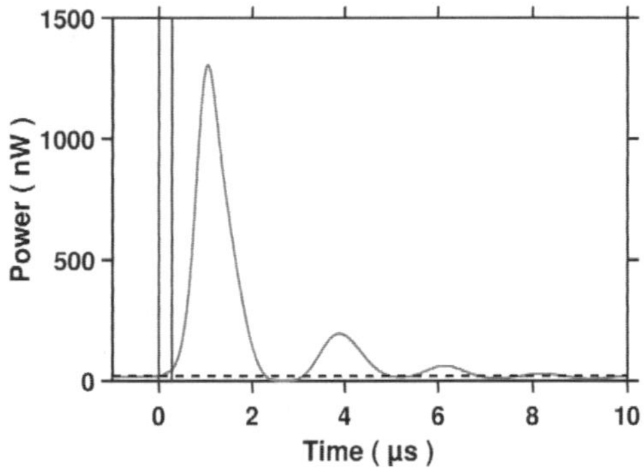

Figure 6. Simulation of a superfluorescent-like burst (red) following a 3π-pulse (blue). The initial decay of the pulse is much faster than the natural decay rate of the transition and is followed by coherent ringing characteristic for superfluorescence. Here no spatial effects of the system are included, resulting in a significant overestimation of the burst intensity. Notice that the initial value of the transmission curve is non-zero due to the seed field (dashed line).

The larger power predicted by the simulation is primarily caused by spatial dependences of the atoms and pump pulse in the system. When spatial distributions are included much better agreement between the predicted and experimentally observed power is expected. The detailed modelling of this will be the subject of future publications. The dynamical behaviour is quite similar and bodes well for a superfluorescent system with thermal atoms. By optimizing the spatial configurations of the setup, thus increased the number of participating atoms, and by better controlling some of the decoherence effects, we expect an improvement of the burst intensity by at least an order of magnitude may be obtained.

5. Conclusion

We have described our approach to and preliminary results from two of the novel methods that are currently receiving broad interest in the laser stabilization and quantum metrology communities. While these approaches rely on technically and experimentally very different approaches they have a great deal in common when one considers the physical systems. Here we have presented the physical systems in a way that underline these similarities.

The passive approach boasts impressive predictions that should be comparable to and beyond the state of the art in laser stabilization technology. In the active approach, the emission line-narrowing effect expected by the single atom cooperativity can provide a laser with a linewidth much narrower than the natural linewidth of the lasing transition making superfluorescence in the bad cavity regime an interesting alternative approach towards an ultra narrow continuous laser source. The narrowing of the linewidth in the superfluorescent light means that such a device could significantly increase the stability and accuracy of reference lasers used in, say, optical atomic clocks.

Both techniques require a continuous system such as a beamline of ultra cold atoms in order to take full advantage of the potential for narrow laser linewidths. We are currently constructing such a beamline system in order to demonstrate both a continuous active laser as well as a passive continuously locked laser.

Acknowledgments
The authors wish to thank Jun Ye, Jingbiao Chen, James K. Thompson, and Matthew A. Norcia for valuable discussions. The research was funded by ESA Contract No. 4000108303/13/NL/PA-NPI272-2012 and the Brødrene Hartmann Foundation.

References
[1] Martin M J, Meiser D, Thomsen J W, Ye J and Holland M J 2011 *Phys. Rev. A* **84** 063813
[2] Meiser D, Ye J, Carlson D and Holland M J 2009 *Phys. Rev. Lett.* **102** 163601
[3] Bloom B J, Nicholson T L, Williams J R, Compbell S L, Bishof M, Zhang X, Zhang W, Bromley S L and Ye J 2014 *Science* **506** 71
[4] Le Targat R et al. 2013 *Nat. Comm.* **4** 2109
[5] Ushijima I, Takamoto M, Das M, Okhubo T and Katori H 2015 *Nat. Phot.* **9** 185
[6] Bishof M, Zhang X, Martin M J and Ye J 2013 *Phys. Rev. Lett.* **111** 093604
[7] Kessler T, Hagemann C, Grebing C, Legero T, Sterr U, Riehle F, Martin M J, Chen L and Ye J 2012 *Nat. Phot.* **6** 687
[8] Cole G D, Zhang W, Martin M J, Ye J and Aspelmeyer M 2013 *Nat. Phot.* **7** 644
[9] Westergaard P G, Christensen B T R, Tieri D, Cooper J, Holland M J, Ye J and Thomsen J W 2015 *Phys. Rev. Lett.* **114** 093002.
[10] Christensen B T R, Henriksen M R, Schäffer S A, Westergaard P G, Tieri D, Ye J, Holland M J and Thomsen J W 2015 *Phys. Rev. A* **92** 053820
[11] Chen J 2009 *Chi. Sci. Bull.* **54** 348
[12] Bohnet J G, Chen Z, Weiner J M, Meiser D, Holland M J and Thompson J K 2012 *Nature* **484** 78
[13] Zhuang W and Chen J 2014 *Opt. Lett.* **39** 21
[14] Norcia M A and Thompson J K 2016 *Phys. Rev. X* **6** 011025
[15] Pan D, Xu Z, Xue X, Zhuang W and Chen J 2014 *2014 IEEE International Frequency Control Symposium (FCS)* **15** 1-4
[16] Ye J, Ma L and Hall J 1998 *JOSA B* **15** 6-15
[17] Wolfe E and Yelin S F 2014 *Phys. Rev. Lett.* **112** 140402
[18] Wolfe E and Yelin S F 2014 *arXiv*:1405.5288v2
[19] Tieri D, Cooper J, Christensen B T R, Thomsen J W, Holland M J 2015 *Phys. Rev. A* **92** 013817
[20] Gross M and Haroche S 1982 *Phys. Rep.* **93** 5 301-396

Ultracold collisions in metastable helium

G Peach[1], D G Cocks[2] and I B Whittingham[2]

[1] Department of Physics and Astronomy, University College London, London WC1E 6BT, UK
[2] College of Science and Engineering, James Cook University, Townsville 4811, Australia

E-mail: g.peach@ucl.ac.uk

Abstract.
Photoassociation processes are studied in ultracold collisions between different isotopes of metastable He(2^3S) and He(2^3P) atoms; Penning and associative ionization rates for collisions between two He(2^3S) atoms are also obtained. Comparisons are made with data from existing experiments.

1. Introduction

Bose-Einstein condensation (BEC) was first observed in alkali atoms in 1995 and a little later in atomic hydrogen. Initially condensates were all formed from atoms in their ground electronic states but subsequently, the search was on to obtain condensates of atoms in excited states. In 2001, a condensate was observed in a dilute gas of ^{4}He(2^3S) at a critical temperature of 4.5 ± 0.5 μK with a maximum number of condensed atoms of about 5 x 10^5.

Ultracold collisions of metastable helium atoms are now widely used to study collision dynamics in dilute quantum gases as the large internal energy can be released during collisions, allowing novel experimental strategies based upon single atom detection to be implemented. Photoassociation of ultracold atoms provides a powerful technique for the study of ultracold collisions in which two interacting cold atoms are resonantly excited by a laser to bound states of the associated molecule. The resulting spectra have a very high resolution of < 1 MHz.

Currently we are investigating elastic and ionizing collisions in spin-polarized metastable helium in the absence of an exciting laser, where the spin-dipole interaction can induce relaxation from the initial spin-polarized state to states from which Penning and associative ionization is highly probable.

2. Theory

2.1. Photoassociation

Photoassociation of ultracold bosonic homonuclear metastable helium has been observed by many groups. The system $^4\mathrm{He}^* + {}^4\mathrm{He}^*$ is resonantly excited by a laser photon with energy hν in the processes

$$^4\mathrm{He}(1s\,2s\,^3S_1)+{}^4\mathrm{He}(1s\,2s\,^3S_1)+h\nu \rightarrow {}^4\mathrm{He}(1s\,2s\,^3S_1)+{}^4\mathrm{He}(1s\,2p\,^3P_j) - E_{rv}\,,$$

where j=0,1,2 and E_{rv} is the binding energy of the rovibrational bound state relative to the limit $^4\mathrm{He}(1s\,2s\,^3S)+{}^4\mathrm{He}(1s\,2p\,^3P_j)$. These states occur in shallow potential wells at interatomic separations $\geq 150\ a_0$. We have carried out detailed calculations for these processes [1] and have extended the theory to consider the similar systems $^3\mathrm{He}^*+{}^3\mathrm{He}^*$ [2] and $^3\mathrm{He}^*+{}^4\mathrm{He}^*$ [3].

Table 1. Helium: Energy levels in MHz

^{4}He fine structure levels	^{3}He hyperfine structure levels
	$2^3\mathrm{P}_0$ $f=\frac{1}{2}$ 34385.941
34061.912 $2^3\mathrm{P}_0$	
	$2^3\mathrm{P}_2$ $f=\frac{3}{2}$ 6961.104
6462.553 $2^3\mathrm{P}_1$	
	$2^3\mathrm{P}_1$ $f=\frac{1}{2}$ 6293.071
2153.072 $2^3\mathrm{P}_2$	
	$2^3\mathrm{P}_1$ $f=\frac{3}{2}$ 1780.880
	$2^3\mathrm{P}_2$ $f=\frac{5}{2}$ 0
	$2^3\mathrm{S}_1$ $f=\frac{1}{2}$ 6739.701177
2246.567 $2^3\mathrm{S}_1$	
	$2^3\mathrm{S}_1$ $f=\frac{3}{2}$ 0

Quite distinct results are obtained for the different isotope combinations because the hyperfine structure of the ^{3}He states cannot be neglected. The relevant energy levels are shown in table 1.

The bound rovibrational levels of the ultracold excited metastable helium system are found by analyzing the eigenvalues of the molecular Hamiltonian

$$\hat{H} = \hat{T} + \hat{H}_{\mathrm{rot}} + \hat{H}_{\mathrm{el}} + \hat{H}_{\mathrm{fs}} + \hat{H}_{\mathrm{hfs}} \,,$$

where $\hat{T}$ is the kinetic energy operator

$$\hat{T} = -\frac{\hbar}{2\mu R^2}\frac{\partial}{\partial R}\left(R^2\frac{\partial}{\partial R}\right) ,$$

and $\hat{H}_{\mathrm{rot}}$ is the rotational operator

$$\hat{H}_{\mathrm{rot}} = \frac{\hat{l}^2}{2\mu R^2} \,,$$

for a system of two atoms with interatomic separation R, reduced mass μ and relative angular momentum $\hat{\mathbf{l}}$.

The total electronic Hamiltonian is

$$\hat{H}_{\mathrm{el}} = \hat{H}_1 + \hat{H}_2 + \hat{H}_{12} \,,$$

where the unperturbed atoms have Hamiltonians $\hat{H}_i$, $i = 1, 2$ and their electrostatic interaction is specified by $\hat{H}_{12}$. The terms $\hat{H}_{\mathrm{fs}}$ and $\hat{H}_{\mathrm{hfs}}$ describe the fine and hyperfine structure of the atoms.

The multichannel equations describing the interacting atoms are obtained from the eigenvalue equation

$$\hat{H}|\Psi\rangle = E|\Psi\rangle$$

by expanding the eigenvector in terms of basis states in the form $|\Phi_a\rangle = |\Phi_a(R,q)\rangle$. Here, a denotes the set of approximate quantum numbers describing the electronic-rotational states of the molecule and q denotes the interatomic polar coordinates (θ,ϕ) and electronic coordinates $(\mathbf{r}_1, \mathbf{r}_2)$.

The expansion

$$|\Psi\rangle = \sum_a \frac{1}{R} G_a(R)|\Phi_a\rangle$$

yields the multichannel equations

$$\sum_a \left\{ T^G_{a'a}(R) + \left[V_{a'a}(R) - E\delta_{a'a}\right] G_a(R) \right\} = 0\,,$$

where

$$T^G_{a'a}(R) = -\frac{\hbar}{2\mu}\langle\Phi_{a'}|\frac{\partial^2}{\partial R^2} G_a(R)|\Phi_a\rangle$$

and

$$V_{a'a}(R) = \langle\Phi_{a'}|\left[\hat{H}_{\rm rot} + \hat{H}_{\rm el} + \hat{H}_{\rm fs} + \hat{H}_{\rm hfs}\right]|\Phi_a\rangle$$

2.2. Potentials

The required Born-Oppenheimer potentials ${}^{1,3,5}\Sigma^+_{g,u}$ and ${}^{1,3,5}\Pi_{g,u}$, see [4], were constructed by matching *ab initio* MCSCF and MRCI short-range potentials onto the long-range dipole-dipole plus dispersion potentials given by

$$V^{\rm long}_\Lambda(R) = -f_{3\Lambda}(R/\lambda) C_{3\Lambda}/R^3 - C_{6\Lambda}/R^6 - C^{\pm}_{8\Lambda}/R^8,$$

where $\Lambda = \Sigma, \Pi$ and $f_{3\Lambda}$ is an R- and Λ-dependent retardation correction. The wavelength for the transition $2\text{s}\,{}^3\text{S}$–$2\text{p}\,{}^3\text{P}$ is λ, where $\lambda/(2\pi) = 3258.17\,a_0$ and the superscripts indicate the sign of $(-1)^{S+w}$, $S = 0,1,2$, $w = 0$ or 1 (gerade/ungerade). The coefficients $C_{n\Lambda}$; $n = 3,6,8$ are well known, see [5] and [6].

2.3. Scattering processes

In the absence of an exciting laser, ionizing collisions in spin-polarized metastable helium can still take place because the spin-dipole interaction can induce relaxation from the initial spin-polarized state to spin states of $[\text{He}(2^3\text{S}_1) + \text{He}(2^3\text{S}_1)]$ from which Penning and associative ionization is highly probable. These processes are:

$$[\text{He}(1\text{s}\,2\text{s}\,{}^3\text{S}_1) + \text{He}(1\text{s}\,2\text{s}\,{}^3\text{S}_1)]\ {}^1\Sigma_g \rightarrow \text{He}(1\text{s}^2\,{}^1\text{S}_0) + \text{He}^+(1\text{s}) + \text{e}^- \quad \text{Penning ionization}$$

and

$$[\text{He}(1\text{s}\,2\text{s}\,{}^3\text{S}_1) + \text{He}(1\text{s}\,2\text{s}\,{}^3\text{S}_1)]\ {}^1\Sigma_g \rightarrow \text{He}_2{}^+ + \text{e}^- \quad \text{associative ionization,}$$

where if $[\text{He}(1\text{s}\,2\text{s}\,{}^3\text{S}_1) + \text{He}(1\text{s}\,2\text{s}\,{}^3\text{S}_1)]$ is prepared in the $\text{He}_2{}^5\Sigma_g$ state the transition rate to the $\text{He}_2{}^1\Sigma_g$ state is small. If this transition does take place, Penning ionization or associative ionization follows rapidly leaving the helium atoms/molecules in their ground electronic states. The collisions between isotopes ^{4}He and ^{3}He are being studied for all three combinations, 4-4, 3-3 and 3-4.

We consider the relative motion of the atoms subject to a central interatomic potential $V(R)$ where

$$V(R) = V_0(R) + V_{\text{ion}}(R)$$

and $V_0(R)$ is the adiabatic potential for the ${}^1\Sigma_g$ state of the He*+He* system. The potential $V_{\text{ion}}(R)$ is included to allow for loss processes. Two ionization models are used, M1 and M2, see [7] and $V_0(R)$ is determined from experimental and theoretical data [8], [9] and [10].

The two models are specified by

$$(M1) \qquad V_{\text{ion}}(R) = +\infty\,; \quad R \le R_b$$
$$= \quad 0\,; \quad R > R_b$$

and

$$(M2) \qquad V_{\text{ion}}(R) = -\frac{1}{2}\mathrm{i}\,\Gamma(R); \quad \Gamma(R) = 0.3a_0 \exp(-R/1.086a_0)\,.$$

In both cases on integrating the radial wave equation we obtain a 2×2 scattering matrix S which must be unitary and symmetric. The cross sections $\sigma(v)$ as a function of the relative velocity v are given by

$$\sigma_{\text{elastic}}(v) = \frac{\pi}{k^2}\sum(2l+1)|1-S_{00}|^2\,; \quad \sigma_{\text{ionization}}(v) = \frac{\pi}{k^2}\sum(2l+1)|S_{01}|^2,$$

where $k = \mu v/\hbar$ and the rates K_e and K_i are obtained by averaging over a Maxwell velocity distribution so that

$$K_e = [v\sigma_{\text{elastic}}(v)]_{\text{av}}; \quad K_i = [v\sigma_{\text{ionization}}(v)]_{\text{av}}\,.$$

The sum $\sum$ over the angular momentum quantum number l for the heteronuclear system ^{3}He + ^{4}He is taken over all values of l, but the systems ^{4}He + ^{4}He and ^{3}He + ^{3}He are symmetric and anti-symmetric respectively with respect to the interchange of the nuclear space and spin coordinates. This means that the sum $\sum$ has to be interpreted as

$$\sum \equiv 2\sum_{l\,even}$$

for the bosonic case and

$$\sum \equiv \frac{1}{2}\sum_{l\,even} + \frac{3}{2}\sum_{l\,odd}$$

for the fermionic system.

The multichannel calculations include magnetic-dipole mediated transitions from the ${}^5\Sigma$ state and represent the ionization loss from the ${}^{1,3}\Sigma$ states by a complex optical potential. The asymptotic solutions of the multichannel equations yield the non-unitary scattering matrix elements $S_{a'a}$. The elastic cross section for channel a is then

$$\sigma(a \to a; v) = \frac{\pi}{k^2}\sum(\ldots)|1 - S_{aa}|^2.$$

The ionization cross section is

$$\sigma(a \to \text{ion}; v) = \frac{\pi}{k^2}\sum(\ldots)|1 - S(a \to \text{ion})|^2$$

where $S(a \to \text{ion})$ is obtained from the non-unitarity of the S-matrix:

$$|S(a \to \text{ion})|^2 = 1 - \sum_{a'}|S_{a'a}|^2.$$

In the above $\sum(\ldots)$ denotes sum/average over unobserved quantum numbers.

Table 2. ^{4}He(2^3S)+^{4}He(2^3P). Photoassociation
Energies in GHz relative to the $j = 2$ asymptote. The third column lists the levels after a 1% variation is applied to the short-range potentials.

Exp	Theor	Variation	assignment
-13.57	-13.259	-13.621	$2_u,\ J=3$
-11.70	-11.434	-11.764	$1_u,\ J=1$
-11.10	-10.930	-11.224	$1_u, 0_u^+\ J=3$
-8.94	-8.705	-8.966	$2_u,\ J=3$
-7.44	-7.262	-7.493	$1_u,\ J=1$
-7.01		-7.103	$1_u, 0_u^+\ J=3$
-5.64	-5.463	-5.647	$2_u,\ J=3$
-4.53	-4.394	-4.551	$1_u, 0_u^+,\ J=1$
-4.26	-4.145	-4.285	$0_u^+, 1_u,\ J=3$
-3.49	-3.438	-3.566	$2_u,\ J=2$
-3.38	-3.251	-3.375	$2_u,\ J=3$

3. Results

3.1. Photoassociation

Three sets of calculations have been undertaken: single-channel, multichannel without Coriolis couplings and full multichannel with Coriolis couplings. At each value of R the single-channel potential is formed by diagonalizing the matrix:

$$V_{a'a}^{\Omega_j} = \langle a'|\hat{H}_{\rm el}|a\rangle + \langle a'|\hat{H}_{\rm fs} + \hat{H}_{\rm hfs}|a\rangle + \frac{\langle a'|\hat{l}^2|a\rangle_{\Omega_j}}{2\mu R^2},$$

where $\langle a'|\hat{l}^2|a\rangle_{\Omega_j}$ is the part of $\langle a'|\hat{l}^2|a\rangle$ diagonal in Ω_j. Ω_j denotes the projection of the total angular momentum quantum number J onto the inter-molecular axis. Results shown in table 2 illustrate the excellent agreement beteen theory and experiment that can be achieved. More details are given in [1].

3.2. Scattering processes

Rates K_e and K_i as a function of temperature T for the elastic scattering and ionization processes obtained using models M1 (with $R_b = 7a_0$) and M2 are listed in table 3, where Model M1 is insensitive to the precise choice of R_b.

The results from the two models agree quite well with each other and show that the rates become increasingly insensitive to the form of $V_{\rm ion}(R)$ as the temperature increases. In table 4 total unpolarized ionization rates are listed for combinations of the different isotopes of helium, where allowance is made for contributions from all the initial spin states of He(2^3S)+He(2^3S), $^1\Sigma_g$, $^3\Sigma_u$ and $^5\Sigma_g$.

Full quantum-mechanical close-coupled calculations are also being carried out and provisional results K(theor) see [11], agree quite well with those of a simple two-stage semi-classical model K(sc) and with experiment K(exp), see [12] and [13].

4. Conclusions

Our analysis of photoassociation processes in the ^{4}He*+^{4}He* system, based upon single-channel and multichannel calculations, permitted criteria to be established for the assignment of the

Table 3. $^4He(2^3S)+^4He(2^3S)$. Scattering
The elastic (K_e) and ionization (K_i) rates for models M1 and M2 are in units of cm^3/s and the data are listed in the form 4.914(-12) = 4.914× 10^{-12}.

$Log_{10}(T(K))$	$K_e(M1)$	$K_e(M2)$	$K_i(M1)$	$K_i(M2)$
-7.0	4.914(-12)	4.216(-12)	1.329 (-9)	1.277 (-9)
-6.0	1.540(-11)	1.322(-11)	1.319 (-9)	1.268 (-9)
-5.0	4.739(-11)	4.072(-11)	1.289 (-9)	1.239 (-9)
-4.0	1.374(-10)	1.185(-10)	1.198 (-9)	1.155 (-9)
-3.0	3.324(-10)	2.895(-10)	9.663(-10)	9.367(-10)
-2.0	5.066(-10)	4.615(-10)	6.980(-10)	6.836(-10)
-1.0	7.424(-10)	7.152(-10)	1.066 (-9)	1.028 (-9)
0.0	1.720 (-9)	1.712 (-9)	1.567 (-9)	1.601 (-9)

Table 4. $He(2^3S)+He(2^3S)$. Scattering
Thermally averaged total unpolarized ionization rates K for the 3-3, 3-4 and 4-4 systems in units of cm^3/s.

System	T(mK)	K(exp)	K(theor)	K(sc)
4-4	1.9(1)	8(2) $\times 10^{-11}$	9.74×10^{-11}	8.0×10^{-11}
3-3	2.0(3)	$1.8(3)\times10^{-10}$	1.30×10^{-10}	1.7×10^{-10}
3-4	1.2(1)	$2.5(8)\times10^{-10}$	2.42×10^{-10}	2.4×10^{-10}

theoretical levels to the experimental observations. Excellent agreement was obtained for the nearly 50 observed levels. We have also investigated photoassociation in the fermionic metastable homonuclear $^3He^*+^3He^*$ system, where the hyperfine structure leads to a very different pattern of energy levels below the $^3He(1s\,2s\,^3S)+^3He(1s\,2p\,^3P_j)$ asymptotes. As no experimental observations are available for this system, we have assessed the observability of the theoretical levels and have identified purely long-range levels and 30 short-range levels suitable for experimental investigation. Very recently we have extended our investigations to photoassociation in the heteronuclear $^3He^*+^4He^*$ system. No purely bound long-range states were found, although several resonances with line widths smaller than 1 MHz were obtained.

Currently we are investigating elastic and ionizing collisions in spin-polarized metastable $^4He^*+^4He^*$, $^3He^*+^3He^*$ and $^3He^*+^4He^*$ systems in the absence of an exciting laser, where the spin-dipole interaction can induce relaxation from the initial spin-polarized state to states from which Penning and associative ionization is highly probable. Two simple models are adopted from which rates for elastic scattering and ionization processes have been calculated for the $^4He^*+^4He^*$ system. The results from the two models agree quite well with each other, and preliminary results from full solutions of the quantum-mechanical coupled equations indicate that the simple models give a good general description of the process.

References

[1] Cocks D G, Whittingham I B and Peach G 2010 *J. Phys.* B: *At. Mol. Opt. Phys.* **43** 135102
[2] Cocks D G, Peach G and Whittingham I B 2011 *Phys. Chem. Chem. Phys.* **13** 18724–33
[3] Peach G, Cocks D G and Whittingham I B 2015 *J. Phys.* B: *At. Mol. Opt. Phys.* **48** 115205

IOP Conf. Series: Journal of Physics: Conf. Series **810** (2017) 012003 doi:10.1088/1742-6596/810/1/012003

[4] Deguilhem B, Leininger T, Gadéa F X and Dickinson A S 2009 *J. Phys.* B: At. Mol. Opt. Phys. **42** 015102
[5] Venturi V, Leo P J, Tiesinga E, Williams C J and Whittingham I B 2003 *Phys. Rev.* A **68** 022706
[6] Zhang J-Y, Yan Z-C, Vrinceanu D, Babb J F and Sadeghour H R 2006 *Phys. Rev.* A **73** 022710
[7] Garrison B J, Miller W H and Schaefer H F 1973 *J. Chem. Phys.* **59** 3193
[8] Müller M W, Merz A, Ruf M-W, Hotop H, Meyer W and Movre M 1991 *Z. Phys.* D: At. Mol. Clusters **21** 89-112
[9] Stärck T and Meyer W 1994 *Chem. Phys. Lett.* **225** 229
[10] Venturi V and Whittingham I B 2000 *Phys. Rev.* A **61** 060703(R)
[11] Cocks D G, Whittingham I B and Peach G (to be published)
[12] Stas R J W, McNamara J M, Hogervorst W and Vassen W 2006 *Phys. Rev.* A **73** 032713
[13] McNamara J M, Stas R J W, Hogervorst W and Vassen W 2007 *Phys. Rev.* A **75** 062715

X-ray spectroscopy of super-intense laser-produced plasmas for the study of nonlinear processes. Comparison with PIC simulations

E Dalimier[1], A Ya Faenov[2,3], E Oks[4], P Angelo[1], T A Pikuz[3,5], Y Fukuda[6], A Andreev[7,8], J Koga[6], H Sakaki[6], H Kotaki[6], A Pirozhkov[6], Y Hayashi[6], I Yu Skobelev[3,9], S A Pikuz[3,9], T Kawachi[6], M Kando[6], K Kondo[6], A Zhidkov[5], E Tubman[10], N M H Butler[11], R J Dance[11], M A Alkhimova[3,9], N Booth[12], J Green[12], C Gregory[12], P McKenna[11], N Woolsey[10], R Kodama[2,5]

[1] LULI - UPMC Paris 06 : Sorbonne Universités ; CNRS, Ecole Polytechnique, CEA : Université Paris-Saclay - F-75252 Paris cedex 05, France

[2] Institute for Academic Initiatives, Osaka University, Suita, Osaka, 565-0871, Japan

[3] Joint Institute for High Temperatures, Russian Academy of Sciences, Moscow 125412, Russia

[4] Physics Department, 206 Allison Lab, Auburn University, Auburn, AL 36849, USA

[5] PPC and Graduate School of Engineering, Osaka University, 2-1, Yamadaoka, Suita, Osaka 565-0871, Japan

[6] Kansai Photon Science Institute, National Institutes for Quantum and Radiological Science and Technology, Kizugawa, Kyoto, 619-0215, Japan

[7] Max Born Institute, Berlin 12489, Max-Born str.2a, Berlin, Germany

[8] ELI-ALPS, Szeged H-6720, Hungary

[9] National Research Nuclear University MEPhI (Moscow Engineering Physics Institute), Moscow 115409, Russia

[10] York Plasma Institute, Department of Physics, University of York, York YO10 5DD, UK

[11] Department of Physics, SUPA, University of Strathclyde, Glasgow G4 0NG, UK

[12] Central Laser Facility, STFC Rutherford Appleton Laboratory, Didcot OX11 0QX, UK

E-mail: elisabeth.dalimier@upmc.fr

Abstract. We present X-ray spectroscopic diagnostics in femto-second laser-driven experiments revealing nonlinear phenomena caused by the strong coupling of the laser radiation with the created plasma. Among those nonlinear phenomena, we found the signatures of the Two Plasmon Decay (TPD) instability in a laser-driven CO_2 cluster-based plasma by analyzing the Langmuir dips in the profile of the O VIII Lyε line, caused by the Langmuir waves created at the high laser intensity 3 10^{18}Wcm^{-2}. With similar laser intensities, we reveal also the nonlinear phenomenon of the Second Harmonic Generation (SHG) of the laser frequency by analyzing the nonlinear phenomenon of satellites of Lyman δ and ε lines of Ar XVII. In the case of relativistic laser-plasma interaction we discovered the Parametric Decay Instability (PDI)-induced ion acoustic turbulence produced simultaneously with Langmuir waves via irradiation of thin Si foils by laser intensities of 10^{21}Wcm^{-2}.

1. Introduction

X-ray spectroscopy in femto-second laser-driven plasmas can be used for a better understanding of intense laser-plasma interaction. Undoubtedly it can also be used for a laboratory modelling of nonlinear physical processes in astrophysical objects. The nonlinear phenomena caused by the strong coupling of the electromagnetic wave "t_l" of the laser at frequency ω_L (and eventually its second harmonic "t_2" at $2\omega_L$) with the created plasma may involve the Langmuir turbulence "l" at the electron plasma frequency ω_{pe} equal to ω_L (or $\omega_L/2$) and the ion acoustic turbulence "s" at the ion plasma frequency ω_{pi}. The processes discovered in the present work, due to high-resolution X-ray spectroscopy, are the Two Plasmon Decay instability TPD [1], the Parametric Decay Instability PDI and the Second Harmonic Generation SHG [2]. For TPD the nonlinear process is summarized by the balance equation $t_l \rightarrow l + l$. For PDI the process involved is $t_l \rightarrow l + s$ and as for SHG it can be explained either by the process $l+l \rightarrow t_2$ or $l + t_l \rightarrow t_2$.

The two super intense femto-second laser facilities involved in these spectroscopic studies of non-linear processes are the Kansai Photon Science Institute (KPSI, National Institutes for Quantum and Radiological Science and Technology, Kyoto, Japan) and the Rutherford Appleton Laboratory RAL (Science and Technology Facilities Council, Didcot, UK).

At KPSI, the lasers JLITE-X (pulse duration 40 fs, pulse energy 60 mJ) and J-KAREN (pulse duration 40 fs, pulse energy 800 mJ) gave access to large intensities $4\ 10^{17}$ Wcm^{-2} and $3\ 10^{18}$ Wcm^{-2} respectively. As for the Vulcan Petawatt laser (pulse duration 500-1500 fs, pulse energy 290 J), relativistic laser-plasma regimes could be reached with intensities as high as $1.4\ 10^{21}$ Wcm^{-2}.

2. Discovery of the Two Plasmon Decay TPD instability

At KPSI, the spectroscopic analysis of the Langmuir dips of O VIII Lyε line revealed the instability process TPD and its characteristics [1].

Let us first recall the mechanism of Langmuir dips formation adding some improvements in their spectral line shape characteristics. The L-dip phenomenon is a resonant coupling between the low frequency ionic field F_i and the quasi-monochromatic electric field E of the high frequency Langmuir wave ω_{pe}. The resonance condition is given by:

$$\omega_F = s\omega_{pe}(N_e) \quad (1)$$

where $\omega_F = 3n\hbar F_i/(2Z_r m_e e)$ is the separation between Stark sublevels caused by F_i (n being the principal quantum number and Z_r the nuclear charge of radiating ion) and $\omega_{pe} = (4\pi e^2 N_e/m_e)^{1/2}$, the electron plasma frequency (N_e being the electron density). s=1 for "one quantum" resonance.

Due to the broad distribution ΔF of the low frequency field, there is always a fraction of radiators, for which the resonance condition is satisfied.

The resonance condition manifests as an L-dip (for each Stark component) located from the unperturbed wavelength λ_0, at the distance $\Delta\lambda_{dip}$ given in equation (2) (see, e.g., [3]):

$$\Delta\lambda_{\delta\iota\pi} = -[\lambda_0^2/(2\pi\chi)]\{\theta\omega_{\pi\varepsilon} + [2\omega_{\pi\varepsilon}^3/(27\nu^3 Z_\rho Z_\pi \omega_{\alpha\tau})]^{1/2}[\nu^2(\nu^2 - 6\theta^2 - 1) + 12\nu^2\theta^2 \pm 6\nu^2\theta]\} \quad (2)$$

The positions of the dips allow an accurate measure of the electronic density N_e as they depend on it as:

$$\Delta\lambda_{dip} = aN_e^{1/2} + bN_e^{3/4} \quad (3)$$

Here a and b are controlled by quantum numbers n and q and by the charges of the radiating Z_r and perturbing Z_p ions (n is the principal quantum-number, $q= n_1-n_2$ the electric quantum number, n_1, n_2 being the parabolic quantum-numbers). $\omega_{\alpha\tau}= 4{,}14\ 10^{16}\ \sigma^{-1}$ is the atomic unit of frequency. The first term in equation (3) is due to the dipole interaction with the micro-field and the second to the quadrupole interaction.

Figure 1 shows the two bump-dip-bump structures in O VIII Lyε line that lead to a spectroscopic diagnostic of TPD. These structures are the one quantum resonance (s=1) dip in the profiles of the two most intense components, originating from the sublevels (311) and (131), (q=2 and q=-2 the electric quantum number values). By comparing the theoretical and experimental separations between these two L-dips

IOP Conf. Series: Journal of Physics: Conf. Series **810** (2017) 012004 doi:10.1088/1742-6596/810/1/012004

$$[\lambda_0^2/(\pi c)]|q|\rceil_{pe}(N_e) = 4(\lambda_0^2/2\pi c)\,\rceil_{pe}(N_e) = 4\lambda_p = 57\text{mA} \quad (4)$$

we determine $\omega_{pe} = 1.26 \times 10^{15}\ s^{-1} = \omega_{pe}$ and then find that it is equal to $\omega_L/2$ with the accuracy 5%.

Since the TPD instability occurs at the quarter of the critical density $N_c = m_e\omega_L^2/4\pi e^2$, the above constitutes the experimental evidence of the TPD. It is important to emphasize that the center of gravity of the structures is shifted to the red by 9 mA, in agreement to the non-uniformity of the ion micro-field effect [4] reflected by the second term in the right side of Eq. (2).

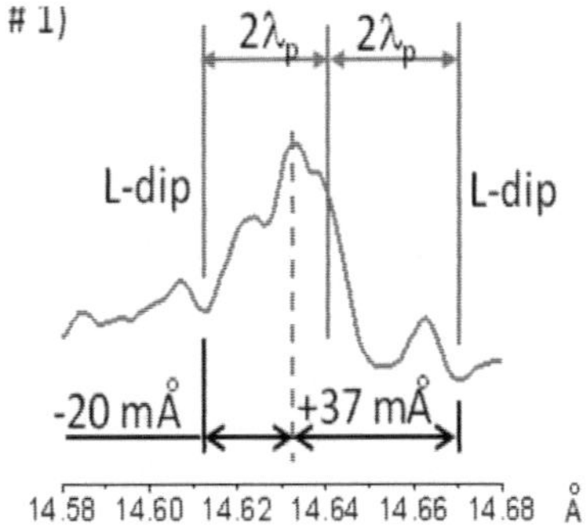

Figure 1. The two "bump-dip-bump" structures in O VIII Lyε line . The experiment was performed at J-Karen in femto-second laser-driven cluster- based experiments with mixture CO_2 and He (laser frequency $\omega_L = 2.4\ 10^{15}\ s^{-1}$, laser intensity $I_L = 3\ 10^{18}\ Wcm^{-2}$). The bumps surrounding the dips are due to partial transfer of the intensity from the wavelength of the dip to adjacent wavelengths.

3. Discovery of the PDI induced ion acoustic turbulence

At RAL, at the relativistic laser intensities, the spectroscopic analysis of the Langmuir-dips in Si XIV lines and of the excessive ("anomalous") Stark broadening of these lines, led to the conclusion of the development of Low-frequency Electrostatic Turbulence (LET) caused by the PDI. Using Vulcan petawatt laser facility (wavelength $\lambda_L = 1.045\ \mu m$; pulse duration 500 to 1500 fs; intensity I_L up to 1.4 $10^{21}\ Wcm^{-2}$), the ultra-intense radiation penetrates the plasma until a region of density N_{rc} higher than the usual critical density N_c [5]. The density N_{rc} is higher than N_c because of the relativistic increase of the mass of electrons, and thus the decrease of the electron plasma frequency.

For the linearly - polarized laser radiation, the relativistic critical density is given by [6]

$$N_{rc} = (\pi a/4)\, N_c \quad (5)$$

with $N_c(\omega) = m_e\omega_L^2/(4\pi e^2)$ and $a = \lambda_L(\mu m)\,[I_L(\omega/cm^2)/(1.37\times10^{18})]^{1/2}$

For the Vulcan laser the critical density N_c can be 1.0 $10^{21}\ cm^{-3}$ and in consequence the relative critical density could be as high as $N_{rc} = 2.3\ 10^{22}\ cm^{-3}$. In fact it is well known that in relativistic regimes the transverse electromagnetic wave intensity is higher than the incident laser radiation intensity due to self-focusing of the beam, Raman and Brillouin backscattering [7], so that the relativistic critical density should be even higher, as we will confirm.

Figure 2 shows typical X-ray spectra of Si XIV Lyβ and Lyγ lines at Vulcan with two laser irradiations of silicon foils. The black trace is for the incident laser radiation 1.01 $10^{21}\ Wcm^{-2}$ at the surface of the target, the blue trace for a laser irradiation 0.24 $10^{21}\ Wcm^{-2}$.

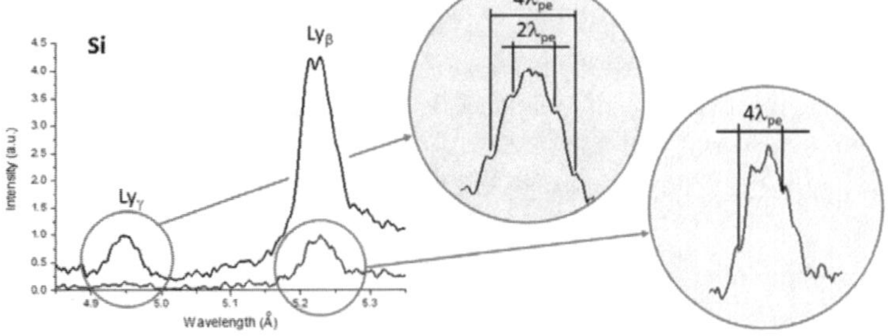

Figure 2. Observation of the Langmuir dips in Lyβ (blue trace) and Lyγ (black trace) corresponding to different incident laser irradiations 0.24 10^{21} Wcm^{-2} (blue trace) and 1.01 10^{21} Wcm^{-2} (black trace).

The identification of the symmetrical L-dips on Lyβ (blue trace) separated by 43mA yields (with s=1 and q=±2) the density N_e=1.74 10^{22} cm^{-3}. The identification of the two couples of symmetrical dips on Lyγ (with s=1, q=±1 for the dips separated by 28mA, and with s=1, q=±2 for the dips separated by 56mA) yields the same density N_e=3.6 10^{22} cm^{-3}.

If the low-frequency (quasistatic) field involved in the formation of L-dips would be represented by the ion microfield, there would be a significant red shift of the center of gravity of all those pairs of L-dips in Fig. 2 (according to the second term in the right side of Eq. (2). However, the experimental profiles do not show any such red shift. This is the experimental evidence that the low-frequency (quasistatic) field involved in the formation of L-dips was dominated by the LET (in which case there is no second term in Eq. (2)).

This is the first (but not the only one) confirmation that the LET was developed at the surface of relative critical density N_{rc}. It is well-known that at this surface the mechanism for the production of the LET is the PDI, i.e., the transformation of the laser wave into the combination of the Langmuir wave and the ion acoustic wave. The density N_{rc} for each shot (blue and black traces) is higher than the density given by (5), as predicted.

There is a second confirmation that the low frequency field is dominated by LET developing at N_{rc}. This confirmation comes from line broadening analysis. Line broadening simulations have been made and compared with two kinds of simulations: the FLYCHK modelling [8] and the Oks's modelling. FLYCHK code takes into consideration the Stark broadening with the ionic micro field and the Doppler, opacity and instrumental broadenings. This code is valid even for high densities $10^{23}cm^{-3}$. In this simulation the Langmuir dips and the LET broadening are excluded. The Oks's modeling valid only for moderate densities has the advantage to take into account the Langmuir dips, and the Stark broadening with both the ionic field and the LET. It takes into consideration the Döppler, opacity and instrumental broadenings, just as FLYCHK.

Figure 3 shows for Si XIV Lyγ the FLYCHK simulation fitting with the experimental result obtained at RAL. The electronic density for the fit is 9 $10^{22}cm^{-3}$ which is obviously higher than 3.6 $10^{22}cm^{-3}$, the density measured from the dips separations and identified to the relativistic critical density. Figure 4 presents Oks' simulation for the same line Si XIV Lyγ. The electron density for the good fit is 3.6 $10^{22}cm^{-3}$, exactly the measured density for the relativistic critical density. Obviously this simulation supports the presence of the LET, i.e. the ion acoustic wave, in the relativistic plasmas.

A third confirmation of the presence of the LET accompanying the PDI in relativistic regimes was obtained by PIC simulations shown in figure 5. 1D PIC simulation for a laser pulse duration 600 fs with intensity 2.7 10^{21} Wcm^{-2} (the estimated intensity of the transverse electromagnetic wave) has been performed. The initial scaled density of Si ions was taken constant over 2μm on the foil. The black line corresponds to the transverse electric field, the blue line to the longitudinal Langmuir wave. The red and green lines represent the evolutions of the electron and ion densities respectively. The

Langmuir wave exists up to N_{rc}= 3.6 $10^{22}cm^{-3}$. The modulation of the ion density is the manifestation of the ion acoustic wave.

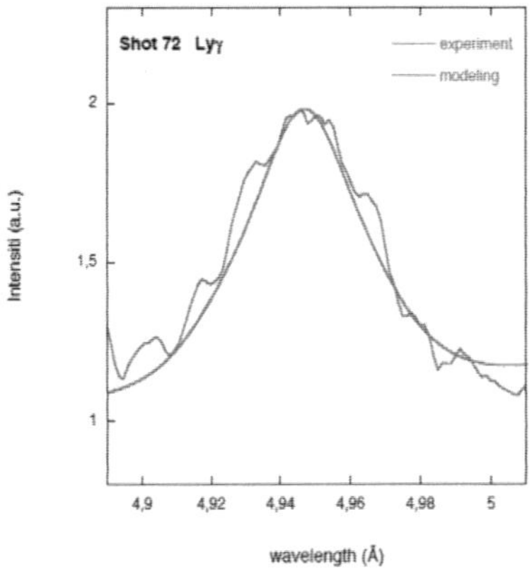

Figure 3. Si Lyγ spectral line shape with FLYCHK (red line) fitted with the experimental results (blue line)

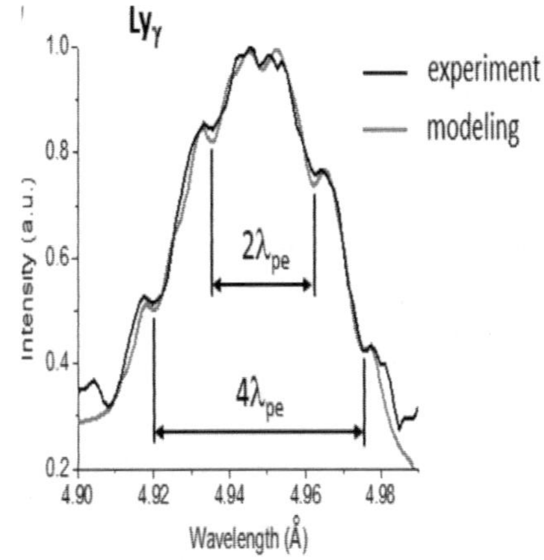

Figure 4. Si Lyγ spectral line shape with Oks's simulation (red line) fitted with the experimental results (black line)

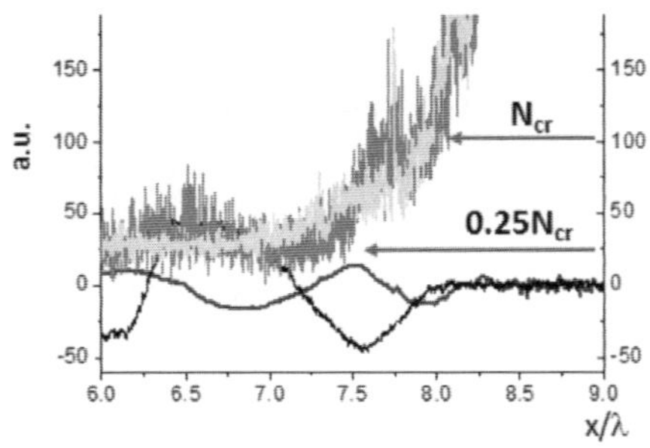

Figure 5. 1D PIC simulation of the PDI in the relativistic regime showing the modulation of the ion density near N_{rc}. Black line: transverse electric field. Blue line: longitudinal Langmuir wave. Red line: electron density. Green line: ion density

4. Discovery of the Second Harmonic Generation SHG of the laser frequency.

At J-KAREN facility, the Second Harmonic Generation SHG was discovered by analyzing the nonlinear phenomenon of satellites on the spectral lines Ar XVII Heδ and Heε. The intensity of the laser is high but not relativistic. In the presence of the laser electromagnetic field ω_L, the combined system "radiator plus laser field" is characterized by quasi-energy levels separated by $\hbar\omega_L$ leading in spectra to satellites $\pm\omega_L$, $\pm 2\omega_L$, ... Due to the high laser intensity various nonlinear processes can occur: for instance the parametric decay instability PDI can be followed by the processes $l+l \rightarrow t_2$ or $l + t_1 \rightarrow t_2$. In the present work we discovered the satellites $\pm 2\omega_L$, $\pm 4\omega_L$, ... of the transverse wave t_2 at $2\omega_L$. At J-Karen the femto second laser-driven cluster-based argon plasma was diagnosed due to a high spectral resolution (3000) and spatial resolution FSSR spectrograph. Figure 6 presents the experimental profiles (dashed blue lines) Heδ and Heε at I_L= 3 10^{18} Wcm^{-2}. The lines exhibit satellites at $2\omega(\lambda_n^2/(2pc))$, λ_n being the unperturbed wavelength, but not at $\omega(\lambda_n^2/(2\pi c))$.

Two important remarks lead to the necessity to develop a new theory of laser satellites under the action of a bi-chromatic field ω and 2ω. First, the second harmonic wave t_2 intensity being much lower than the pump wave t_l intensity, the satellites at ω_L should be stronger than those at $2\omega_L$. This is not the case in the experiment. Moreover the second harmonic is supposed to penetrate the plasma until the density $4N_c$=7.2 10^{21} cm^{-3} (N_c being the critical density for ω_L), but the average density deduced from the spectra is lower. As a consequence a new theory had to be developed improving Blochinzev's [9] and Oks's [10, 3] theories of satellites valid only for the lateral Stark components of hydrogen lines under a monochromatic field. Let us remark that the highly excited lines Heδ,ε are hydrogenic for the densities considered. The line Heε has a central Stark component, so that the theory of satellites was advanced by extending it to the central Stark components. Also a further advance was made by developing a theory of satellites in a bi-chromatic field $E_1\cos(\omega t) + E_2\cos(2\omega t)$ in the plasma region diagnosed by spectroscopy.

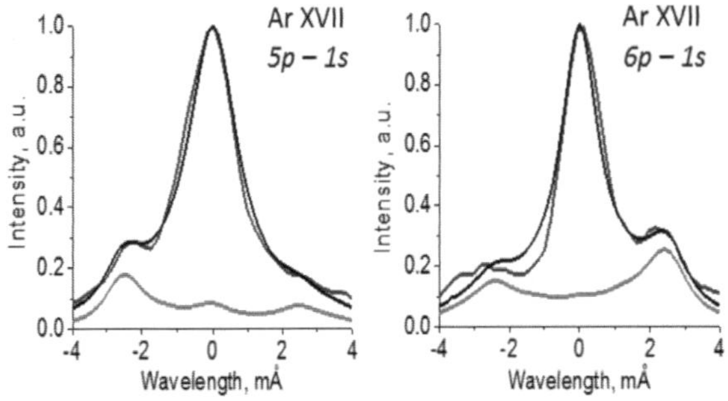

Figure 6. Ar XVII Heδ and ε profiles obtained at J-KAREN at I_L= 3 10^{18} Wcm^{-2}. The blue lines are for the experimental profiles, the solid red lines for the theoretical profiles from a bi-chromatic region and the solid black lines for the total theoretical profiles including a region of no-periodic-field. The black lines exhibit satellites at $2\omega_L$.

The argon Heδ and ε profiles obtained at J-KAREN at I_L= 3 10^{18} Wcm^{-2} are shown in figure 6.

The best agreement of the experimental profiles was obtained with simulations (solid black lines) involving a space integration over a bi-chromatic field region at the electronic density 3 10^{20} cm^{-3} (solid red lines), explaining the broad pedestals of the final profiles, and a no-periodic field region at the electronic density 10^{20} cm^{-3} reproducing the intense central part of the profiles. The simulations confirm the laser satellites at $2\omega_L$. The intensity of the SHG in the plasma versus the incident laser field in vacuum has been evaluated at 2% from the above spectroscopic analysis.
It is important to emphasize that at lower laser intensity 4 10^{17} Wcm^{-2} (JLITE-X laser at KPSI) the Second Harmonic Generation was not detected.

Finally 2D PIC simulations to support SHG diagnostics yielded fruitful results. They were performed for a 40 fs pulse of 3 10^{18} W cm^{-2} linearly polarized laser on to Ar XVII cluster 500 nm diameter. The figure 7 (a) gives the 2D spatial distribution of electric field E_y(y,z) at 42 fs (z-axis laser pulse propagation from bottom to the top, cluster center at z=y=10 mm); the transformation of the plane waves into spherical waves is obvious. The distribution of the electric field along the z-axis is given in figure 7 (b) confirming the conversion efficiency into $2\omega_L$: 2%.

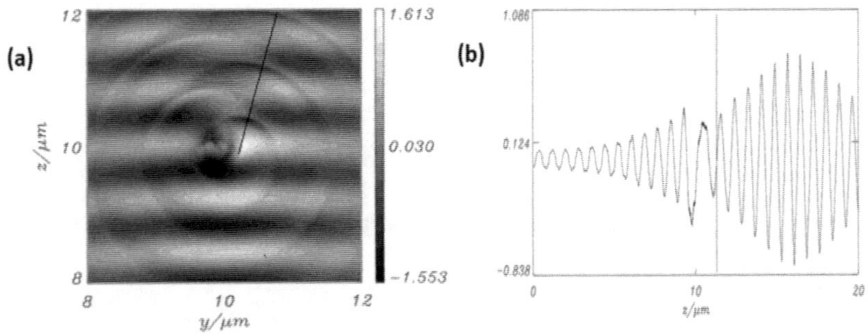

Figure 7. 2D PIC simulations for the experimental conditions, taking care of all non-linear processes for SHG (non restricted to parametric processes). (a) the 2D spatial distribution of the electric field $E_y(y,z)$; (b) the distribution of the electric field along the z-axis.

5. Conclusion

We demonstrated that X-ray spectroscopy in femto-second laser plasmas leads to a better understanding of nonlinear physical processes, such as TPD, PDI, SHG, in these plasmas. No less important is that it can be also used in a laboratory modelling of nonlinear physical processes in astrophysical objects because in both situations the ratio of the energy density of the turbulent field to the thermal energy density is of the same order. This non-perturbing method opens up an avenue for diagnosing these processes, measuring their thresholds and testing theories. The nonlinear process responsible of the PDI has been totally identified by X-ray spectroscopic diagnostic. The PIC simulations supported the discovery of the ion acoustic turbulence and the conversion of the short, intense laser light into the second harmonic generation.

Acknowledgments

We thank the Vulcan (CRL, UK) and J-KAREN and JLITE-X (KPSI, Japan) technical preparation teams for their support during the experiments. The research leading to these results has received funding from the Science and Technology Facilities Council, and the Engineering and Physical Science Research Council (Grant No.EP/J003832/1) of the United Kingdom. This work was partly supported by the Funding Program for Next Generation World-Leading Researchers (NEXT Program) from JSPS, a Grant-in-Aid for Scientific Research (A) No. 26247100 by JSPS, and the Consortium for Photon Science and Technology (CPhoST) program by MEXT. The work is partly supported by Russian Foundation for Basic Research via grants #15-32-21121 and #14-22-02089 and RAS Presidium Program for Basic Research #13. A.A. gratefully acknowledges the computing time granted by NIC and provided on the supercomputer JUROPA at JSC.

References

[1] Oks E, Dalimier E, Faenov A Ya et al. 2014 *J. Phys. B: At. Mol. Opt. Phys. Fast Track Communications* **47** 221001
[2] Oks E, Dalimier E, Faenov A Ya et al. 2015 *Optics Express* **23** 31991
[3] Oks E 1995 *Plasma Spectroscopy: The Influence of Microwave and Laser Fields (Springer Series on Atoms and Plasmas)* vol 9 (Berlin Springer)
[4] Oks E, Bödeeker S and Kunze H-J 1991 *Phys. Rev.* A **44** 8338
[5] Lünow W 1968 *Plasma Phys.* **44** 879
[6] Guerin S, Mora P, Adam JC, Laval G 1996 *Phys. Plasmas* **3** 2693
[7] Pukhov A 2003 *Rep. Prog. Phys.* **66** 47, Mourou GA, Tajima T, Bulanov SV 2006 *Rev. Modern Phys.* **78** 309, Sauvan P et al. 2009 *J. Phys. B: At. Mol. Opt. Phys.* **42** 195501
[8] Chung HK, Chen MH, Morgan WL, Ralchenko Yu, Lee RW 2005 *High Energy Density Physics* **1** 3
[9] Blochinzew D I 1933 *Phys. Z. Sow. Union* **4** 501
[10] Oks E 1984 *Sov. Phys. Doklady* **29** 224

Excitation of atoms and ions in plasmas by ultra-short electromagnetic pulses

V A Astapenko[1], S V Sakhno[1] , S Yu Svita[1] and V S Lisitsa[1,2,3]

[1]Moscow Institute of Physics and Technology, Dolgoprudnyi, Russia
[2]Russian Research Center 'Kurchatov Institute', Moscow, Russia
[3]National Research Nuclear University MEPhI (Moscow Engineering Physics Institute), Moscow, Russia

vlisitsa@yandex.ru

Abstract. The problem of atoms and ions diagnostics in rarefied and dense plasmas by ultra-short laser pulses (USP) is under consideration. The application of USP provides: 1) excitation from ground states due to their carrier frequency high enough, 2) penetration into optically dense media due to short pulses duration. The excitation from ground atomic states increases sharply populations of excited atomic states in contrast with standard laser induced fluorescence spectroscopy based on radiative transitions between excited atomic states. New broadening parameter in radiation absorption, namely inverse pulse duration time $1/\tau$ appears in addition to standard line-shape width in the profile $G(\omega)$. The Lyman-beta absorption spectra for USP are calculated for Holtsmark static broadening mechanism. Excitation of highly charged H-like ions in hot plasmas is described by both Gaussian shapes for Doppler broadening and pulse spectrum resulting in analytical absorption line-shape. USP penetration into optically thick media and corresponding excitation probability are calculated. It is shown a great effect of USP duration on excitation probabilities in optically thick media. The typical situations for plasma diagnostics by USP are discussed in details.

1. Introduction

The standard line broadening theory deals with the investigations of emitted or absorbed atomic/ions spectra on the frequency detuning from unperturbed frequency of specific atomic transition. So the line shapes are described by their dependences on the dimensionless ratio of frequency shift to the typical line widths depending on specific broadening mechanism (Doppler, Stark, collisions, etc). There are several problems in investigations of atomic spectra under plasma conditions. The first one is connected with small populations of excited atomic levels resulting in difficulties of registration signals of laser induced fluorescence (LIF). Really the most laser experiments deal with local excitations of atomic system from one to another excited atomic energy states where the transition frequency belongs to visible spectral range. At the same time the population difference between excited states are not so different to increase the difference sharply under the action of laser pulse. So the fluorescence signal is blended strongly by plasma background radiation. The second problem is connected with penetration of laser radiation inside optically dense medium when one would try to excite atomic systems from ground states.

Both problems pointed above can be solved by application of ultrashort laser pulses for plasma diagnostics by laser radiation. Really modern technology of ultrashort pulses generations provides the movement of carrier laser frequency in VUV and even soft X-ray spectral range, see for example reference [1] where pulse duration is of order of 67 as for the carrier frequency near 200 eV. It opens possibilities to excite atomic systems not only from excited but also from ground atomic states. As far as the populations of ground states are 3-4 order of magnitude larger than excited states that means a sharp increase of atomic states populations by their excitations from ground states. The second problem of large optical thickness for radiative transitions from ground states is also simplified by application of ultrashort laser pulses. Such simplification is achieved due to broad spectra of the ultrashort pulses which make it possible to avoid strong absorption in the far wings of the broadening spectral lines. Both problems will be considered in details in present paper.

So in the case of ultrashort pulses the new parameter appears in the absorption line spectra connected with the pulse duration τ in addition with standard broadening parameters connected with standard line shapes. The connection with the new parameters follows from the general expression for the absorption probability of the ultrashort laser pulse given by the equation [2]:

$$W(\tau)=\frac{c}{4\pi^2}\int_0^\infty \sigma(\omega')\frac{|E(\omega',\omega,\tau)|^2}{\hbar\omega'}d\omega', \tag{1}$$

here absorption cross section is given by the formula

$$\sigma(\omega')=\frac{2\pi^2 e^2}{mc}f_0\,G(\omega'),$$

$G(\omega')$ is standard line shape of a spectral line, f_0 is oscillator strength, $E(\omega',\omega,\tau)$ is the Fourier transform of USP electric field which for Gaussian envelope takes the form

$$|E(\omega')|^2 \approx \frac{\pi}{2}\tau^2 E_0^2 \exp\{-(\omega-\omega')^2\tau^2\}, \tag{2}$$

here ω, τ are the carrier frequency and the duration of an electromagnetic pulse, E_0 is the amplitude of the electric field in a pulse.

2. Transition probabilities from ground atomic states for different broadening mechanisms your paper

We will consider below the most famous mechanisms of spectral line broadening by plasmas, namely Holtsmark and Doppler broadening.

2.1. Holtsmark static broadening

Holtsmark broadening of spectral lines is due to the Stark effect on the atomic structure under the action of static ion plasma microfield. It typical value for the linear Stark effect is equal to the broadening by the averaged value of electric field determined by ion density by the well known relationship

$$\Delta = C_{st}\,N^{2/3}$$

where C_{st} is the Stark constant for a specific radiative transitions in neutral hydrogen or hydrogen-like ions, N being the ion density.

Well known line shape in static theory is described by Holtsmark distribution function $H(\beta)$, $\beta = F/F_0$.

$$H(\beta) = \frac{2}{\pi}\int_0^\infty x \cdot \sin(\beta x) \cdot \exp\left(-x^{3/2}\right) dx$$

Substituting the Holtsmark line shape into general equation (1) we obtain the absorption line shape for USP in Holtsmark plasma microfield:

$$W_{Hol} \cong \frac{\sqrt{\pi}}{4} f_0 E_0^2 \frac{\tau}{\omega_0 \Delta} J_{Hol}\left(\Delta\tau, \frac{\omega - \omega_0}{\Delta}\right), \quad (3)$$

$$J_{Hol}(\alpha, \beta) \cong \int_0^\infty x \exp\left(-x^{3/2} - x^2/4\alpha^2\right)\left[x \frac{\cos(\beta x)}{2\alpha^2} + \beta \sin(\beta x)\right] dx, \quad (4)$$

here the parameters are equal to

$$\alpha = \Delta\tau, \ \beta = (\omega - \omega_0)/\Delta$$

The general absorption line shapes for different pulse durations is presented on figure1 for Ly-beta line.

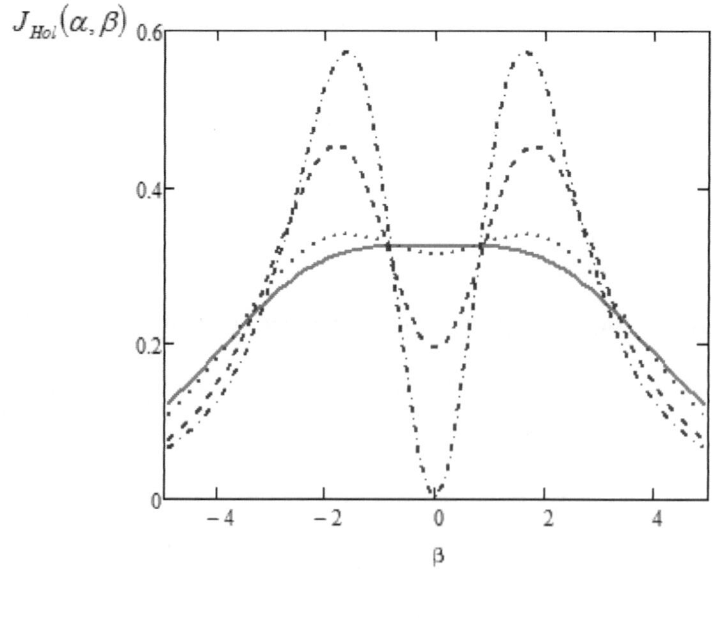

Figure 1. Ly_β excitation probability vs dimensionless frequency shifts β for different values of pulse duration parameter $\alpha = \Delta\tau$: solid line – $\alpha = 0.4$, dotted line – $\alpha = 0.5$, dashed line – $\alpha = 1$, dotted-dashed line $\alpha = 10$.

One can see that the well-known dips in the line centers disappear with the decrease of pulse duration.

The specific example of the Hydrogen Ly-beta line shape in the plasma with standard parameters $N_e=10^{18}$ cm^{-3} and electron temperature $T_e=1$ eV is presented on figure 2 for different pulse durations (cycles number N).

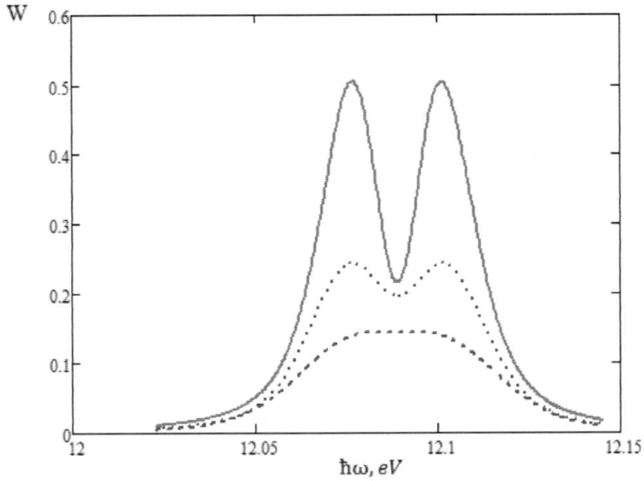

Figure 2. Excitation probability spectrum for various pulse duration: solid line – N=500 cycles, dotted line – N=300 cycles, dashed line – N=200 cycles of carrier frequency.

The effect of pulse duration on the absorption probabilities is shown on figure 3 where the probability W is presented as a function of pulse duration. It is clear that for the long pulse durations the probability follows to the linear dependence on pulse duration corresponding to the standard representation for the absorption per unit time.

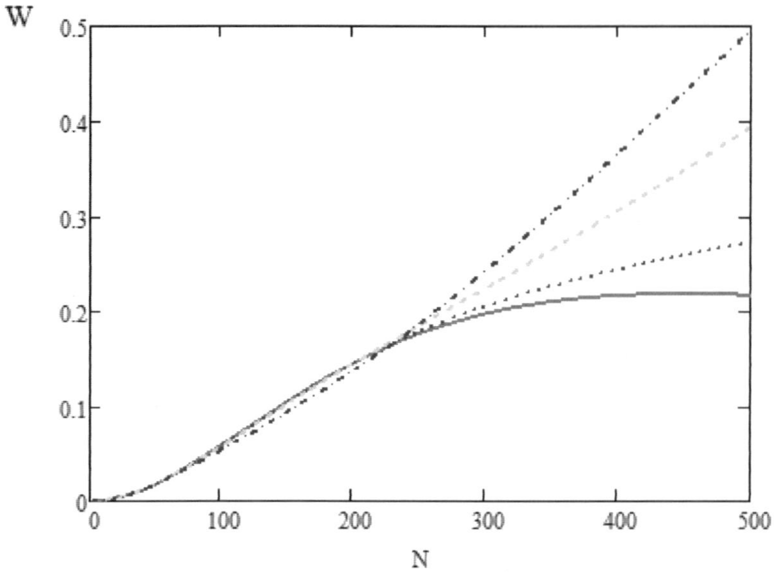

Figure 3. Dependence of Ly_β excitation probability on pulse duration (cycles number N) for different detuning of carrier frequency. Solid line – β=0, dotted line – β=0.5, dashed line – β=1, dotted-dashed line – β=2.

2.2. Doppler broadening

Doppler broadening mechanism is typical for highly charged ions in high temperature plasmas. The line shapes of H-like ions in their radiation transitions to the ground states depend also on the thin structure of H-like ions satisfying following relationships:

$$\Delta_{fs} \approx \frac{\alpha^2 Z^4}{4n^3}, \tag{4}$$

where $\alpha \cong 1/137$ is the fine structure constant, n is the principal quantum number of the upper level.

The Doppler broadening is of specific interest for USP absorption shapes due to the identical dependencies between Fourier components of Doppler line shape and Gaussian line shape of the USP electric field. It results in analytical expression for the absorption line shape following from the general equation (1) [3]:

$$W_{nj} \cong \frac{\pi}{8} f_{nj} \frac{E_0^2}{\omega_{nj}\,\Delta\omega_D^2} F(y,\rho_{nj}),\quad F(y,\rho) = \frac{y^2}{\sqrt{1+y^2}} \exp\left\{-\rho^2 \frac{y^2}{1+y^2}\right\} \tag{5}$$

Where the following parameters are introduced:

$$y = \sqrt{2}\,\Delta\omega_D \tau,\quad \rho_{nj} = \frac{\omega_c - \omega_{nj}}{\sqrt{2}\,\Delta\omega_D} \tag{6}$$

here ω_{nj} is the own frequency of the transition $1s \rightarrow np_j$.

IOP Conf. Series: Journal of Physics: Conf. Series **810** (2017) 012005 doi:10.1088/1742-6596/810/1/012005

It is obvious that the total probability of excitation of the np-state is

$$W_n = W_{n,1/2} + W_{n,3/2}. \tag{7}$$

Hereafter we will analyze the probability of excitation of a hydrogen-like ion normalized to the squared amplitude of the electric field strength E_0^2 as a function of the duration and the carrier frequency of a pulse: $\tilde{W} = W/E_0^2$.

The absorption spectra of H-like ion are shown on figure 4 with account of thin structure. It follows that the dip between fine structure components disappeared with the decrease of pulse duration.

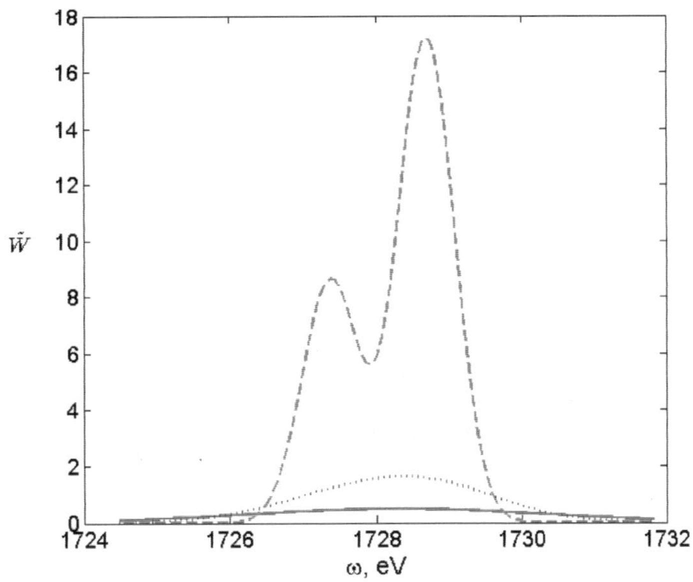

Figure 4. Normalized spectrum of excitation of the transition in a hydrogen-like ion with the nucleus charge Z = 13 by pulses of different durations: solid curve - τ = 0.24 fs, dotted curve - τ = 0.48 fs, dashed curve - τ = 2.4 fs.

The effect of pulse duration on absorption probabilities is shown on figure 5 for different ion charges for carrier frequency equal to: $\omega = (\omega_{j=1/2} + \omega_{j=3/2})/2$. It is seen a strong dependence of the function $\tilde{W}(\tau)$ on the ion charge.

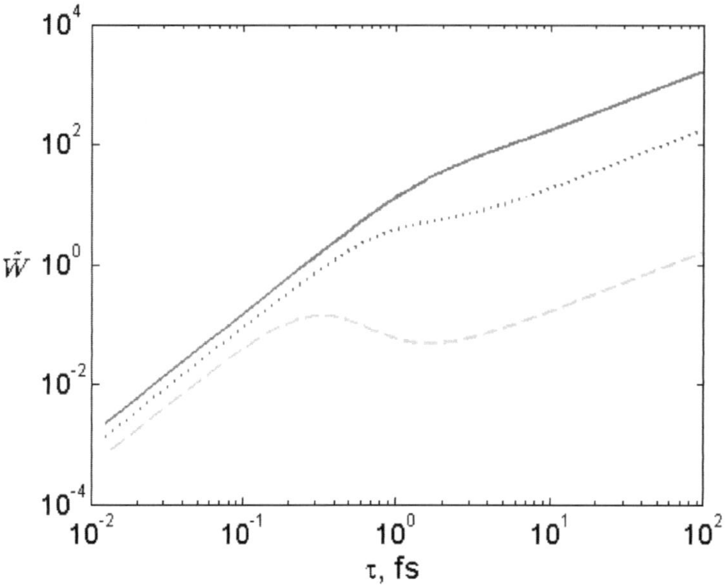

Figure 5. Dependence of the normalized probability of excitation of a hydrogen-like ion at the transition on the pulse duration for different nucleus charges: solid curve – Z = 10, dotted curve - Z = 13, dashed curve - Z = 18.

3. Absorption probabilities for USP in optically dense media

Let us consider the effect of pulse duration on the UPS penetration into optically thick media. The optical thickness is described by the standard equation:

$$\Lambda(L) \approx \kappa_0 L \approx 2\pi N_0 \lambda\!\!\!^{-2} L \frac{A}{\Gamma} \tag{8}$$

here Δ is line-width ($\Delta \approx \mathrm{v}\omega_c/c$ – Doppler; $\Delta \approx N \mathrm{v} \sigma$ – Lorenz), ω_c is the central frequency of transition, ω is carrier frequency of the pulse, $\beta = (w - \omega_c)/\Delta$, w is current frequency, $\delta = (\omega_c - \omega)/\Delta$, $\alpha = \Delta\tau$.

By substitution of the optical thickness (8) into general expression for the reduction of the pulse intensity in the dense media one can arrive to the expression of the electric field intensity in the dense media as following:

$$E^2(\Lambda, \alpha, \beta, \delta) = \alpha^2 \exp\left\{-(\beta + \delta)^2 \alpha^2 - \Lambda J_{L,D}(\beta)\right\}, \tag{9}$$

here

$$J_D(\beta) = \exp(-\beta^2/2)/\sqrt{2\pi}, \; J_L(\beta) = 1/\left[\pi(1 + \beta^2)\right].$$

The general results for the spectra of field squared in the dense media are illustrated by figure 6.

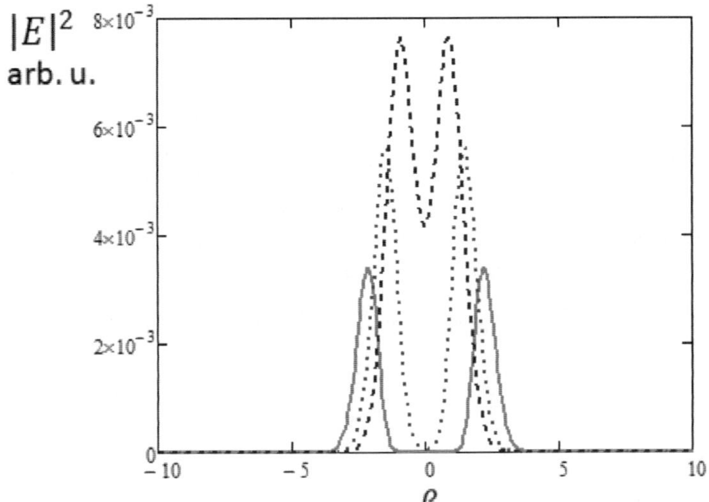

Figure 6. Spectra of field squared inside dense medium for Lorenz lineshape for α=1, δ=0 and different optical thickness Λ: solid line – Λ=100, dotted line – Λ=30, dashed line – Λ=10.

It is simple to obtain general expression for the USP absorption probability as a function of media optical depth following equation (8) and general expression (1). The result has the form:

$$W(\Lambda,\alpha,\delta)\propto\alpha^2\int\exp\{-(\beta+\delta)^2\alpha^2-\Lambda J_{L,D}(\beta)\}J_{L,D}(\beta)d\beta \qquad (10)$$

The dependence of the USP absorption probability on the medium optical depths is presented on figure7 for different pulse durations.

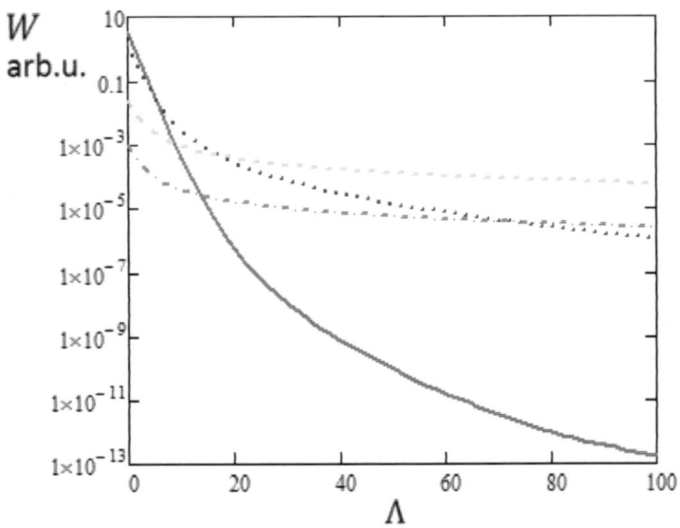

Figure 7. Excitation probability vs optical thickness for δ=0 and different pulse duration parameter α=Δτ: solid line – α = 2, dotted line – α = 1, dotted-dashed line – α = 0.02, dashed line – α = 0.1.

It is clear seen a strong effect of pulse duration on the absorption probability due to its penetration into the media.

4. Conclusion

General scheme of USP absorption spectroscopy looks as following: (1) excitation from ground atomic states by USP can be considered as ground state population transfer to the excited state (practically redefinition of ground state); (2) typical situation: large optical thickness for radiative transitions from ground states and transparency for ones between excited states; (3) the excitation from ground states by USP is accompanied by electron collisions excitation transfer between excited atomic states, the effect providing increase of radiation emission in optically thin spectral lines corresponding to transitions between excited states. Combined actions of USP and electron collisions results in the transfer of effective ground state to the excited atomic levels resulting in sharp increase of latter populations. It is especially clear for atomic systems in thermodynamic equilibrium. Here the Boltzman exponential factor and, as a result, levels populations increase sharply after transfer of the ground energy level to the excited one under the action of USP.

References

[1] Zhao K, Zhang Q, Chini M, Wu Y, Wang X, Chang Z. 2012 *Optics Lett.* **37** 3891
[2] Astapenko V A 2010 *Phys. Lett. A* **374** 1585
[3] Astapenko V A and Svita S.Yu. 2015 *Rus. Phys. J.* **57** 1503

Latest advances in the semiclassical theory of the Stark broadening of spectral lines in plasmas

Eugene Oks

Physics Department, 206 Allison Lab, Auburn University, Auburn, AL 36849, USA

goks@physics.auburn.edu

Abstract. Spectral lines continue to be used as an important practical tool for diagnosing various laboratory and astrophysical plasmas. As experiments move to very high electron densities of plasmas or to strongly magnetized plasmas, new physical phenomena have to be taken into account while calculating Stark broadening of hydrogen spectral lines. As for spectral lines having dipole-forbidden components (such as, He-, He-like-, Li-, Li-like-lines), it is shown that they provide the best way for diagnosing Langmuir solitons in plasmas.

1. Stark broadening of hydrogen lines in plasmas of densities up to or more than $N_e \sim 10^{20}$ cm^{-3}

Kielkopf & Allard (KA) recently presented benchmark measurements of the Full Width at Half Maximum (FWHM) of the H_α line, reaching the electron densities N_e by two orders of magnitude greater than the corresponding previous benchmark experiments, namely up to $N_e = 1.4 \times 10^{20}$ cm^{-3} [1]. Their experiment was performed in a laser-produced hydrogen plasma of the average T = 28000 K. Their experimental FWHM are presented by dots in figure 1.

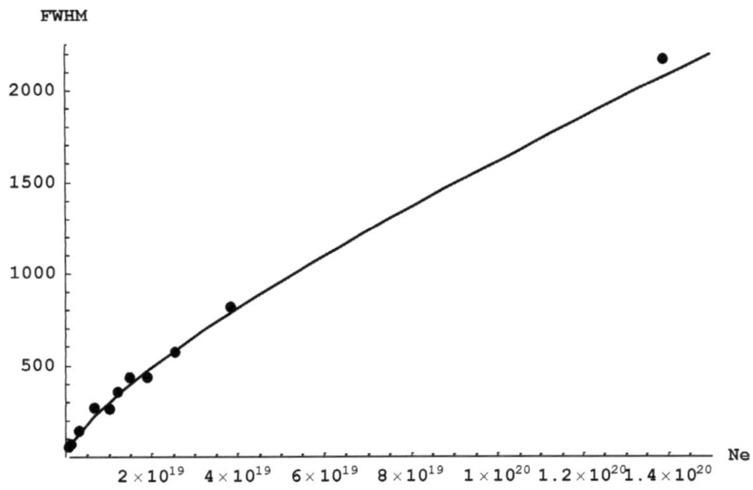

Figure 1. Comparison of the experimental FWHM of the H_α line from Kielkopf-Allard experiment [1] (dots) with the corresponding FWHM yielded by our analytical theory from paper [2] (solid line). The FWHM is measured in Angstrom, while N_e – in cm^{-3}.

At this range of N_e, no theoretical calculations of the FWHM of the H_α line existed, except the calculation by KA; however, the latter is incomplete because of the neglect of the contribution by plasma electrons (as noted by KA). In paper of 2015 [2] we developed a consistent analytical theory that is relevant to the range of the electron densities reached in KA experiment. At this range of N_e, a new factor becomes significant – the factor never taken into account in any previous simulations or analytical theories. This new factor is a rising contribution of the electrostatic plasma turbulence at the *thermal* level of its energy density. This contribution becomes comparable to the corresponding contribution by electron and ion microfields at this range of N_e.

According to Bohm and Pines [3], the number of collective degrees of freedom in a unit volume of a plasma is $N_{coll} = 1/(6\pi^2 r_D^3)$, where r_D is the Debye radius. Therefore the energy density of the oscillatory electric fields at the thermal level is $E_t^2/(8\pi) = N_{coll}T/2$, so that $E_t^2 = 16\pi^{1/2}e^3N_e^{3/2}/(3T^{1/2})$. At the absence of the magnetic field, the thermal energy density of the collective degrees of freedom is distributed in equal parts between two branches: $F_0^2 = E_0^2 = E_{tot}^2/2$, where F_0, E_0, and E_{tot} are the root-mean-square thermal electric fields of the ion acoustic turbulence, the Langmuir turbulence, and the total turbulence, respectively.

The contribution of the thermal ion acoustic turbulence was taken into account by calculating the convolution of Sholin-Oks' distribution of the turbulent field $\mathbf{F}_t$ [4] with the APEX distribution of the ion microfield $W_i(u)$. The thermal Langmuir turbulence was allowed for as an additional contribution to the dynamical Stark width according to Oks-Sholin paper [5]. (More details can be found in [2].) The resulting FWHM of the H_α line yielded by this analytical theory is shown in figure 1 by the solid line. It is seen that the agreement is very good. Even at the highest density point of KA experiment [1] ($N_e = 1.39\times10^{20}$ cm^{-3}, T = 34486 K), the theoretical FWHM differs by just 4.5% from the most probable experimental value and is well within the experimental error margin. This theory can be also used for calculating Stark profiles and the FWHM of other hydrogen spectral lines.

2. Influence of magnetic-field-caused modifications of trajectories of plasma electrons on spectral line shapes: application to magnetic fusion and white dwarfs

A general framework for calculating shapes of hydrogen (or deuterium) spectral lines in strongly-magnetized plasmas with the allowance for spiraling trajectories of perturbing electrons has been developed in paper [6] of 2016. It has been shown that in this situation the first order term $\Phi^{(1)}(B)$ of the Dyson expansion of the electron broadening operator does not vanish – in distinction to the case of rectilinear trajectories.

For the atomic electron to experience the spiraling nature of the trajectories of perturbing electrons, it requires $\rho_{De}/r_{Bp} > 1$, where ρ_{De} is the electron Debye radius. It is equivalent to $B > B_{cr} = c(4m_eN_e)^{1/2}$, B_{cr}(Tesla) $= 1.81\times10^{-7}[N_e(\text{cm}^{-3})]^{1/2}$. This condition is fulfilled modern tokamaks, in many DA white dwarfs, as well as in capacitor-produced plasmas and in plasmas produced by high-intensity lasers.

We consider the situation where the temperature T of the radiators satisfies the condition $T \ll (11.12 \text{ keV}/n)(M/M_H)^2$, where M is the radiator mass, M_H is the mass of hydrogen atoms, n is the principal quantum number of the energy levels, from which the spectral line originates. For plasmas of tokamaks and DA white dwarfs this condition is fulfilled for the head lines (i.e., the most intense lines) of spectral series. Under this condition, the Lorentz field effects can be disregarded compared to the "pure" magnetic field effects.

The first order of the nondiagonal elements of the electron broadening operator $\Phi_{\alpha\alpha'}{}^{(1)}$ can be represented in the form (the diagonal elements of $\Phi^{(1)}$ vanish):

$$\Phi_{\alpha\alpha'}{}^{(1)} = -\,i\, X_{\alpha\alpha'}\,(e^2/\hbar)\rho_{De}N_e \int_0^\infty dv_p W_p(v_p) \int_{-\infty}^{\infty} dv_z W_z(v_z)\, f_0(u), \qquad u = B/B_{cr.} \qquad (2.1)$$

The universal function f(u) in equation (2.1) that controls the phenomenon under consideration is expressed in [6] analytically via elliptic integrals. Figure 2 presents its plot.

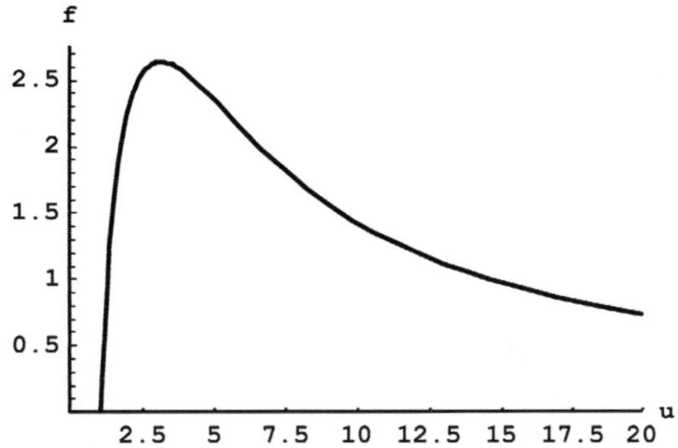

Figure 2. The universal function f, controlling the effect of spiraling trajectories of plasma electrons on hydrogen/deuterium spectral lines, versus the scaled magnetic field $u = B/B_{cr}$, $B_{cr} = c(4m_eN_e)^{1/2}$.

Examples related to several tokamaks, such as T-10 in Russia, EAST in China, and the future ITER in France, show the following. The primary effect of spiraling trajectories of plasma electrons is that the ratio of the intensity of the central peak in the Zeeman triplet to the intensity of any of the two lateral peaks increases by up to a factor of two. For DA white dwarfs this ratio increases by over 20%. The secondary effect – both for tokamaks and DA white dwarfs – is an additional shift of the lateral components of the Zeeman triplet. More details can be found in [6].

3. Role of Lorentz-Doppler Broadening of Hydrogen/Deuterium Spectral Lines in Magnetized Plasmas: Applications to Magnetic Fusion and Solar Physics

Under the condition $T > (11.12\ \mathrm{keV}/n)(M/M_H)^2$, which is the opposite to the situation considered in the previous section, the Lorentz field effects dominate over the "pure" Zeeman effect. This condition is fulfilled for highly-excited hydrogen/deuterium lines observed in tokamaks ($n \sim 10$) and in the quiet Sun ($n \sim 30$). Also the Lorentz field exceeds the most probable ion microfield when the magnetic field B exceeds the following critical value: $B_c(\mathrm{Tesla}) = 4.69\mathrm{x}10^{-7}N_e^{2/3}/[T(K)]^{1/2}$. This condition is also met in tokamaks and in the quiet Sun. In this situation the broadening is controlled by the Lorenz(field)- and Doppler-effects, and the combination of these two effects cannot be taken into account via a convolution, as noted in paper [7]: they entangle in a more complicated way.

In papers [8] of 2013 and [9] of 2015 were obtained, in particular, analytical expressions for Lorentz(field)-Doppler profiles of hydrogen/deuterium lines – for arbitrary values of the scaled magnetic field b = (Lorentz shift)/Doppler shift. For observations along the magnetic field **B** these expressions were obtained for the first time. For observations perpendicular **B** these expressions were obtained for the first time for $b > 1$ and $b = 1$. For $b < 1$ we corrected errors in analytical results from paper [9].

For the lack of space we present here only one result from [8, 9]: for observation perpendicular to **B**, the width of the Lorentz-Doppler profiles is a *non-monotonic function of the magnetic field.* As b increases from zero, the width first decreases, then reaches a minimum at $b = 1$, and then increases.

This counterintuitive result is illustrated in figure 3. The narrowing effect is the most pronounced when the Lorentz-field shift is equal to the Doppler shift.

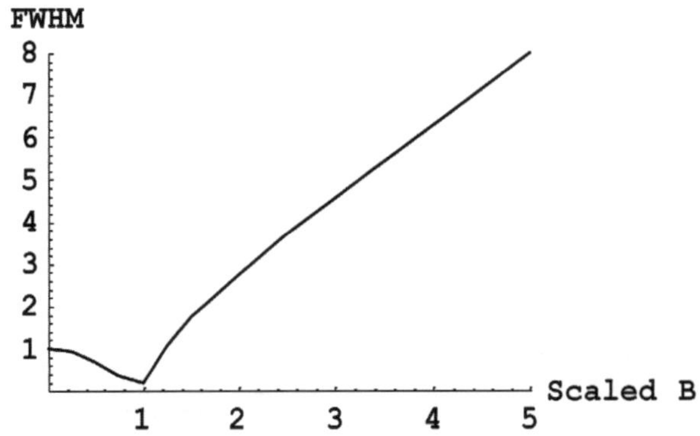

Figure 3. FWHM of the Ly-beta line observed perpendicular to the magnetic field **B** with a polarizer along **B**. The scaled magnetic field is (Lorentz shift)/Doppler shift. FWHM is in units of the Doppler FWHM.

4. Inhibition of the ion dynamical broadening by a strong magnetic field: questionable notions in the literature

In magnetic fusion machines, a strong magnetic field **B** partially inhibits the ion dynamical broadening compared to the B = 0 case. Physically this is because a relatively large Zeeman splitting $\omega_B/2$ diminishes the range of ion impact parameters ρ, for which the characteristic frequency of the variation of the ion field v_i/ρ exceeds $\omega_B/2$. This effect was described analytically in year 1994 by Derevianko and Oks [11] using an advanced Stark broadening formalism (called GT). Figure 4 shows the corresponding Stark width reducing factor for the Ly_β line versus B at $N_e=10^{14}$ cm^{-3} and T=5 eV. At B=0, the exact analytical solution for Ly_β was given by Derevianko and Oks in paper [12]

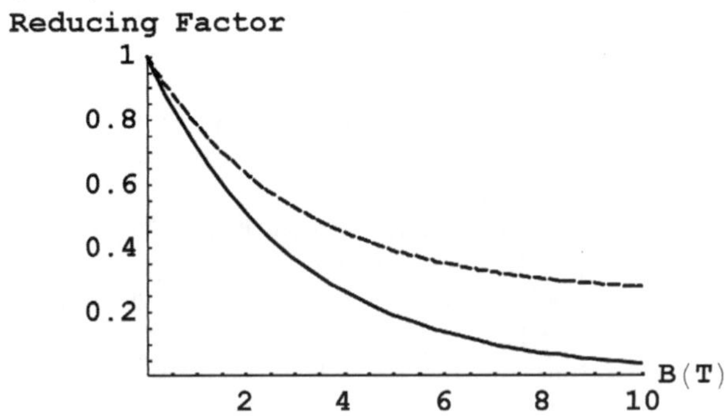

Figure 4. Stark width reducing factor for the Ly_β line versus the magnetic field B at $N_e=10^{14}$ cm^{-3} and T=5 eV: for σ-components (solid line) and for the π-component (dotted line).

15 years after Derevianko-Oks paper [11], Rosato et al published paper [13] describing the same effect using a less advanced Stark broadening formalism (called ST). Essentially Rosato et al repeated Sholin-Demura-Lisitsa's version of the ST [14] with the substitution of the magnetic splitting ω_B of hydrogen/deuterium energy levels instead of the electric splitting ω_F.

In our paper [15] and in our review [16] it was shown that the analytical results by Rosato et al [13] yield a very significant inaccuracy – up to orders of magnitude. In a later paper [17], Rosato et al made the second attempt to show that for magnetized plasmas the ST is allegedly better than the GT. The results from Rosato et al 2012 [17] cannot be trusted (just like those from Rosato et al 2009 [13]) for the following reasons.

First and foremost, it is known since the publication of paper by Sholin-Demura-Lisitsa in 1973 [14] that the impact shift of the spectral components of hydrogen/deuterium lines of the Ly series is proportional to the following combination of the parabolic quantum numbers $n(n_1-n_2)$, so that for the σ-components of the Ly_α line $n_1-n_2=0$ and this shift vanishes, which is a rigorous analytical result for rectilinear trajectories of the perturbers. Despite this, computations of Rosato et al 2012 [17] yielded a non-zero impact shift of the σ-components of the Ly_α line (for rectilinear trajectories of the perturbers), which is obviously incorrect.

Second, it is very important that for the magnetic fields typical for the edge plasmas of tokamaks, used as an example by Rosato et al both in 2009 and in 2012, the so-called nonadiabatic contribution to the broadening becomes much smaller than the *primary, adiabatic* contribution (i.e., the contribution of the component of the ion microfield parallel to the magnetic field). The primary, adiabatic contribution is calculated in the GT *exactly* (not by perturbation theory), while in the ST by Rosato et al – only by the order of magnitude (using the 2nd order of the perturbation theory). This is an undisputable fact. As a result, the primary, adiabatic contribution was significantly overestimated by Rosato et al both in 2009 [13] and in 2012 [17]. Some details are presented below.

In our paper [15] and in our review [16], we calculated the ratio of the adiabatic broadening cross-section by Rosato et al [13] to the adiabatic broadening cross-section by the GT: $\kappa = \sigma_{aRos}/\sigma_{aGT}$. While calculating σ_{aRos}, we had set to unity a certain coefficient of the order of unity. After Rosato et al pointed this out in 2012 [17], we have calculated the ratio of the adiabatic broadening cross-sections $\kappa = \sigma_{aRos}/\sigma_{aGT}$ more accurately and found the following [6] (below the components of a particular line are identified by the parabolic quantum numbers: $(n_1n_2m)_\alpha - (n_1n_2m)_\beta$; the ratio κ is calculated at T = 4 eV and $N_e = 10^{13}$ cm^{-3}).

For the Paschen-alpha line, specifically its component (102) – (101), which is one of the two most intense σ-components: κ = 14. (Here and below the components are labeled by the parabolic quantum numbers $(n_1\, n_2\, m)$.) For the Paschen-beta line, specifically its component (121) – (020), which is the σ-component closest to the line center: κ = 22. For the Balmer-alpha line, specifically the component (101) – (100), which is one of the two most intense σ-components: κ = 9. Thus, more accurate calculations lead qualitatively to the same conclusion: *Rosato et al overestimated the primary, adiabatic contribution to the dynamical Stark broadening by ions in magnetic fusion plasmas by up to an order of magnitude*.

After producing the above erroneous results for the primary, adiabatic contribution, Rosato et al 2012 [17] focused at the much smaller nonadiabatic contribution. After attempting to numerically calculate the nonadibatic contribution by the GT, they claimed that it violates unitarity at small impact parameters. However, the violation of the unitarity at small impact parameters for both the adiabatic and nonadiabatic contributions is actually the characteristic feature of the ST by Rosato et al for *all* hydrogen/deuterium lines (resulting, after some rather arbitrary cutoff, in the *inability of the ST from Rosato et al to calculate the adiabatic contribution better than by the order of magnitude*). As for the GT, Touma et al [18] rigorously showed analytically that for the overwhelming majority of hydrogen/deuterium spectral lines, the *nonadiabatic contribution does not violate unitarity* at small impact parameters, as illustrated also in figures A.1 – A.3 from [6].

As for the Ly_α line, where the secondary, nonadiabatic contribution of GT does require the unitarity cutoff at small impact parameters ρ (which was mentioned by us already in 1995 [19]), the *width function* for the σ-components $A_-(\rho)$ is *proportional to ρ at small ρ*, which is a *rigorous analytical result*: $A_-(\rho) = \text{const}\ \rho$. In distinction, the corresponding *attempted "GT computations" by Rosato et al 2012 for the Ly_α line yielded that $A_-(0)$ is infinite (?), which is again obviously incorrect.*

Thus, all of the above shows that the results from Rosato et al 2012 [17] are incorrect. One of the possible reasons was actually noted by Rosato et al 2012 [17]: they wrote that "*the broadening function used in the GT involves a multiple integral of strongly oscillating functions, not suitable for fast numerical evaluation*". So, their computations failed to properly handle the multiple integral of strongly oscillating functions. This by itself already shows that the results by Rosato et al 2012 [17] cannot be trusted – just like the results from Rosato et al 2009 [13].

5. Manifestations of Langmuir solitons in satellites of dipole-forbidden spectral lines of helium, lithium, and of the corresponding ions

In our paper [20] there was calculated analytically the shape of satellites of dipole-forbidden lines in a spectrum *spatially-integrated* through a Langmuir soliton (or through a sequence of Langmuir solitons separated by a distance L). The Langmuir solitons have the following form in space according to book [21]: $F(x, t) = E(x) \cos \omega_p t$, $E(x) = E_0/\mathrm{ch}(x/\lambda)$, $\lambda << L$. Here $\omega_p = \omega_{pe} - 3T_e/(2m_e\omega_{pe}a^2)$, where ω_{pe} is the plasma electron frequency. For diagnosing solitons it is necessary not only to find experimentally an electric field oscillating at the frequency $\sim \omega_{pe}$, but also to make sure that the spatial distribution of the amplitude corresponds to the formfactor E(x).

Under any quasimonochromatic electric field, dipole-forbidden spectral lines of helium, lithium, and of the corresponding ions can exhibit satellites. In cases of a relatively large separation between the forbidden and allowed lines, or a relatively weak amplitudes, the intensities of the far (+) and near (–) satellites are $S_\pm = a_\pm E^2(x)$, where $a_\pm$ does not depend on x.

The spatially-integrated profile of the satellites, calculated with the allowance for the quadratic shift of their frequencies $bE^2(x)$, where b does not depend on x, has the form:

$$S_\pm(\Delta\omega) = (1/L)\int_{L/2}^{L/2} dx\, a_\pm E^2(x)\, \delta[f(x)], \qquad f(x) = \Delta\omega \pm \omega_p - bE^2(x). \tag{5.1}$$

After calculating the integral in Eq. (5.1), in paper [20] it was obtained:

$$S_\pm(\Delta\omega) = (\lambda/L)[a_\pm/(2|b|)]/[1 - (\Delta\omega \pm \omega_p)/(bE_0^2)]^{1/2}. \tag{5.2}$$

The profiles $S_\pm(\Delta\omega)$, formally calculated by Eq. (5.2), have singularities at $\Delta\omega \pm \omega_p = bE_0^2$. From the physical point of view, for obtaining a finite result at $\Delta\omega \pm \omega_p = bE_0^2$, it is necessary to replace the δ-function in Eq. (5.1) by a real profile, e.g., by the Lorentz profile representing the dynamical Stark broadening by electrons and by some ions:

$$\delta[f(x_0)] \rightarrow (1/\pi)\gamma/\{\gamma^2 + [f(x_0)]^2\}, \tag{5.3}$$

where x_0 is the root of the equation $f(x) = 0$. Using the Taylor expansion of f(x) at $x = x_0$ and taking into account the first derivative df/dx vanishes at $x = x_0$, the right side of Eq. (5.3) can be approximated as $(1/\pi)\gamma/\{\gamma^2 + [d^2f(x_0)/dx^2]^2(x - x_0)^4\}$. Then using the integral

$$(1/\pi)\int_{-\infty}^{\infty} dz\, \gamma/(\gamma^2 + z^4) = 1/(2\gamma)^{1/2}, \tag{5.4}$$

we find

$$S_{soliton}^{max} = S_\pm(\Delta\omega = \pm\, \omega_p - bE_0^2) = [\lambda a_\pm/(2^{1/2}\pi L b^{1/2})]E_0^2/\gamma^{1/2}, \tag{5.5}$$

where $S_{soliton}^{max}$ is the peak intensity of the satellites in the case of solitons.

In the case of non-solitonic Langmuir waves, the peak intensity of the satellites is $S_{nonsoliton}^{max} = a_\pm E_0^2/(\pi\gamma)$. So, for the ratio of the peak intensity of the satellite under solitons to the peak intensity of the same satellite under nonsolitons we get

$$S_{soliton}^{max}/S_{nonsoliton}^{max} = [\lambda/(2^{1/2}L)][\gamma/(bE_0^2)]^{1/2} >> 1 \tag{5.6}$$

for the typical situation where $\gamma >> bE_0^2$, i.e., where the dynamical Stark width γ is much greater than the quadratic shift bE_0^2 of the satellites.

Thus, in the case of Langmuir solitons, the peak intensity of the satellites can be significantly enhanced – by orders of magnitude – compared to the case of non-solitonic Langmuir waves. This distinctive feature of satellites under Langmuir solitons allows distinguishing them from non-solitonic Langmuir waves.

We note that for a more general case, where both the dynamical Stark broadening and the Doppler broadening are taken into account, the δ-function in Eq. (5.1) should be substituted by the Voigt profile resulting in

$$S_{soliton}^{max}/S_{nonsoliton}^{max} \sim \Delta\omega_{Voigt}/(bE_0^2)]^{1/2} >> 1, \quad (5.7)$$

where $\Delta\omega_{Voigt}$ is the halfwidth of the Voigt profile. Again, in the typical situation where $\Delta\omega_{Voigt} >> bE_0^2$, the peak intensity of the satellites is significantly enhanced – by orders of magnitude – compared to the case of non-solitonic Langmuir waves.

Finally we mention that Hannachi et al [22] performed simulations for finding the effect of Langmuir solitons on the hydrogen Ly_α line. The effect was an additional broadening. However, even at the low electron density $N_e = 10^{14}$ cm^{-3}, the effect was very small compared to the Stark broadening by plasma microfields. Moreover, the additional broadening rapidly diminished with the increase of N_e, so that there would be practically no additional broadening at $N_e > 10^{15}$ cm^{-3}. Therefore, it seems that the results by Hannachi et al [22] could not be useful for the experimental diagnostics of Langmuir solitons.

References

[1] Kielkopf J F and Allard N F 2014 *J. Phys. B: At. Mol. Opt. Phys.* **47** 155701
[2] Oks E 2015 *J. Phys. B: At. Mol. Opt. Phys.* **49** 065701
[3] Bohm D and Pines D 1953 *Phys. Rev.* **92** 609
[4] Sholin G V and Oks E 1976 *Sov. Phys. Tech. Phys.*, 21, 144
[5] Oks E and Sholin G V 1975 *Sov. Phys. JETP* **41** 482
[6] Oks E 2016 *J. Quant. Spectr. Rad. Transfer* **171** 15
[7] Galushkin Yu I 1970 *Sov. Astron.* **14** 301
[8] Oks E 2013 *Intern. Rev. Atom. Mol. Phys.* **4** 105
[9] Oks E 2015 *J. Quant. Spectr. Rad. Transfer* **156** 24
[10] Breton C, De Michelis C, Finkental M, and Mattioli M 1980 *J. Phys. B* **13** 1703
[11] Derevianko A and Oks E 1994 *Phys. Rev. Letters* **73** 2079
[12] Derevianko A and Oks E 1998 *14th Int. Conf. Spectral Line Shapes*, AIP Conf. Proc. **467** 148
[13] Rosato J, Marandet Y, Capes H, Ferri S, Mosse C, Godbert-Mouret L, Koubiti M, and Stamm R 2009 *Phys. Rev. E* **79** 046408
[14] Sholin G V, Demura A V, and Lisitsa V S 1973 *Sov. Phys. JETP* **37** 1057
[15] Oks E 2010 *Intern. Rev. Atom. Mol. Phys.* **1** 169
[16] Oks E 2012 In: *Atomic Processes in Basic and Applied Physics*, Shevelko V & Tawara H edits. Springer Series on Atomic, Optical and Plasma Physics **68** (Springer, New York) 393
[17] Rosato J, Capes H, Godbert-Mouret L, Koubiti M, Marandet Y, and Stamm R 2012 *J. Phys. B: At. Mol. Opt. Phys.* **45** 165701
[18] Touma J, Oks E, Alexiou S, and Derevianko A 2000 *J. Quant. Spectr. Rad. Transfer* **65** 543
[19] Oks E, Derevianko A, and Ispolatov Ya 1995 *J. Quant. Spectr. Rad. Transfer* **54** 307
[20] Oks E *Sov. Phys. Doklady* 1984 **29** 224
[21] Kadomtsev B B 1982 *Collective Phenomena in Plasma* (Pergamon, Oxford)
[22] Hannachi I, Stamm R, Rosato J, and Marandet J 2016 *Europhys. Lett.* **114** 23002

IOP Conf. Series: Journal of Physics: Conf. Series **810** (2017) 012007 doi:10.1088/1742-6596/810/1/012007

Spectral line-shape study by cavity-enhanced complex refractive index spectroscopy

D. Lisak*, A. Cygan, S. Wójtewicz, P. Wcisło, M. Zaborowski, G. Kowzan, P. Masłowski, R. Ciuryło

Institute of Physics, Faculty of Physics, Astronomy and Informatics, Nicolaus Copernicus University in Toruń, Grudziadzka 5, 87-100 Toruń, Poland

E-mail: *dlisak@fizyka.umk.pl

Abstract. We present a unique experimental setup that enables measurement of high-resolution complex refractive index spectra of gas sample using several cavity-enhanced methods: frequency-stabilized cavity ring-down spectroscopy, cavity mode-width spectroscopy and one-dimensional cavity mode-dispersion spectroscopy. Presented examples of applications cover molecular spectral line-shape investigations, line intensities and unperturbed line center frequency measurements and high-precision detection of systematic instrumental errors of various spectroscopic methods.

1. Introduction

High-accuracy measurements of spectral line shapes are important in a growing number of fields including basic research, metrology and many gas-sensing applications. This is associated with development of new ultra-accurate spectroscopic techniques using ultra-stable and narrow laser systems. For very weak absorption lines, the cavity ring-down spectroscopy (CRDS) yields excellent performance and spectral line shapes with extremely high signal-to-noise ratios [1, 2] were demonstrated using frequency-stabilized version of CRDS (FS-CRDS) [3, 4]. The FS-CRDS is also one of the well-known and most accurate techniques that provide line-shape parameters with sub-percent combined standard uncertainties [5, 6, 7]. On the other hand, achieving 0.1% agreement between line intensities determined independently in different laboratories is very challenging [7] and this problem is still open. Apart from the determination of the physical conditions in a sample, the accuracy of most often used absorptive intensity-based spectroscopic techniques is limited by the linearity and repeatability of the detection system. Even the accurate FS-CRDS method may be prone to systematic distortions of exponential light decays caused e.g. by too low bandwidth and poor linearity of the detection system [8], frequency stability of the excited mode and multimode excitation [9], or incomplete light extinction in case of continuous-wave CRDS systems [10].

The exponential light decay measured in the CRDS is directly related to the spectral width of the cavity mode. This phenomenon was exploited in cavity mode-width spectroscopy (CMWS) [11, 12, 13] in which the absorption spectrum can be retrieved by precise measurements of half-widths of cavity modes. Analysis of the CRDS and CMWS methods revealed their complementarity in terms of achievable accuracies at different levels of intra-cavity absorption [11]. For low absorptions the best precision is expected with the CRDS technique, where ring-down decays are long and not affected by non-exponential distortions. In the opposite case of high absorption and short ring-down decays, the cavity modes become broad and, consequently, the precision obtained by the CMWS should be higher.

Measurement of dispersive shifts of cavity modes provides another possibility for very accurate quantitative spectroscopy. Since frequency is a physical quantity that can be measured with the highest accuracy, direct measurement of absolute or differential cavity modes frequencies to obtain a dispersive spectrum should eliminate potential problems associated with nonlinearities of a detection system as well as minimize systematic instrumental errors. The recently developed one-dimensional cavity mode-dispersion spectroscopy (1D-CMDS) [13] is so far the only method that provides one-dimensional spectra and contrary to any absorptive methods both axes in the 1D-CMDS spectrum can be linked to a primary frequency standard [14]. Since the 1D-CMDS is a light intensity-independent technique, it can be considered as a reference tool for validation of other spectroscopic techniques [15]. Typical cavity mode widths used in a high-sensitivity absorption spectroscopy are of the order of a few kHz. Therefore, precise measurement of cavity mode shapes, crucial for the CMWS and 1D-CMDS methods, requires the use of extremely narrow spectrally and finely tunable lasers.

Described above new cavity-enhanced spectroscopy techniques may be especially useful in applications that require the high accuracy of weak absorption measurements, e.g. gas spectroscopy for atmospheric monitoring and gas metrology, Doppler thermometry, as well as fundamental study of spectral line shapes. Moreover, direct comparison of spectra obtained with various cavity-enhanced techniques enables identification of potential instrumental errors at sub-percent level of accuracy required in many modern applications.

Here we present a unique experimental setup that enables measurements of absorptive and dispersive spectra at the same time and with the same gas sample by three independent cavity-enhanced techniques: FS-CRDS, CMWS and 1D-CMDS. Validation of the FS-CRDS method with respect to the 1D-CMDS is presented in a broad range of pressures. A complex-fit line-shape analysis [15] is used for simultaneously recorded absorptive and dispersive spectra. Moreover, absolute frequency of an unperturbed $^{12}C^{16}O$ transition from the near-infrared (3-0) band measured with an uncertainty five-fold lower than that of previously available data is presented [14]. Both frequency axes of dispersive spectrum were linked to the primary frequency standard. Three techniques (FS-CRDS, CMWS, 1D-CMDS) and various approaches to data analysis were used to estimate potential systematic instrumental errors.

2. Experimental setup

An experimental setup enabling measurement of absorptive and dispersive spectral line shapes simultaneously by the FS-CRDS, CMWS and 1D-CMDS methods is presented in Fig. 1. A continuous-wave ECDL laser is incorporated into a two-beams configuration scheme with a careful polarization separation of the probe and locking beams. The vertical polarization is used for Pound-Drever-Hall locking [16] of the laser to one of the modes of a high-finesse optical cavity filled with the gas sample. The horizontally polarized beam, detuned up or down from the locking point center by about one cavity free spectral range, can be finely scanned across the cavity mode to record its shape. Since the laser frequency is tightly locked to the cavity, any instabilities of the comb of cavity modes caused by acoustic and mechanical vibrations of the cavity are equally transferred to absolute frequency stabilities of both beams. In this way, cavity mode shapes measured with respect to the locking point can result in high-precision determination of modes half-widths and positions. To prevent the thermal drift of the comb of cavity modes, the cavity length is actively stabilized to a Nd:YAG laser having a long-term stability of 10 kHz. A dual-AOM frequency scanning system of the probe beam is used to make dispersive measurements insensitive to the temporal drift of the laser locking point [15]. In this configuration, two acousto-optic modulators (AOMs) driven by the same RF signal are used to

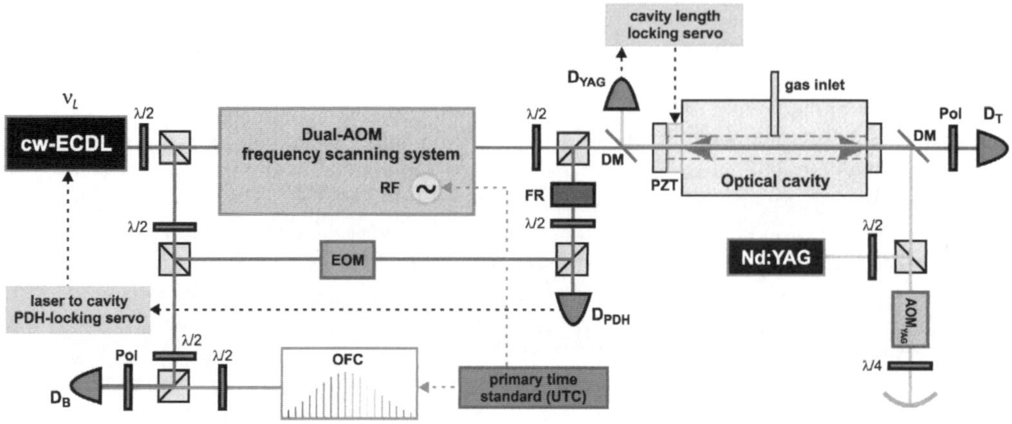

Figure 1. Schematic of an optical frequency comb (OFC)-assisted experiment for realization of three cavity-enhanced spectroscopy methods: FS-CRDS, CMWS and 1D-CMDS. Shown elements are a continuous-wave external cavity diode laser (cw-ECDL), an electro-optic modulator (EOM) for PDH locking of the cw-ECDL frequency, Faraday rotator (FR), a piezoelectric transducer (PZT), an acousto-optic modulator for the Nd:YAG laser (AOM_{YAG}), photodiodes (D_T, D_{PDH}, D_B, D_{YAG}), dichroic mirrors (DM), a quarter-wave plate ($\lambda/4$), half-wave plates ($\lambda/2$) and polarizers (Pol). The dashed lines represent electrical connections between locking electronic elements as well as reference frequency signal for the OFC and dual-AOM frequency scanning system.

point by point scan the probe laser beam alternately around the upper and lower cavity mode in relation to the locking point with relatively short switching time between them, less than 20 ms. More details about the experimental setup can be found in Refs. [13, 14, 15].

In general, a CMWS spectrum is obtained directly from determined cavity modes widths, whereas frequency intervals measured between cavity modes centers provide a differential dispersive 1D-CMDS spectrum. Ring-down decays measured in the FS-CRDS are initiated by fast switching off one of two acousto-optic modulators. As usually, the inverse of determined decay time constants form the absorptive CRDS spectrum. The absolute frequency of the ECDL laser is measured with respect to the optical frequency comb (OFC). Both the OFC as well as the RF signal for driving AOMs are referenced to the primary frequency standard, UTC(AOS), provided via the OPTIME network [17, 18] from AstroGeodynamic Observatory in Borowiec (Poland) [19, 20].

3. Results and discussion

Experimental demonstration of the cavity-enhanced complex refractive index spectroscopy was made on the P3 transition from the $^{13}C^{16}O$ (3 − 0) band as an example [15]. The recorded complex spectra (absorption from FS-CRDS and dispersion from 1D-CMDS) are shown in Fig. 2 for two gas pressures (20 and 60 Torr). It should be noted that both components of the complex spectra were measured with the same gas sample at the same time. Experimental spectra were fitted using the speed-dependent Nelkin-Ghatak profile (SDNGP) [21] which takes into account two line-shape effects beyond the Voigt profile approximation: the speed-dependence of collisional broadening and shift, with quadratic speed-dependence model, and the velocity-changing collisions, which leads to Dicke narrowing. Two schemes of data analysis were used to demonstrate the advantages of the complex spectra measurements. First, individual

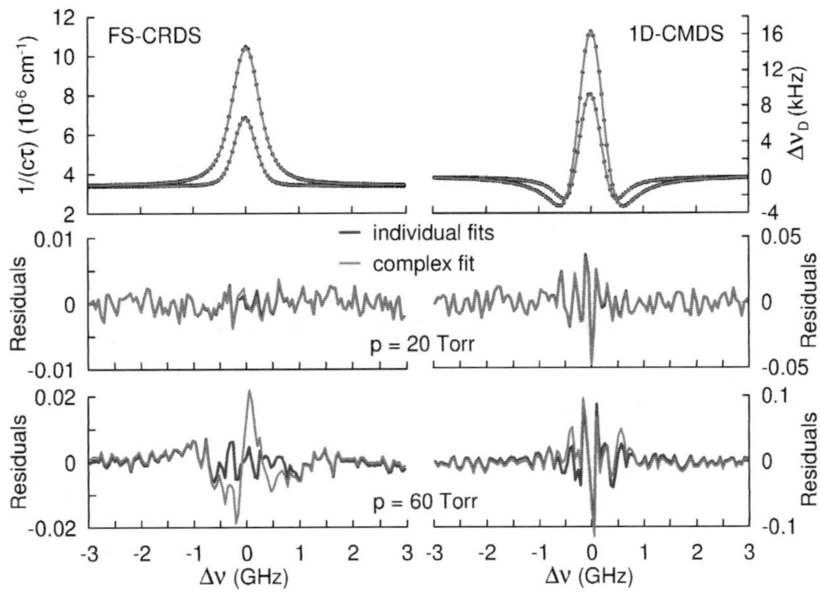

Figure 2. Experimental spectra of the $^{13}C^{16}O$ $(3-0)$ band P3 line measured at pressures of 20 and 60 Torr. Absorption and differential dispersion spectra were measured with the FS-CRDS and 1D-CMDS, respectively. Below residuals of the SDNGP fits are shown for two cases: blue and red residuals correspond to individual fits of absorption or dispersion spectra and a complex line shape fits to the complex (absorption and dispersion) spectra, respectively.

fits of the real and imaginary parts of the complex SDNGP were fitted to absorption (FS-CRDS) and dispersion (1D-CMDS) spectra, respectively. Next, the complex SDNGP was fitted simultaneously to absorption and dispersion spectra. From the fit residuals shown in Fig. 2 it is clear that a perfect agreement between absorption and dispersion spectra was obtained for low-pressure spectrum (20 Torr). For the high-pressure spectrum (60 Torr) systematic distortions of the fit residuals are observed for the complex fit, but not for individual fits. This is clear indication of systematic instrumental error introduced by one of used experimental methods. In this case the ring-down time in FS-CRDS was too short to be measured properly around the line center of the 60 Torr absorption spectrum. This kind of systematic instrumental errors is difficult to detect using only one experimental technique. Especially at the accuracy level when non-Voigt line-shape effects must be taken into account, the instrumental line-shape distortions may be incorrectly interpreted as a part of the physical line-shape effects.

Another confirmation of the source of systematic line-shape distortion can be obtained by an analysis of the pressure dependence of fitted line-shape parameters which should depend linearly on gas pressure. In Fig. 3 a ratio of integrated line area A to pressure p is plotted vs pressure. Assuming that fitted SDNGP properly describe measured line shape, this A/p ratio should be constant. In Fig. 3 it is constant within the fit uncertainties only for dispersion 1D-CMDS spectra. For the FS-CRDS the instrumental effect of too short ring-down times is manifested as underestimated A/p for high absorption conditions at pressures above 20 Torr.

Similar analysis of A/p ratios obtained from fits of the Voigt profile, also shown in Fig. 3, demonstrates high sensitivity of differential dispersion spectra to a systematic error related to a wrong line-shape model. The observed nonlinearity of the line area with pressure for the

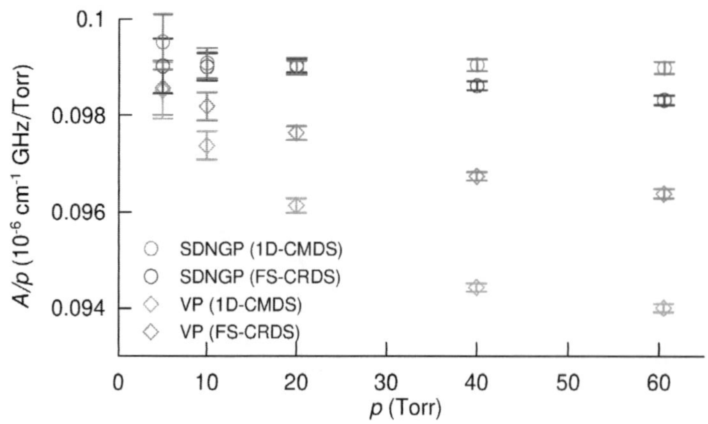

Figure 3. Dependence of a pressure scaled integrated line area of $^{13}C^{16}O$ (3 – 0) band P3 line on pressure obtained from fits of the VP and SDNGP to absorption (FS-CRDS) and differential dispersion (1D-CMDS) spectra.

1D-CMDS is twice as large as that for the FS-CRDS. This feature predestinates dispersion spectroscopy as a very sensitive tool for spectral line shape investigations. Different sensitivity of absorption and dispersion spectra to the line-shape model was also observed in the NICE-OHMS experiment [22].

The three spectroscopic methods, presented above FS-CRDS, CMWS and 1D-CMDS, were used for a high accuracy measurement of unperturbed line center frequencies and pressure shifts of $^{12}C^{16}O$ lines from the (3 – 0) band [14]. The frequency axes of the spectra were linked to the primary frequency standard UTC(AOS). It should be noted that in the case of the 1D-CMDS method both axes of the spectrum were linked to this frequency standard. In Fig. 4(a) the dependence of R24 line centers on CO pressure is shown for three experimental methods. Taking into account that the three experimental methods are not sensitive to the same kind of instrumental errors, systematic differences between FS-CRDS, CMWS and 1D-CMDS results enables estimation of systematic errors affecting determined line positions, which could not be properly estimated from results of only one method. The unperturbed line center frequencies obtained from each method by linear extrapolation of line positions to zero pressure are shown in Fig. 4(b) together with their average and a literature value of Mondelain et al. [23].

4. Conclusions

We have shown an experimental setup that enables measurements of the complex refractive index spectra using FS-CRDS in the time domain and CMWS in the frequency domain for absorption, and 1D-CMDS in the frequency domain for dispersion. This unique setup enables high-precision comparisons between absorption and dispersion spectrum of the same sample, which is tremendously useful to detect and quantify potential systematic instrumental distortions of spectra.

The purely frequency-based 1D-CMDS method is also very attractive for metrology applications because, unlike any absorption spectroscopy techniques, it is based only on frequency measurements. Therefore both axes of experimental spectra can be linked to the most accurate among standards of physical quantities.

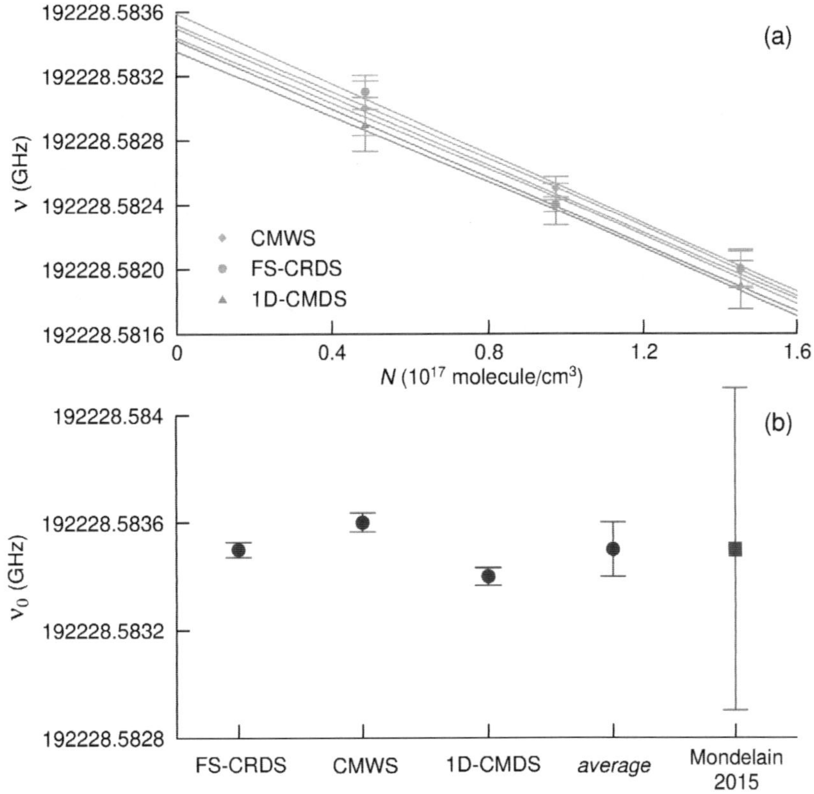

Figure 4. Graph (a): dependence of the fitted center frequencies of the $^{12}C^{16}O$ (3 − 0) band R24 line on pressure in the case of SDNGP fits to experimental spectra obtained with FS-CRDS, CMWS and 1D-CMDS (graph a). Lines indicate expanded (2σ) uncertainties and error bars correspond to scatter of individual results given by expanded (2σ) standard deviations. Graph (b): unperturbed line center frequency ν_0 obtained from the linear extrapolation of plots in graph (a) to zero pressure for three experimental methods. Additionally our average ν_0 and a literature value by Mondelain et al. [23] are shown. The error bars correspond to combined expanded (2σ) standard uncertainties.

Acknowledgments
The research is part of the program of the National Laboratory FAMO in Toruń, Poland and is supported by the National Science Centre, Poland, Project Nos. DEC-2013/11/D/ST2/02663, DEC-2012/05/D/ST2/01914, and 2015/17/B/ST2/02115, by the Foundation for Polish Science TEAM Project co-financed by the EU European Regional Development Fund and by MOLIM COST action. P.W. is supported by the Foundation for Polish Science START program.

References
[1] Cygan A *et al* 2012 *Phys. Rev. A* **85** 022508
[2] Lin H *et al* 2015 *J. Quant. Spectrosc. Radiat. Tr.* **161** 11
[3] Hodges J T *et al* 2004 *Rev. Sci. Instrum.* **75**, 849
[4] Long D A *et al* 2012 *Chem. Phys. Lett.* **536**, 1
[5] Lisak D *et al* 2009 *Phys. Rev. A* **79** 052507

[6] Wójtewicz S *et al* 2014 *J. Quant. Spectrosc. Radiat. Tr.* **144** 36
[7] Polyansky O L *et al* 2015 *Phys. Rev. Lett.* **114** 243001
[8] Wójtewicz S *et al* 2011 *Phys. Rev. A* **84** 032511
[9] Lehmann K K and Romanini D 1996 *J. Chem. Phys.* **105** 10263
[10] Long D A *et al* 2013 *Proc. SPIE* **8726** 87260O-6
[11] Cygan A *et al* 2013 *Opt. Express* **21** 29744
[12] Long D A *et al* 2014 *Appl. Phys. B* **114** 489
[13] Cygan A *et al* 2015 *Opt. Express* **23** 14472
[14] Cygan A *et al* 2016 *J. Chem. Phys.* **144** 214202
[15] Cygan A *et al* 2016 *Meas. Sci. Technol.* **27** 045501
[16] Drever R W P *et al* 1983 *Appl. Phys. B* **31** 97
[17] Śliwczyński Ł *et al* 2013 *Metrologia* **50** 133
[18] Morzyński P *et al* 2015 *Sci. Rep.* **5** 17495
[19] Azoubib J *et al* 2003 *Metrologia* **40** S245
[20] Jiang J *et al* 2015 *Metrologia* **52** 384
[21] Lance B *et al* 1997 *J. Mol. Spectrosc.* **185**, 262
[22] Wang J Y, *et al* 2014 *J. Quant. Spectrosc. Radiat. Tr.* **136**, 28
[23] Mondelain D. *et al* 2015 *J. Quant. Spectrosc. Radiat. Tr.* **154**, 35

Fourier Transform Spectroscopy of two trace gases namely Methane and Carbon monoxide for planetary and atmospheric research application

R Hashemi [1,2], A S Dudaryonok [3], N N Lavrentieva [3], A C Vandaele [4], J Vander Auwera [2], AV Nikitin [5,6], Vl G Tyuterev [7], K Sung [8], M A H Smith [9], V M Devi [10], A Predoi-Cross [1]

[1] Department of Physics and Astronomy, University of Lethbridge, 4401 University Drive, T1K 3M4 Lethbridge, AB, Canada
[2] Service de Chimie Quantique et Photophysique, C.P. 160/09, Université Libre de Bruxelles, B-1050 Brussels, Belgium
[3] V.E. Zuev Institute of Atmospheric Optics, Siberian Branch of the Russian Academy of Sciences, 1 Akademishian Zuev square, 634021 Tomsk, Russia
[4] Belgian Institute for Space Aeronomy, Planetary Aeronomy Division, 3 avenue circulaire, B-1180 Brussels, Belgium
[5] Laboratory of Theoretical Spectroscopy, V.E.Zuev Institute of Atmospheric Optics, Russian Academy of Sciences, Tomsk State University, Tomsk, Russian Federation
[6] QUAMER Laboratory, Tomsk State University, Tomsk, Russian Federation
[7] Groupe de Spectrométrie Moléculaireet Atmosphérique, UMRCNRS 6089, Universitéde Reims, U.F.R. Sciences, Reims Cedex 2, France
[8] Jet Propulsion Laboratory, California Institute of Technology, Pasadena, California, USA
[9] Science Directorate, NASA Langley Research Center, Hampton, Virginia, USA
[10] Department of Physics, College of William & Mary, Williamsburg, Virginia, USA

Abstract. Two atmospheric trace gases, namely methane and carbon monoxide have been considered in this study. Fourier transform absorption spectra of the 2–0 band of $^{12}C^{16}O$ mixed with CO_2 have been recorded at total pressures from 156 to 1212 hPa and at 4 different temperatures between 240 K and 283 K. CO_2 pressure-induced line broadening and line shift coefficients, and the associated temperature dependence have been measured in an multi-spectrum non-linear least squares analysis using Voigt profiles with an asymmetric profile due to line mixing. The measured CO_2-broadening and CO_2-shift parameters were compared with theoretical values, calculated by collaborators. In addition, the CO_2-broadening and shift coefficients have been calculated for individual temperatures using the Exponential Power Gap (EPG) semi-empirical method. We also discuss the retrieved line shape parameters for Methane transitions in the spectral range known as the Methane Octad. We used high resolution spectra of pure methane and of dilute mixtures of methane in dry air, recorded with high signal to noise ratio at temperatures between 148 K and room temperature using the Bruker IFS 125 HR Fourier transform spectrometer (FTS) at the Jet Propulsion Laboratory, Pasadena, California. Theoretical calculations for line parameters have been performed and the results are compared with the previously published values and with the line parameters available in the GEISA2015 [1] and HITRAN2012 [2] databases.

1. Introduction

The success of spectroscopic planetary atmospheric remote sensing studies often relies on having highly accurate sets of laboratory data for line parameters of the majority of atmospheric trace constituents, to interpret them. Advances in remote sensing instrumentation are pushing the requirements for accuracy of spectroscopic line parameters included in databases such as GEISA [1] or HITRAN2012 [2]. Today's state of the art line shape models go beyond the Voigt model, and include subtle effects such as line mixing, correlation between collisional narrowing, the velocities of molecules, the ro-vibrational relaxation mechanisms [3-7].

Besides being an active greenhouse gas present in the terrestrial atmosphere, carbon monoxide is also present as trace constituent in the atmosphere of Venus [8]. Understanding its sources, sinks and distribution in the Venusian atmosphere can help scientists better understand the dynamics taking place in the mesosphere and the lower thermosphere (from 70 to 120 km). The Venusian atmosphere is probed by the "Solar Occultation in the InfraRed" (SOIR) instrument on board of the ESA "Venus Express" spacecraft [9]. Precise spectroscopic line shape parameters of this species perturbed by CO_2 are required to retrieve CO concentrations from SOIR or NOMAD spectra on the first overtone (2–0) band of CO, located near 4260 cm^{-1} (2.3 μm). Our study, published in Ref. [10] is intended to extend the knowledge for line parameters of CO broadened by CO_2, over a range of temperatures.

For this propose, we have used number of 21 spectra of CO/CO_2 mixtures recorded using Fourier transform spectrometer at University of Brussels. Different line shape models have been used to reproduce the observed spectra: The Voigt profile (VP) [11], the Rautian profile (RP) [12] to account for the Dicke narrowing effects [13], and the quadratic Speed Dependent Voigt profile (qSDV) [14,15] to account for the dependence of line shape parameters on the relative speed of the molecules. CO_2 pressure induced broadening and shift coefficients of the lines of $^{12}C^{16}O$ have been determined, together with the temperature dependence of the CO_2 broadening parameters. We were able to retrieve the line mixing (LM) coefficients [16], as well. The retrieved parameters were then compared with CO_2 broadening and shift coefficients calculated using a semi-empirical method.

Due to the importance of methane as a trace atmospheric gas and a greenhouse gas and its contributions to the terrestrial Carbon Cycle, we have carried out a precise line shape study to obtain the self- and air half-width coefficients, self- and air-shift coefficients and the off-diagonal relaxation matrix element coefficients for methane transitions in the spectral range known among spectroscopists as the "methane octad". In addition, the temperature dependences of these coefficients have been measured. The results have been published in Ref. [17].

2. Experimental details

The experimental sessions for the two experiments presented here were carried out in different locations. The high resolution absorption spectra of the 2–0 band of carbon monoxide mixed with carbon dioxide at total pressures ranging from 156 to 1212 hPa and at 4 different temperatures between about 240 K and 283 K have been recorded using a Bruker IFS 125 HR Fourier transform spectrometer. The instrument was fitted with a Tungsten source, a CaF_2 beamsplitter, a band-pass filter limiting the transmission to the 3980–5100 cm^{-1} range, and an InSb detector cryogenically cooled to 77 K. The gas mixtures were prepared in the 19.7(2) cm long double jacketed stainless steel cell, closed by CaF2 windows and located inside the evacuated spectrometer.

The 14 high signal to noise ratio spectra of pure Methane and of dilute mixtures of Methane in dry air with 0.005 cm^{-1} resolution, have been recorded at temperatures from 148 K to room temperature using the Bruker IFS 125 HR Fourier transform spectrometer (FTS) at the Jet Propulsion Laboratory, Pasadena, California. The coolable absorption cell used in this experiment has an optical path of 20.38 cm and has been built to fit inside the sample compartment of the Bruker FTS [18] and [19]. The spectrometer setup consisted of a Globar source, a CaF_2 beam-splitter and a liquid-N_2 cooled InSb detector. Of the 14 spectra, 8 spectra were recorded with pure methane samples and 6 of the spectra were obtained with dilute mixtures of $^{12}CH_4$ in dry air.

3. Analysis of both set of the spectra and results

The analysis of both sets of spectra was done using the multispectrum nonlinear least-squares curve fitting technique. The CO_2 broadening and shift coefficients of the P(22) to R(21) lines and their temperature dependences and first-order line mixing coefficients, have been retrieved. For methane we have focused on the 4300–4500 cm^{-1} spectral region for the range of terrestrial atmospheric temperatures. The off-diagonal relaxation matrix element coefficients and quadratic speed dependence were taken into account for several transitions.

The following expressions were used to determine the half-width and shift coefficients:

$$b_L = P_{tot}[b_L^0(foreign)(1-\chi)\left(\frac{T_0}{T}\right)^{n_1} + b_L^0(self)\chi\left(\frac{T_0}{T}\right)^{n_2}$$

$$\Delta\tilde{\nu} = P_{tot}\left[d^0(foreign)(1-\chi) + d\delta^0(self)\chi\right]$$
$$\delta^0(T) = d\delta^0(T_0) + d\acute{\delta}(T - T_0)$$

where P_{tot} is the total sample pressure, T_0 = 296 K, χ is the mole fraction, and b_L is the Lorentz half-width at pressure P and temperature T, b^0_L is the Lorentz half-width coefficient of the line at the reference pressure P_0 and temperature T_0, δ^0 represent the shift coefficient (in $cm^{-1}atm^{-1}$). n_1 and n_2 are the temperature dependence exponents of the foreign and self broadening coefficients, respectively. δ' quantifies the temperature dependence of the shift coefficients.

Figure 1 presents the sample fit obtained for CO spectra corresponding to the high total pressure achieved in this work. The residuals show that, at high pressure, the qSDV line shape model, with or without first-order line mixing, best reproduces the observed line profiles.

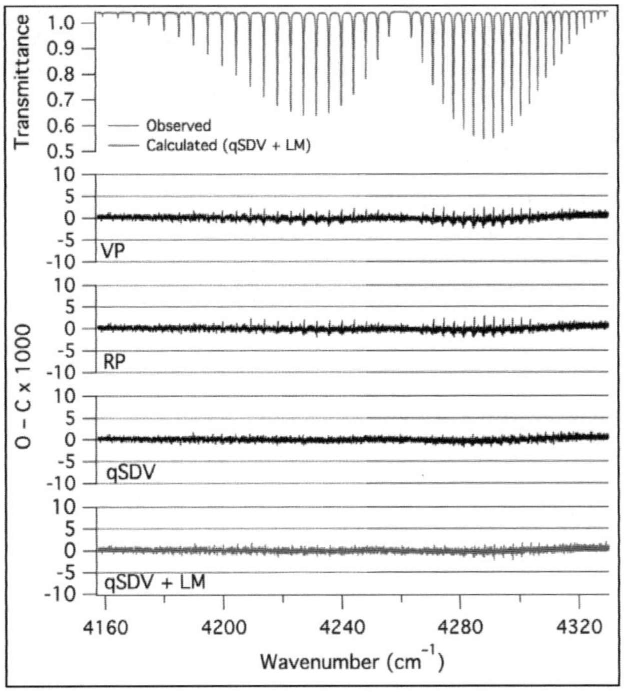

Figure 1. The transmittance spectrum of the 2–0 band of $^{12}C^{16}O$ broadened by CO_2: observed and calculated at the same conditions using the qSDV line shape model with weak line mixing. The lower panels present the residuals for the displayed spectrum of the multi-spectrum analysis of the 21 spectra with the VP, RP and qSDV line shape models, and the qSDV line shape model with weak line mixing. (Figure taken from [10].)

3.1 Sample of results obtained for Carbon monoxide

We measured the CO_2-pressure broadening and CO_2-pressure shift coefficients. The temperature dependence of the CO_2 broadening coefficients measured in this work using the 3 line shape models are presented in figure 2. Interestingly, the present measurements exhibit an intensity dependence, similar for the 3 line shape models. The results reported by Nakazawa and Tanaka [20] for the R branch of the fundamental band do exhibit a minimum for the same m value as in the present work, but strongly disagree at higher J. These comparisons may suggest that the temperature dependence of the CO_2 broadening coefficients are the same for the fundamental and first overtone bands.

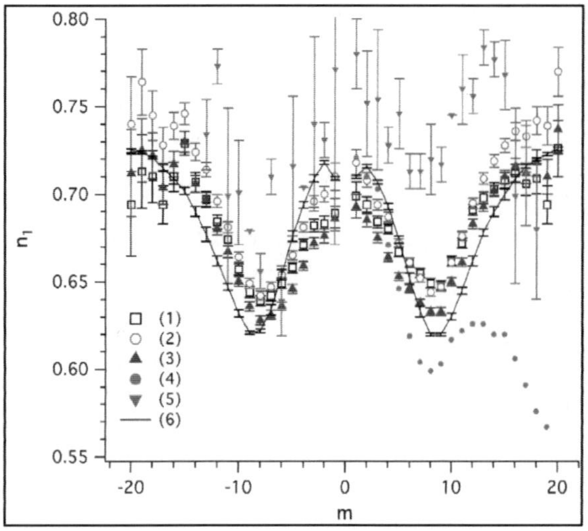

Figure 2. The temperature dependence exponent of CO_2 broadening coefficients for the 2–0 band of $^{12}C^{16}O$ measured in this work with the VP (1), RP (2) and qSDV (3) line shape models. The error bars associated with these measurements represent the uncertainty of measurement, estimated by the least squares fitting algorithm. Measurements reported for the fundamental band of $^{12}C^{16}O$ by Nakazawa and Tanaka using the Lorentz profile [20] (4) and by Sung and Varanasi [21] using the VP (5), and values calculated using a semi-classical method are also shown. (Figure taken from [10].)

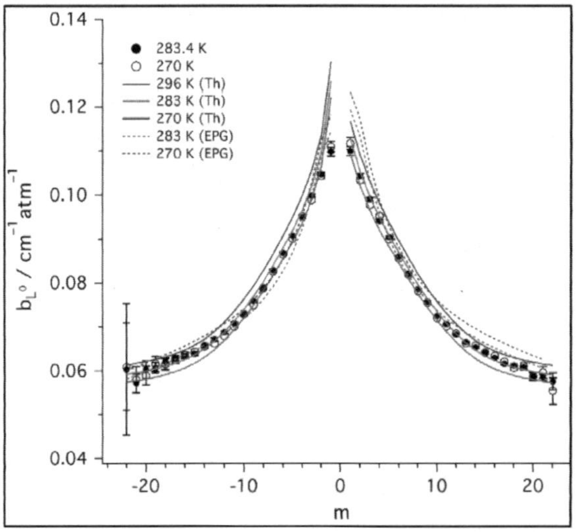

Figure 3. Comparison of CO_2 broadening coefficients measured at 2 temperatures in the 2–0 band of $^{12}C^{16}O$ with semi-classical calculations. (Figure taken from [10].)

Theoretical calculations were also performed to obtain line CO_2-broadening and CO_2-shift coefficients [22]. To allow comparisons with the theoretical calculations, the CO_2 broadening coefficients were also measured separately at each of the 4 temperatures, using the qSDV line shape model, including first-order line mixing. The retrieved parameters were then compared with CO_2 broadening (figure 3) coefficients calculated using a semiempirical method. The semi-classical method used is based on the impact theory of broadening and includes correction factors, the parameters of which being determined by fitting the calculated broadening or shift coefficients to the experimental data.

The narrowing coefficients measured using the RP line shape model and the dynamic friction coefficient also deduced. At the lowest (240.5 K) and highest (283.4 K) temperatures the dynamic friction coefficient obtained to be 0.0368 and 0.0320 cm^{-1}atm^{-1}, respectively. The measured narrowing coefficients indicated that the narrowing effect induced by velocity changing collisions decreases as the molecule rotates faster. However, except at high J, the measured narrowing coefficients are larger than the theoretical value, which indicates that the broadening coefficients may be affected by speed dependence [23] and that the Rautian line shape model with line mixing does not reproduce our spectra

IOP Conf. Series: Journal of Physics: Conf. Series **810** (2017) 012008 doi:10.1088/1742-6596/810/1/012008

completely. The speed dependence of the CO_2-broadening coefficients and the first-order line mixing coefficients obtained in this work using the qSDV profile. The first-order line mixing coefficients determined for the 3 line shape models are presented in figure 4. Even though the obtained values are small, taking the line mixing effects into account was important as they affect the measured CO_2 shift coefficients.

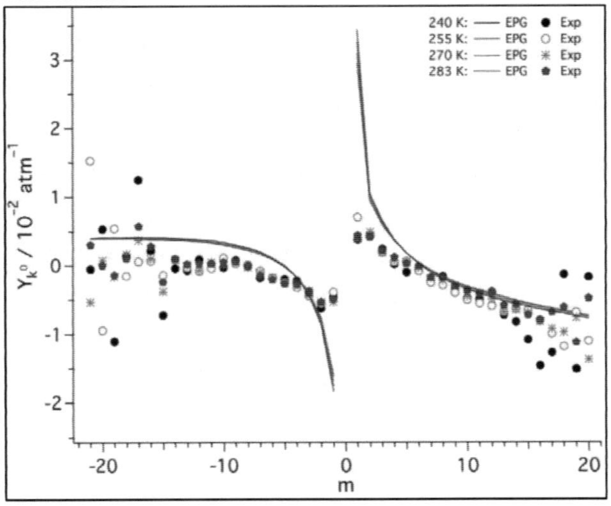

Figure 4. Weak line mixing coefficients in the 2–0 band of $^{12}C^{16}O$, measured and calculated using the EPG law at the same temperatures. (Figure taken from [10].)

3.2 Sample of obtained results for Methane study

For methane, short spectral intervals (2-5 cm^{-1}) were fitted at a time using the multi-spectrum analysis program of Benner *et al.* [24]. We added the spectra one by one starting from the ones recorded at the lowest pressure until all 14 spectra were included in the fits. We retrieved line positions and intensities, the Lorentz CH_4–CH_4 and CH_4–air-half-width and shift coefficients, and their temperature dependences. The initial value for speed dependence parameter was fixed to 0.08. Our goal was to minimize the difference between the observed spectra and the simulated spectra by adjusting different line parameters. The background also was fitted to a first order Chebyshev polynomial and there was no need to add channel spectra.

The line positions and intensities have been measured and compared to our theoretically calculated values and to the values in the databases [1,2]. The comparisons of positions and intensities suggest that our results agree well with these datasets. For line positions, the agreement is in the order of 10^{-4} cm^{-1}. For the intensities, we have obtained the percentage difference and it is less than 4% for most of the cases compared to HITRAN2012 [2] and GEISA [1] databases.

The temperature dependences of CH_4–CH_4 and CH_4–air half-width coefficients in the 4300–4500 cm^{-1} region are plotted in Figure 5 as a function of |m|. We used the power law model from Eq. (1) to obtain the temperature exponents of CH_4–CH_4 and CH_4–air half-width coefficients. In the top panel, the temperature exponents for CH_4–CH_4 half-width coefficients are shown and in the bottom panel are shown the results of temperature dependence of air-width coefficients.

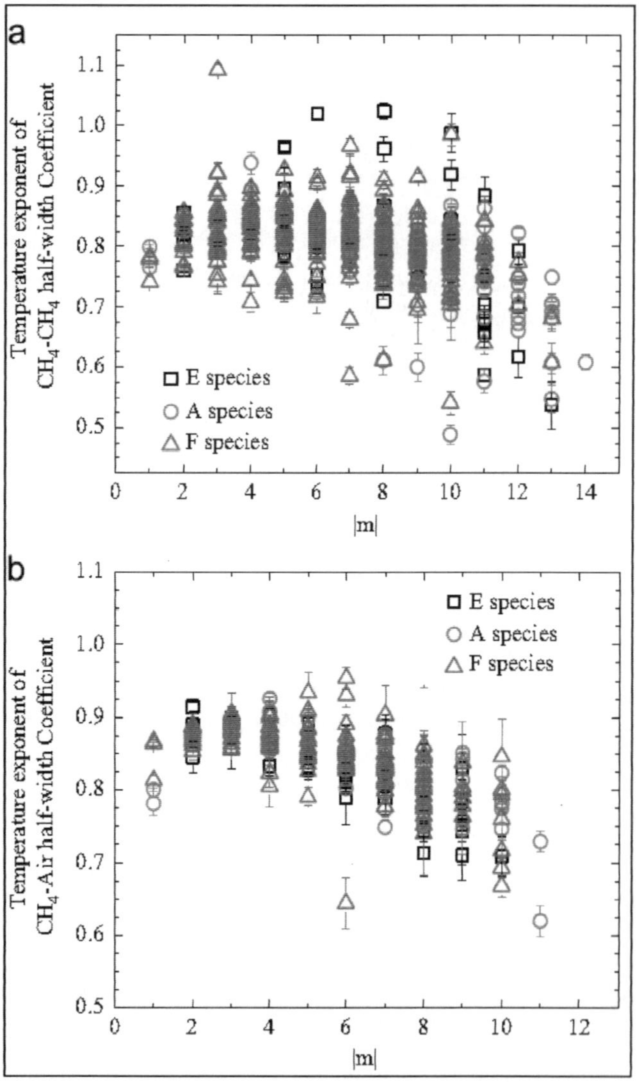

Figure 5. Temperature dependence exponents of CH_4–CH_4 and CH_4–air half-width coefficients in the $\nu_3+\nu_4$ band of methane. (Figure taken from [17].)

The Voigt profile was not able to fit the spectra within the experimental noise levels at higher pressures. This was also the case where the selected interval included pairs of transitions involved in collisional line mixing. To remedy the fault, line mixing for both CH_4–CH_4 and CH_4–air half-width coefficients was modeled using the off-diagonal relaxation matrix formalism [25]. Figure 6 shows the graph of the CH_4–air and CH_4–CH_4 coefficients from the present work, and corresponding values from the ν_4 and ν_2 bands [26] and [27] are shown for comparison.

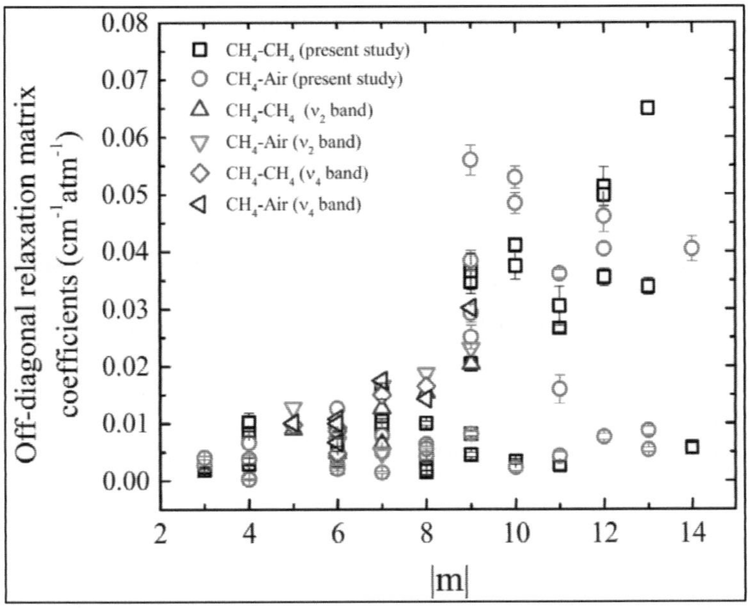

Figure 6. Off-diagonal relaxation matrix element coefficients in the $\nu_3+\nu_4$ band of methane compared to Refs. [26,27]. (Figure taken from [17].)

4. Conclusions

In the first part of this study, we performed analysis of high-resolution Fourier Transform Spectra of CO mixed with CO_2 using a multispectrum least squares fitting technique. The quadratic speed-dependent Voigt profile with weak line mixing properly modeled the observed molecular line shapes, to within the noise level. We measured the CO_2 broadening and shift coefficients for transitions in the 2–0 band of $^{12}C^{16}O$, up to J=22. The temperature dependence of the broadening parameters, as well as the weak line mixing coefficients were also obtained. Theoretical semi-empirical calculations were carried out to obtain CO_2 broadening and CO_2 shift parameters. Comparison with the corresponding observed coefficients showed that the calculated values are comparable at 296 K. Also, the weak CO_2 line mixing coefficients were obtained using the EPG method.

Then, the accurate line shape parameters such as CH_4–CH_4 and CH_4–air half-width coefficients, CH_4–CH_4 and CH_4–air shift coefficients, their temperature dependence and line mixing coefficients were determined in the $\nu_3+\nu_4$ band of methane. For several transitions the temperature exponents for self half-width coefficients were slightly higher than those for air half-width coefficients. The quadratic SDV profile including line mixing through the off-diagonal relaxation matrix elements formalism was used in the retrievals. The results of this laboratory study addressed the broad objectives of the Atmospheric Composition Laboratory Research Program to develop further the understanding of terrestrial atmospheric and planetary composition [28] and to provide improved prognostic capability for environmental issues.

5. Acknowledgments

This work was supported by funding from the Natural Sciences and Engineering Research Council of Canada through the Discovery Grant and CREATE programs, and "Alberta Innovates – Technology Futures" (AITF). Financial support from the Fonds de la Recherche Scientifique (F.R.S.-FNRS, Belgium, contract FRFC) and from the Russian Fund for Basic Research (grant 16-52-16016 NCNIL) are gratefully acknowledged. Methane research performed at the Jet Propulsion Laboratory, California Institute of Technology. We are grateful to Dr. Arlan Mantz of the Department of Physics, Astrophysics and Geophysics, Connecticut College for his contributions to the design and building of the cryogenic cell. Dr. D. Chris Benner of the Department of Physics, College of William and Mary is thanked for allowing us to use his Labfit software. The support of the CNRS (France) and RFBR (Russia) in the frame of Laboratoire International Associé (LIA) SAMIA and of Tomsk State University Academic D. Mendeleev funding Program are acknowledged. The research performed at the Jet Propulsion Laboratory, California Institute of Technology, the College of William and Mary and NASA Langley Research Center is supported by grants and cooperative agreements with the National Aeronautics and Space Administration.

6. References

[1] Jacquinet-Husson N, Armante R, Scott NA, Chédin A, Crépeau L et al 2016 The 2015 edition of the GEISA spectroscopic database", *J. Mol. Spectrosc.* **327** 31-72

[2] Rothman LS, Gordon IE, Babikov Y, Barbe A, Benner DC, P.F. Bernath, Birk M, Izzocchi L, Boudon V, Brown LR, Campargue A et al. 2013 *J. Quant. Spectrosc. Radiat. Transf.* **130** 4–50

[3] Hartmann JM, Boulet C, Robert D 2008 Elsevier Science

[4] Ciurylo R and Szudy J, 1997 *J. Quant. Spectrosc. Radiat. Transf.* **57** p 411-423

[5] Boone CD, Walker KA, Bernath PF 2007 *J. Quant. Spectrosc. Radiat. Transf.* **105** 525-532

[6] Ciurylo R and Szudy J 2001 *Phys. Rev. A.* **63** 042714

[7] Mondelain D, Payan S, Camy-Peyret C et al. 2007 *J. Mol. Spectrosc.* **244** 130-137

[8] Benner DC, Rinsland CP, Devi VM, Smith MAH, Atkins D 1995 *J. Quant. Spectrosc. Radiat. Transf.* **53** 705-721

[9] Bertaux JL et al. 2007, Nature **450** 646-649

[10] Hashemi R, Predoi-Cross A, Dudaryonok AS, Lavrentieva NN, Vandaele AC, Vander Auwera J 2016 *J. Mol. Spectrosc.* **326** 60-72

[11] Voigt W, 1912 Sitzber. *Bayr Akad. München Ber.* 603–620

[12] Rautian SG, Sobel'man I 1967 *Sov. Phys. Uspekhi.* **9** 701–716

[13] Dicke RH 1953 *Phys. Rev.* **89** 472–473

[14] Rohart F, Mäder H, Nicolaisen HW 1994 *J. Chem. Phys.* **101** 6475–6486

[15] Tran H, Boulet C, Snels M, Piccioni G 2011 *J. Quant. Spectrosc. Radiat. Transf.* **112** 925–936

[16] Rosenkranz PW 1975 IEEE. *Trans. Anten. Propag.* **23** 498–506

[17] Hashemi R, Predoi-Cross A, Nikitin AV, Tyuterev VI G, Sung K, Smith MAH, Devi VM 2016 *J. Quant. Spectrosc. Radiat. Transf.* in press

[18] Sung K, Mantz AW, Smith MAH, Brown LR, Crawford TJ, Devi VM, et al. 2010 *J. Mol. Spectrosc.* **262** 122–34

[19] Mantz AW et al. 2010 *65th International Symposiumon Molecular Spectroscopy*, Ohio State University, Columbus, OH

[20] Nakazawa T, Tanaka M 1982 *J. Quant. Spectrosc. Radiat. Transf.* **28** 471–480

[21] Sung K and Varanasi P 2005 *J. Quant. Spectrosc. Radiat. Transf.* **91**, p 319–332

[22] Bykov AD, Lavrentieva NN, Sinitsa LN, 2004 *Mol. Phys.* **102** 1653–1658

[23] Lisak D, Masłowski P, Cygan A, Bielska K, Wójtewicz S, Piwin´ ski M, Hodges JT, Trawin´ ski RS, Ciuryło R 2010 *Phys. Rev. A.* **81** 042504

[24] Benner DC, Rinsland CP, Devi VM, Smith MAH, Atkins D, 1995 *J. Quant. Spectrosc. Radiat. Transf.* **53** 705-721

[25] Levy A, Lacome N, Chackerian Jr C. 1992 Nar-ahari Rao K,Weber A, editors. Spectroscopy of the earth's atmosphere and interstellar medium. *Boston, MA: Academic Press.* 261–337

[26] Smith MAH, Benner DC, Predoi-Cross A, Devi VM. 2011 *J Quant Spectrosc RadiatTransf.* **112** 952–68

[27] Smith MAH, Benner DC, Predoi-Cross A, Devi VM. 2014 *J. Quant. Spectrosc. Radiat. Transf.* **133** 217–34

[28] Brown LR, Sung K, Benner DC, Devi VM, BoudonV, Gabard T et al. 2013 *J. Quant. Spectrosc. Radiat. Transf* **130** 201–19

Ionization-potential depression and other dense plasma statistical property studies - Application to spectroscopic diagnostics.

Annette Calisti, Sandrine Ferri, Caroline Mossé, Bernard Talin
Aix Marseille University, CNRS, PIIM, UMR7345, Marseille, France
E-mail: annette.calisti@univ-amu.fr

Abstract. The radiative properties of an emitter surrounded by a plasma, are modified through various mechanisms. For instance the line shapes emitted by bound-bound transitions are broadened and carry useful information for plasma diagnostics. Depending on plasma conditions the electrons occupying the upper quantum levels of radiators no longer exist as they belong to the plasma free electron population. All the charges present in the radiator environment contribute to the lowering of the energy required to free an electron in the fundamental state. This mechanism is known as ionization potential depression (IPD). The knowledge of IPD is useful as it affects both the radiative properties of the various ionic states and their populations. Its evaluation deals with highly complex n-body coupled systems, involving particles with different dynamics and attractive ion-electron forces. A classical molecular dynamics (MD) code, the BinGo-TCP code, has been recently developed to simulate neutral multi-component (various charge state ions and electrons) plasma accounting for all the charge correlations. In the present work, results on IPD and other dense plasma statistical properties obtained using the BinGo-TCP code are presented. The study focuses on aluminum plasmas for different densities and several temperatures in order to explore different plasma coupling conditions.

1. Introduction

For various purposes, it is necessary to simulate virtual plasma composed of electrons and ions in different ionization states. Depending on the plasma conditions, e.g. when they are such that the electron-ion interactions are strong or those generated by direct target interaction with intense laser beams, it is required to account, in the simulations, for bound states as well as quantum mechanical processes to create and destroy them. This permits to avoid non-physical Coulomb collapse and to extend the applicability of classical MD simulations, which are ideally suited for fully-ionized, non-degenerate plasma, to a widest range of plasma conditions. Various methods have been developed and proposed by different groups. Among them, one can mention the methods based on the density functional theory (DFT) such as the quantum Molecular Dynamics method (QMD) [1] which is widely used to provide benchmark data but is limited to low temperature or the orbital free density functional theory molecular dynamics (OF-DFT-MD) [2] which has been developed to extend QMD to higher temperatures. Other methods exist which do not use DFT but simulate interacting classical ions and electrons incorporating some quantum mechanical effects through the interaction pseudo-potential between ions and electrons [3, 4].

We have also chosen to simulate the ions and electrons as classical particles and to incorporate a minimum of quantum information through a regularized potential allowing to model collisional ionization and recombination processes. The BinGo-TCP code has been designed to deal with neutral mixtures composed of ions of the same atom with different charge states and electrons [5, 6]. Within the limits of classical mechanics, all charge-charge interactions are accounted for in the particle motion. In this article, after a brief recall of the modeling basis, the importance of such a model to get a good understanding in correlation effects will be emphasized through a study of the ionization potential depression (IPD) and some other statistical properties in dense plasmas. In particular, some results obtained on aluminum plasmas at various conditions of densities and temperatures, corresponding to different conditions of coupling parameter, will be presented and analyzed.

2. Modeling

A classical multi-component plasma molecular dynamics (MD) code, the BinGo-TCP code, has been developed to follow the evolution of plasmas involving ions of various charge states and electrons. The overall structure of the BinGo-TCP code is consistent with that of a standard code of classical MD simulation. Non-relativistic point-like charged particles are simulated in a cubic box. The system inside the simulation box is neutral. The periodic boundary conditions are used together with the minimum image convention and the Newton's equations are solved using a Velocity-Verlet algorithm. A molecular dynamics simulation requires the definition of the potential function that governs interactions between the particles inside the simulation box.

2.1. Soft potentials

The electron-electron or ion-ion interactions are taken to be screened Coulomb:

$$V_{ii,ee} = Z_{i,e}^2 e^2 e^{-r/\lambda}/r, \qquad (1)$$

where the screening length, $\lambda \simeq s/2$, with s the box size, is chosen large enough not to affect the simulated results. Simulating ions and electrons together can be performed provided that an ion-electron potential finite at short distances avoid electron-ion collapse. Such effective regularized potentials are designed to approximately account for known quantum properties appropriate for the kind of investigated plasma. Here, the ion-electron regularized potential depends on Z_i:

$$V_{ie}(r) = -Z_i e^2 e^{-r/\lambda}(1 - e^{-r/\delta(Z_i)})/r, \qquad (2)$$

with the regularization distance δ chosen to reproduce the ionization energy E_i of the unperturbed ion of charge Z_i in the ground state when the electron is located at the ion (r = 0).

$$\delta(Z_i) = -Z_i e^2/E(Z_i) \qquad (3)$$

In other words, an electron located at an ion ($r = 0$) occupies the fundamental state of the ion whose charge is Z with a nucleus charge $Z + 1$.

2.2. Ionization/recombination protocol.

In the simulation, the particle motion is followed with very small time-steps $\sim 10^{-20}$s appropriate for the description of the micro motion of electrons around ions. These time steps required by the simulations are small compared to typical collision rates, thus any statistics on collisional events are expensive. The setting up of the population of electrons temporary trapped in the ion wells, i.e. the reaching of the required equilibrium state, depends on collisional events between electrons, and thus, is a very slow process. The choice of the previous soft ion-electron potential allows to implement a ionization/recombination protocol to control the plasma ion charge distribution and the trapping of electrons in the ion wells. Usually, the ionization-recombination mechanisms rely on an approximate analysis of collisional events between one ion and 1 or 2 electrons. In dense plasmas, the concept of collisions is not straightforward as the interactions involve all the particles within the screening length. To design a ionization/recombination protocol the definition of a collisional process is crucial but necessarily empirical. An analysis of the local environment of a given ion is performed to evaluate if the ion is susceptible to be ionized or recombined. The location of the two nearest neighbor electrons together with the sign of their total energy is used to evaluate if locally the plasma, at one step of its evolution, is favorable to an ionization (in case of positive energy) or a recombination (in case of negative energy) of the ion. When the local environment is favorable to an ionization, the ion of charge Z is changed to an ion of charge $Z+1$ with an electron located on it. This pre-ionization state will give rise to an effective ionization if the interactions of the trapped electron with the surrounding plasma, permit it.

A preparation phase of the particle set (into the simulation box) before extracting any sampling from simulations, is necessary. At the end of the preparation phase the system follows a quasi stable evolution with stationary trapped and free electron populations. The ionization/recombination process implemented in the code has two fundamental functions. It allows the evolution of the charge state population towards a stationary state depending on temperature, density and composition of the plasma and it favors the setting up of a population of electrons temporary trapped in the ion wells.

With this model, one gains the ability to describe the ion-electron coupling accounting for mixtures of ions undergoing changes of their ionization stages. It is important to note that the coupling of electrons with radiation is ignored and that the notion of discrete energy for the ionic excited states is here replaced by its continuous equivalent.

3. Some plasma properties.

Depending on its total energy and its nearest neighbor ion, an electron is either trapped or free. An illustration is given by the electronic total energy distribution functions measured in aluminum plasmas for various conditions.

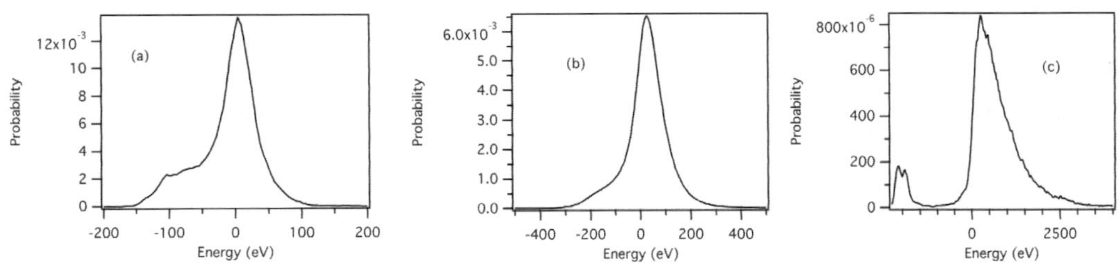

Figure 1. Electronic total energy distribution function measured in aluminum plasmas at (a) $T_e = T_i = 20$eV and $\rho = 0.65 g.cm^{-3}$, (b) $T_e = T_i = 50$eV and $\rho = 2.7 g.cm^{-3}$ and (c) $T_e = T_i = 550$eV and $\rho = 6.5 g.cm^{-3}$.

The analysis of these distribution functions together with the knowledge of the electron number in the simulation box, allows to infer the proportion of free electrons and thus the mean charge value. During a simulation run, various charge states of the same ion exist. The distribution function of these charge states is not representative of the plasma equilibrium as some electrons are trapped in the ion wells (revealed by the negative part of the energy distribution functions, cf. Figure 1). Therefore, to estimate the equilibrium charge states distribution function, a detailed examination of the electrons around each charge state has to be performed. The ionic charge state distribution function obtained for an aluminum plasma at

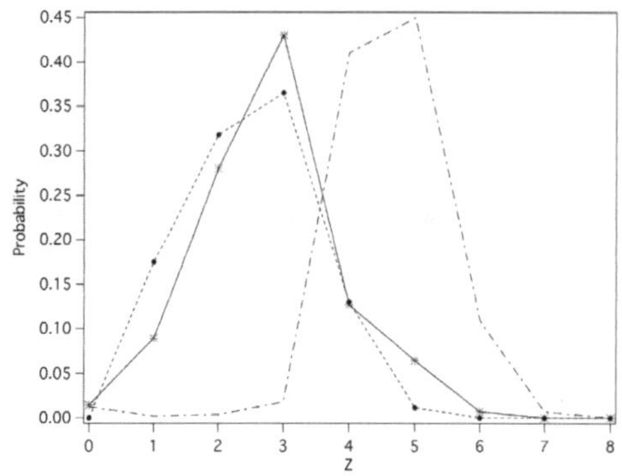

Figure 2. Charge state distribution function in an aluminum plasma at T_e=T_i=20eV and $\rho = 0.65g.cm^{-3}$ measured with (full line and stars) and without (dot-dash line) accounting for trapped electrons compared with the Flychk result (dash line and dots).

T_e=T_i=20eV and $\rho = 0.65g.cm^{-3}$, is shown in Figure 2. It can be seen that the result which accounts for the trapped electrons, compares quite well with the one obtained with the Flychk code [7].

4. The ionization potential depression.

Due to its environment, an ion embedded in a dense plasma will undergo a potential significantly different from the one experienced by the same isolated ion. This plasma-dependent potential will lower the energy required to free a bound electron. This mechanism which is known as the ionization potential depression (IPD), will affect the number of bound states and consequently the radiation emitted by the system.

The evaluation of the IPD has important implications for dense plasma physics and in particular for the detailed prediction of the dense plasma equation of state and radiative opacity in stellar interiors, inertial confinement fusion research, or planetary interiors. Recent experiments [8, 9] leading to IPD measurements in situ has renewed interest for this issue (see for example [10, 11, 12, 13, 14, 6]). Taking advantage of the particular design of the ionization protocol in the BinGo-TCP code, it is possible, when the ion is in a pre-ionization state, to measure the necessary energy to free an electron in the ground-state of a given ion and this, by accounting for all the interactions with the surrounding plasma. Figure 3 shows the distribution of this energy for four charge states existing in an aluminum plasma at T_e=T_i=50eV and $\rho = 2.7g.cm^{-3}$. The curves are not normalized and their relative intensities reflect the number of pre-ionization events for each ion of charge Z. It can be seen that due to a fluctuating local environment of the ions, the ionization energy is characterized by a distribution function. If one compares the mean energies deduced from these distribution functions with the corresponding energies of the isolated ions, it is possible to infer the ionization potential depression due to the interactions with the environment. The results are shown in Figure 4 with the results obtained with two theoretical IPD models, the Stewart-Pyatt (SP) [15] and the Ecker-Kröll (EK) [16] model, which apply across a wide range of densities. In those conditions of plasma, the electron

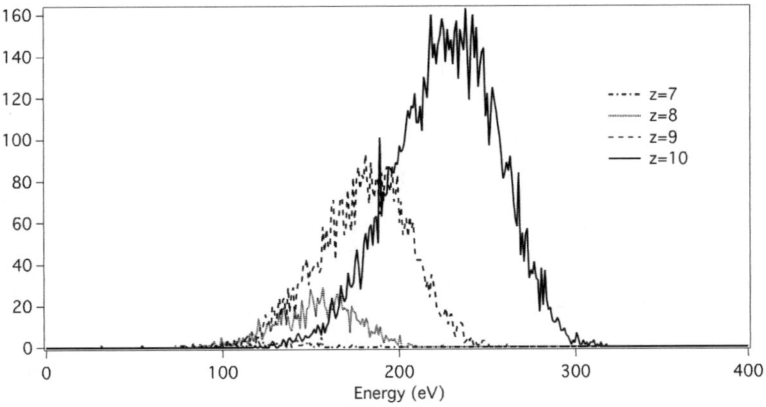

Figure 3. Distribution of the ionization energy of an electron in the ground state of an ion of charge Z in an aluminum plasma at $T_e = T_i = 50$eV and $\rho = 2.7 g.cm^{-3}$.

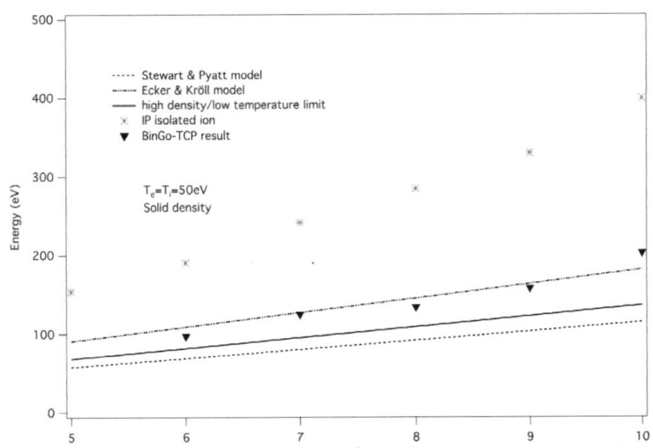

Figure 4. The ionization potential depression obtained with the BinGo-TCP code (down triangles) for an aluminum plasma at $T_e = T_i = 50$eV and $\rho = 2.7 g.cm^{-3}$, compared with the IPD calculated with the SP (dot line) and EK (dot-dash line) models for the same conditions. The high density/low temperature limit of the SP model is also plotted (full line) for comparison.

coupling plasma parameter is $\Gamma_e = 0.35$ and the measured mean charge is $Z_b = 7.4$ leading to a ion coupling plasma parameter, $\Gamma_i = 9.96$. The effects of the charge correlations on the lowering of the necessary energy to free an electron are very strong and cannot be ignored. The comparisons with the theoretical models show that for these conditions, our results are in better agreement with the EK model than with the SP model. Other results obtained for other conditions of plasma (aluminum plasmas at $T_e = T_i = 20$eV and $\rho = 0.65 g.cm^{-3}$ and $T_e = T_i = 550$eV and $\rho = 6.5 g.cm^{-3}$), not presented here, show a good agreement with the SP model and thus, one cannot conclude on the relevance of one or the other model.

5. Discussion.

Due to the specific implementation of the ionization/recombination protocol, our classical MD method gives access to the potential surrounding an ion accounting for the influence of the free electrons and neighboring ions. By comparison with the ionization potential of the equivalent isolated ions, the ionization potential depression can be estimated. Our simulations provide data for further discussion on IPD models.

Moreover, the local electric fields produced by the electrons and ions of the plasma, which derived from the force calculations on individual charges, can be measured by accounting for all the interactions between charges. Characterizing properly these electric fields is very important for spectroscopic diagnostics as they interact with the emitters and modify the radiated spectra. The TCP simulation is well designed to obtain time sequences of electric field. These can be used in the resolution of the Schrödinger equation describing the time evolution of the emitter wave functions in the time-dependent field, or to infer, via a statistical analysis, the field distribution and correlation functions, characterizing the contributions of ions or electrons, or both [17]. As

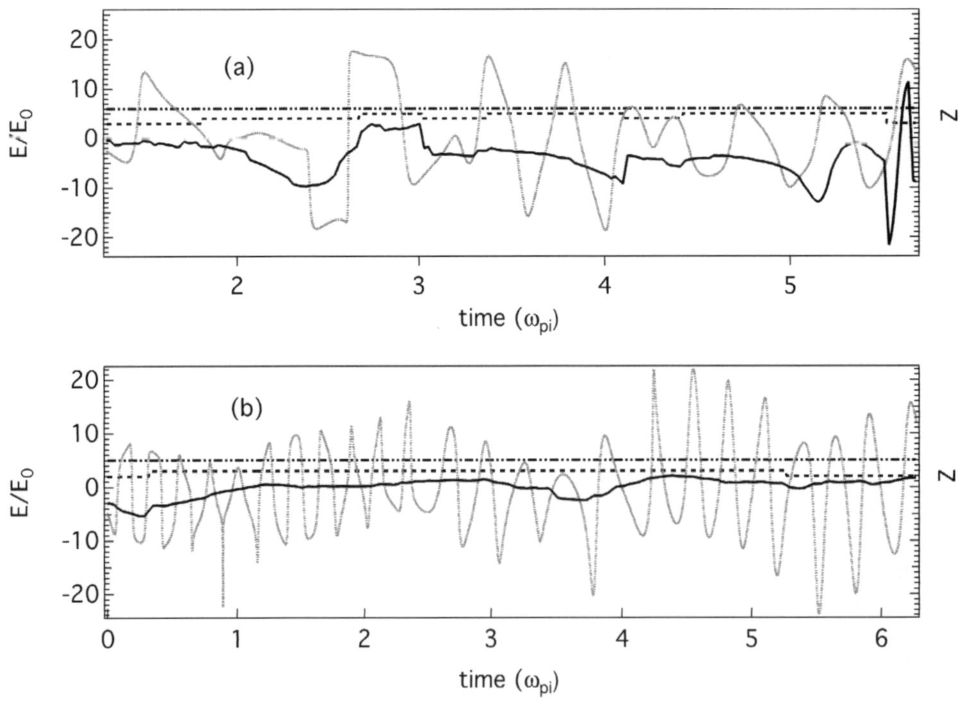

Figure 5. Ionic charge and field histories in an aluminum plasma at T_e=T_i=20eV and $\rho = 0.65g.cm^{-3}$. Bare ionic charge - dot dash line -; Ionic charge accounting for trapped electrons (effective charge)- dash line-; Electronic electric field measured at the bare ionic charge - dot line -; Electronic electric field seen by effective charges - full line -

it has been mentioned previously, it is possible to identify, during a simulation run, the trapped and free electrons and thus to measure the electric fields at a given charge. Examples of two electronic electric field histories measured at two different bare ionic charges ($Z = 6$ and $Z = 5$) are shown in Figure 5(a) and (b), respectively, for the same plasma conditions as Figure 2. The dot line corresponds to the total electronic electric field at the ion and the full line corresponds to the electronic electric fields without the trapped electrons then accounted in the ion charge. It is also shown the value of the bare ionic charge (dot-dash line) and the ionic charge accounting for trapped electrons, the effective charge (dash line). A statistical analysis of the field histories at the effective charges will give access to statistical properties of electric fields, such as field distribution or correlation functions in plasmas composed of various ionic charges and electrons. The same field histories used in the resolution of the emitter evolution equation will permit to infer the influence of correlations on spectral line shapes. And finally, a statistical analysis of the ionization and recombination events will be performed to evaluate the corresponding rates. This work is in progress.

References

[1] Car R., Parrinello M., Phys. Rev. Lett. 55 (1985) 2471.
[2] Zérah G., Clérouin J., and Pollock E. L., Phys. Rev. Lett. 69 (1992) 446; Lambert F., Clérouin J., Zérah G., Phys. Rev. E 73 (2006) 016403.
[3] Graziani F. R. et al., High Energy Density Phys. 8 (2012) 105-131.
[4] Belkacem M., Megi F., Reinhard P.-G., Suraud E., and Zwicknagel G., Eur. Phys. J. D 40 (2006) 247255.
[5] Calisti A., Ferri S., Marciante M., Talin B., High Energy Density Physics 13 (2014) 1-8.
[6] Calisti A., Ferri S., Talin B., J. Phys. B : At Mol. Opt. Phys. 48 (2015) 224003.
[7] Chung H.-K., Chen M.H., Morgan W.L., Ralchenko Yu., and Lee R.W., HEDP 1 (2005) p.3.
[8] Ciricosta O. et al., Phys. Rev. Lett. 109 (2012) 065002; Ciricosta O. et al., Nat. Commun. 7:11713 (2016).
[9] Hoarty D.J. et al., Phys. Rev. Lett. 110 (2013) 265003.
[10] Crowley B. J. B., High Energy Density Phys. 13 (2014) 84-102.
[11] Iglesias C. A., High Energy Density Phys. 12 (2014) 5-11.
[12] Son S. K. et al., Phys. Rev. X 4 (2014) 0311004.
[13] Fletcher L. B. et al., Phys. Rev. Lett. 112 (2014) 145004.
[14] Stransky M., Phys. Plasmas 23 (2016) 012708.
[15] Stewart J. and Pyatt K., Jr., Astrophys. J. 144, 1203 (1966).
[16] Ecker G. and Kröll W., Phys. Fluids 6, 62 (1963).
[17] Calisti A. et al., High Energy Density Phys. 7 (2011) 197-202

Molecular line shape parameters for exoplanetary atmospheric applications

S N Yurchenko, J Tennyson and E J Barton

Department of Physics and Astronomy, University College London, Gower Street, WC1E 6BT London, UK

E-mail: s.yurchenko@ucl.ac.uk

Abstract.
We describe the recent updates to the ExoMol database regarding the molecular spectral line shapes. ExoMol provides comprehensive molecular line lists with a special emphasis on the applications involving characterization of hot atmospheres such as those found in exoplanets and cool stars. Among important requirements of such applications are (i) the broadening parameters for hydrogen and helium dominating atmospheres and (ii) very broad ranges of temperature and pressures. The current status of the available line shape data in the literature, demands from the exoplanetary community and their specific needs are discussed.

1. Introduction

Molecules in the atmospheres of exoplanets, and similar astronomical objects such as brown dwarfs, are observed at pressures approaching atmospheric and therefore display pronounced pressure effects in their spectra [1, 2, 3]. There is, however, a severe lack of data appropriate for reproducing these pressure effects.

The ExoMol project provide extensive line lists for a large range of molecules, mostly related to the atmospheric retrievals for exoplanets and cool stars [4]. The selection of molecules and the spectroscopic coverage in ExoMol are dictated by the requirements from these applications. A major requirement is that the line lists must be sufficiently complete to be applicable for high temperatures specific for atmospheres of most of these objects [5]. Spectroscopic data for almost 40 molecules are currently covered by the ExoMol database, including diatomic, triatomic, tetratomic and two pentatomic species. The available line lists were generated both by ExoMol and other groups, see Table 1.

Recently we have undertaken a major upgrade of the ExoMol database [6]. The upgrade was motivated by the growing role of ExoMol as a major provider of the hot spectroscopic line lists for atmospheric retrievals from exoplanetary observations. The most important modifications include: (i) introduction of application programming interface (API); (ii) line shape parameters; (iii) cooling functions; (iv) radiative life times [7] and Landé g-factors [8]. In this contribution the new line shape data provided by ExoMol are discussed.

2. Requirements for the line shape parameters from the exoplanetary applications.

It is widely recognized that Voigt profiles only provide an approximate solution, especially for high resolution atmospheric studies [116]. However Voigt profiles are in widespread use and

Table 1. Molecular line lists included in the ExoMol database and Voigt line shape parameters when available.

Molecule	Reference	Broadener	Molecule	Reference	Broadener
AlO	[9]		HD^+	[10]	
BeH	[11]		HeH^+	[12]	
CaH	[11]		HNO_3	[13]	Air
CaH	[14]		KCl	[15]	
CaO	[16]		LiH	[10]	
CH	[17]		LiH^+	[10]	
CH_4	[18]	Air, He, H_2, CH_4	MgH	[11]	
CN	[19]		MgH	[20]	
CP	[21]		NaCl	[15]	
CrH	[22]		NaH	[23]	
CS	[24]	Air, CS	NH	[25]	
FeH	[26]		NH_3	[27]	Air, H_2, He, NH_3
H_2CO	[28]	Air, He, H_2, H_2CO	PH_3	[28]	Air, He, H_2, PH_3
H_2O	[29]	Air, H_2, He, H_2O	PN	[30]	
H_3^+	[31]		ScH	[32]	
HCl	[33]	Air, H_2, He, HCl	SiO	[34]	
HCN/HNC	[35]	Air, He, H_2, HCN	TiH	[36]	
SO_2	[37]	Air, He, H_2	SO_3	[38]	
H_2S	[39]		H_2O_2	[40]	
VO	[41]				
H_3^+	[42]				

are easily computed [117, 118, 119]; they are therefore uniformly used to represent pressure broadening effects in exoplanetary models and hence in the ExoMol database.

Table 1 illustrates the coverage of the broadening parameters in ExoMol, which is also a reflection of the current status in the field. Below we list the main factors affecting the development of the line shapes for the exoplanetary applications.

(i) The dominant species in gas giant planets such as hot Jupiter atmospheres are H_2 and He, often at high pressure [120]. However most of the line shapes parameters available in the literature are for the air (or N_2) as the main broadener and motivated by the terrestrial applications. Even though the needs for the line shapes broadened by H_2 and He have been recently recognized by HITRAN [121], the corresponding line shape data is still incomplete [122, 2].

(ii) The large size of the line lists in ExoMol, which consist of tens of billions of lines for larger molecules, make it practically impossible to populate the whole database with accurate, line-by-line line shape parameters, experimental or theoretical. Only simple and very approximate models (e.g. based only on one quantum number J'') can afford productions of the data on such a large scale. However even simple models do not exist for most of the species in question, especially for H_2 and He as broadeners.

(iii) The high temperatures ($T > 1000K$) of hot Jupiters, and other exoplanets of current interest [123], require line profiles for a large range of rotation and vibration excitations over extended temperature ranges. It is considered to be important to model at least the rotational (J) dependence of the line widths broadened by H_2 [120].

(iv) There is no agreement on the value of the line wing cut-off to be used when computing Voigt profiles. On one hand, it is imperative to use a reasonable cut-off distance in order to speed

Table 2. Voigt parameters available in the literature and used in the ExoMol database.

Molecule	Broadener	Reference
H_2O	H_2	[43, 44, 45, 46, 47, 48, 49, 50, 51, 52, 53, 54, 55]
H_2O	He	[56, 57, 58, 59, 60, 43, 44, 45, 48, 49, 61, 52, 53, 54]
H_2O	Air	[62, 63, 64, 65, 66, 67]
H_2O	H_2O	[62, 63, 68, 69, 70]
CH_4	H_2	[71, 72, 73, 74]
CH_4	He	[75, 76, 77, 73]
CH_4	Air	[78, 79, 80]
CH_4	CH_4	[81, 82, 79]
NH_3	H_2	[83, 84, 85, 86]
NH_3	He	[83, 84, 85, 86]
NH_3	Air	[87, 88, 86]
NH_3	NH_3	[87, 89, 90, 91, 92, 93]
PH_3	H_2	[94, 89, 90, 92]
PH_3	He	[95, 89, 90, 92]
PH_3	Air/N_2	[95, 96, 97]
PH_3	PH_3	[96, 97, 98, 99]
H_2CO	H_2	[100]
H_2CO	He	[100]
H_2CO	N_2	[101]
H_2CO	H_2CO	[101]
SO_2	H_2	[102]
SO_2	He	[102, 103]
SO_2	Air	[102, 103]
HCN	Air/N_2	[104, 105, 106]
HCN	HCN	[105, 107]
HCN	H_2	[108]
HCN	He	[108]
HCl	N_2	[109]
HCl	HCl	[110]
HCl	H_2	[111]
HCl	He	[112, 109]
CS	Air	[113]
CS	CS	[114]
HNO_3	Air	[115]

up the line-by-line calculations of the molecular opacities. On the other hand, the choice of the cut-off is known to effect the results at high pressures. Common practices include a fixed cut-off (e.g. 25 cm^{-1} or 100 cm^{-1} [120]), variable cut-off depending on pressure P (e.g. min(25P; 100) cm^{-1} [85]) or proportional to the line (half-)width (e.g. 500 widths [3]). The importance of folding back the truncated wings to ensure that the strength of the profile is conserved [85] is also considered in certain circumstances.

(v) Efficient computational algorithms for the Voigt evaluations [117, 124] as well as the sampling method [125] are required when billions of lines are involved.

Some progress in estimating H_2 and He pressure-broadening parameters for water at elevated temperatures [50, 126] which are assumed to appropriate not only for our $H_2{}^{16}O$ line lists [29, 127] but also those for $H_2{}^{17}O$ and $H_2{}^{18}O$ [128].

In practice it is computationally prohibitive to perform radiative transport calculations on hot exoplanets line-by-line. Therefore the models either use precomputed cross sections, such as is done by τ-Rex [28, 129], or tables of k-coefficients, as done by the NEMISIS [130]. This means that appropriate values are precomputed on a grid of temperatures and pressures as inputs to such codes. Progress has been made recasting very large line lists, such as that for methane, into a background quasi-continuous pressure-independent cross section which can be combined with a full pressure-dependent treatment of the stronger lines [131, 132]. This approach would appear to provide a practical way forward.

3. Conclusion

We are in the advanced stages of developing a diet for pressure broadening of molecules present in exoplanetary atmospheres and those of other hot astronomical objects. A full discussion of this problem will be given elsewhere [133].

Pressure broadening data is urgently needed for atmospheric studies (retrievals) of exoplanets and cool stars. There is huge demand on the comprehensive solutions of the line shape problems for most of the molecules important for exoplanetary studies. The ExoMol database is arguably the main source of opacities for hot species important for modeling these atmospheres, where we have created a structure for depositing and curating any molecular data important for spectroscopic properties of hot atmospheric and other gaseous environments. We invite the molecular data producers to contribute to this database. The line shape data is especially important as the line parameters are missing or incomplete for the H_2-rich atmospheres even for the most important absorbers. Exoplanetary atmospheric retrieval is a hot topic at the moment with a lot of interest from the society, which makes it a good place to be for an expert in the molecular line profiles. The lack of data, a strong demand from the field and interest from the public is a very attractive mixture for work in this direction to be properly recognized and rewarded.

We also invite the community to visit and test the new ExoMol database at www.exomol.com. Any feedback will be greatly appreciated.

Acknowledgements

This work is supported by ERC Advanced Investigator Project 267219. We thank the support of the COST action MOLIM CM1405.

References

[1] Tinetti G, Tennyson J, Griffiths C A and Waldmann I 2012 *Phil. Trans. Royal Soc. London A* **370** 2749–2764
[2] Freedman R S, Marley M S and Lodders K 2008 *Astrophys. J. Suppl.* **174** 504
[3] Hedges C and Madhusudhan N 2016 *Mon. Not. R. Astron. Soc.* **458** 1427–1449
[4] Tennyson J and Yurchenko S N 2012 *Mon. Not. R. Astron. Soc.* **425** 21–33
[5] Yurchenko S N, Tennyson J, Bailey J, Hollis M D J and Tinetti G 2014 *Proc. Nat. Acad. Sci.* **111** 9379–9383
[6] Tennyson J, Yurchenko S N, Al-Refaie A F, Barton E J, Chubb K L, Coles P A, Diamantopoulou S, Gorman M N, Hill C, Lam A Z, Lodi L, McKemmish L K, Na Y, Owens A, Polyansky O L, Rivlin T, Sousa-Silva C, Underwood D S, Yachmenev A and Zak E 2016 *J. Mol. Spectrosc.* **327** 73–94
[7] Tennyson J, Hulme K, Naim O K and Yurchenko S N 2016 *J. Phys. B: At. Mol. Opt. Phys.* **49** 044002
[8] Laporta V, Tennyson J and Celiberto R *Plasma Sources Sci. Technol.*
[9] Patrascu A T, Tennyson J and Yurchenko S N 2015 *Mon. Not. R. Astron. Soc.* **449** 3613–3619
[10] Coppola C M, Lodi L and Tennyson J 2011 *Mon. Not. R. Astron. Soc.* **415** 487–493
[11] Yadin B, Vaness T, Conti P, Hill C, Yurchenko S N and Tennyson J 2012 *Mon. Not. R. Astron. Soc.* **425** 34–43
[12] Engel E A, Doss N, Harris G J and Tennyson J 2005 *Mon. Not. R. Astron. Soc.* **357** 471–477
[13] Pavlyuchko A I, Yurchenko S N and Tennyson J 2015 *Mon. Not. R. Astron. Soc.* **452** 1702–1706
[14] Li G, Harrison J J, Ram R S, Western C M and Bernath P F 2012 *J. Quant. Spectrosc. Radiat. Transf.* **113** 67 – 74 ISSN 0022-4073 URL `http://www.sciencedirect.com/science/article/pii/S0022407311003323`

[15] Barton E J, Chiu C, Golpayegani S, Yurchenko S N, Tennyson J, Frohman D J and Bernath P F 2014 *Mon. Not. R. Astron. Soc.* **442** 1821–1829
[16] Yurchenko S N, Blissett A, Asari U, Vasilios M, Hill C and Tennyson J 2016 *Mon. Not. R. Astron. Soc.* **456** 4524–4532
[17] Masseron T, Plez B, Van Eck S, Colin R, Daoutidis I, Godefroid M, Coheur P F, Bernath P, Jorissen A and Christlieb N 2014 *Astron. Astrophys.* **571** A47
[18] Yurchenko S N and Tennyson J 2014 *Mon. Not. R. Astron. Soc.* **440** 1649–1661
[19] Brooke J S A, Ram R S, Western C M, Li G, Schwenke D W and Bernath P F 2014 *Astrophys. J. Suppl.* **210** 23
[20] GharibNezhad E, Shayesteh A and Bernath P F 2013 *Mon. Not. R. Astron. Soc.* **432** 2043–2047
[21] Ram R S, Brooke J S A, Western C M and Bernath P F 2014 *J. Quant. Spectrosc. Radiat. Transf.* **138** 107–115
[22] Burrows A, Ram R S, Bernath P, Sharp C M and Milsom J A 2002 *Astrophys. J.* **577** 986–992
[23] Rivlin T, Lodi L, Yurchenko S N, Tennyson J and Le Roy R J 2015 *Mon. Not. R. Astron. Soc.* **451** 5153–5157
[24] Paulose G, Barton E J, Yurchenko S N and Tennyson J 2015 *Mon. Not. R. Astron. Soc.* **454** 1931–1939
[25] Brooke J S A, Ram R S, Western C M, Li G, Schwenke D W and Bernath P F 2014 *Astrophys. J. Suppl.* **210** 23 URL `http://stacks.iop.org/0067-0049/210/i=2/a=23`
[26] Wende S, Reiners A, Seifahrt A and Bernath P F 2010 *Astron. Astrophys.* **523** A58
[27] Yurchenko S N, Barber R J and Tennyson J 2011 *Mon. Not. R. Astron. Soc.* **413** 1828–1834
[28] Waldmann I P, Tinetti G, Barton E J, Yurchenko S N and Tennyson J 2015 *Astrophys. J.* **802** 107
[29] Barber R J, Tennyson J, Harris G J and Tolchenov R N 2006 *Mon. Not. R. Astron. Soc.* **368** 1087–1094
[30] Yorke L, Yurchenko S N, Lodi L and Tennyson J 2014 *Mon. Not. R. Astron. Soc.* **445** 1383–1391
[31] Neale L, Miller S and Tennyson J 1996 *Astrophys. J.* **464** 516–520
[32] Lodi L, Yurchenko S N and Tennyson J 2015 *Mol. Phys.* **113** 1559–1575
[33] Li G, Gordon I E, Bernath P F and Rothman L S 2011 *J. Quant. Spectrosc. Radiat. Transf.* **112** 1543 – 1550
[34] Barton E J, Yurchenko S N and Tennyson J 2013 *Mon. Not. R. Astron. Soc.* **434** 1469–1475
[35] Barber R J, Strange J K, Hill C, Polyansky O L, Mellau G C, Yurchenko S N and Tennyson J 2014 *Mon. Not. R. Astron. Soc.* **437** 1828–1835
[36] Burrows A, Dulick M, Bauschlicher C W, Bernath P F, Ram R S, Sharp C M and Milsom J A 2005 *Astrophys. J.* **624** 988–1002
[37] Underwood D S, Tennyson J, Yurchenko S N, Huang X, Schwenke D W, Lee T J, Clausen S and Fateev A 2016 *Mon. Not. R. Astron. Soc.* **459** 3890–3899
[38] Underwood D S, Tennyson J, Yurchenko S N, Clausen S and Fateev A 2016 *Mon. Not. R. Astron. Soc.* **462** 4300–4313
[39] Azzam A A A, Yurchenko S N, Tennyson J and Naumenko O V 2016 *Mon. Not. R. Astron. Soc.* **460** 4063–4074
[40] Al-Refaie A F, Polyansky O L, I R, Ovsyannikov, Tennyson J and Yurchenko S N 2016 *Mon. Not. R. Astron. Soc.* **461** 1012–1022
[41] McKemmish L K, Yurchenko S N and Tennyson J 2016 *Mon. Not. R. Astron. Soc.* (in press)
[42] Mizus I I, Alijah A, Zobov N F, Tennyson J and Polyansky O L 2016 *Mon. Not. R. Astron. Soc.*
[43] Lavrentieva N N, Voronin B A, Naumenko O V, Bykov A D and Fedorova A A 2014 *Icarus* **236** 38
[44] Sousa-Silva C, Yurchenko S N and Tennyson J 2016 *J. Chem. Phys.* **145** 091102
[45] Steyert D W, Wang W, Sirota J, Donahue N M and Reuter D C 2004 *J. Quant. Spectrosc. Radiat. Transf.* **83** 183–191
[46] Brown L R and Plymate C 1996 *J. Quant. Spectrosc. Radiat. Transf.* **56** 263 – 282
[47] Brown L R, Benner D C, Devi V M, Smith M A H and Toth R A 2005 *J. Mol. Struct.* **742** 111–122
[48] Gamache R R, Lynch R and Brown L R 1996 *J. Quant. Spectrosc. Radiat. Transf.* **56** 471 – 487
[49] Dick M J, Drouin B J and Pearson J C 2009 *J. Quant. Spectrosc. Radiat. Transf.* **110** 619 – 627
[50] Faure A, Wiesenfeld L, Tennyson J and Drouin B J 2013 *J. Quant. Spectrosc. Radiat. Transf.* **116** 79–86
[51] Langlois S, Birbeck T P and Hanson R K 1994 *J. Mol. Spectrosc.* **167** 272 – 281
[52] Dutta J M, Jones C R, Goyette T M and Lucia F C 1993 *Icarus* **102** 232–239
[53] Golubiatnikov Y G 2005 *J. Mol. Spectrosc.* **230** 196–198
[54] Zeninari V, Parvitte B, Courtois D, Lavrentieva N N, Ponomarev Y N and Durry G 2004 *Mol. Phys.* **102** 1697–1706
[55] Drouin B and Wiesenfeld L 2012 *Phys. Rev. A* **86** 022705
[56] Petrova T M, Solodov A M, Solodov A A and Starikov V I 2016 *J. Mol. Spectrosc.* **321** 50–58
[57] Petrova T M, Solodov A M, Solodov A A and Starikov V I 2013 *J. Quant. Spectrosc. Radiat. Transf.* **129** 241–253

[58] Petrova T M, Solodov A M, Solodov A A and Starikov V I 2012 *Mol. Phys.* **110** 1493–1503
[59] Solodov A M and Starikov V I 2008 *Opt. Spectrosc.* **105** 14–20
[60] Solodov A M and Starikov V I 2009 *Mol. Phys.* **104** 43–51
[61] Lazarev V V, Ponomarev Y N, Sumpf B, Fleischmann O, Waschull J, Kronfeldt H and Stroinova V N 1995 *J. Mol. Spectrosc.* **173** 177–193
[62] Mérienne M F, Jenouvrier A, Hermans C, Vandaele A C, Carleer M, Coheur P F, Colin R, Fally S and Bach M 2003 *J. Quant. Spectrosc. Radiat. Transf.* **82** 99–117
[63] Gamache R R and Hartmann J M 2004 *Can. J. Chem.* **82** 1013–1027
[64] Gasster S D, Townes C H, Goorvitch D and Valero F P J 1988 *J. Opt. Soc. Am. B* **5** 593–601
[65] Payne V, Delamere J, Cady-Pereira K, Gamache R, Moncet J L, Mlawer E and Clough S 2008 *IEEE Transactions on Geoscience and Remote Sensing* **46** 3601–3617
[66] Gamache R 2005 *J. Mol. Spectrosc.* **229** 9–18
[67] Gamache R and Laraia A 2009 *J. Mol. Spectrosc.* **257** 116–127
[68] Markov V 1994 *J. Mol. Spectrosc.* **164** 233–238
[69] Golubiatnikov G, Koshelev M and Krupnov A 2008 *J. Quant. Spectrosc. Radiat. Transf.* **109** 1828–1833
[70] Cazzoli G, Puzzarini C, Buffa G and Tarrini O 2008 *J. Quant. Spectrosc. Radiat. Transf.* **109** 2820–2831
[71] Pine A S 1992 *J. Chem. Phys.* **97** 773
[72] Margolis J S 1993 *J. Quant. Spectrosc. Radiat. Transf.* **50** 431
[73] Fox K, Jennings D E, Stern E A and Hunnard R 1988 *J. Quant. Spectrosc. Radiat. Transf.* **39** 473 – 476
[74] Strong K, Taylor F W, Calcutt S B, Remedios J J and Ballard J 1993 *J. Quant. Spectrosc. Radiat. Transf.* **50** 363 – 429
[75] Varanasi P and Chudamani S 1990 *J. Quant. Spectrosc. Radiat. Transf.* **43** 1
[76] Gabard T, Grigoriev I M, Grigorovich N M and Tonkov M V 2004 *J. Mol. Spectrosc.* **225** 123 – 131
[77] Grigoriev I M, Filippov N N, Tonkov M V, Gabard T and Le Doucen R 2001 *J. Quant. Spectrosc. Radiat. Transf.* **69** 189 – 204
[78] Smith M A H, Benner D C, Predoi-Cross A and Malathy Devi V 2009 *J. Quant. Spectrosc. Radiat. Transf.* **110** 639–653
[79] Brown L R, Benner D C, Champion J P, Devi V M, Fejard L, Gamache R R and et al 2003 *J. Quant. Spectrosc. Radiat. Transf.* **82** 219–238
[80] Antony B K, Niles D L, Wroblewski S B, Humphrey C M, Gabard T and Gamache R R 2008 *J. Mol. Spectrosc.* **251** 268–281
[81] Predoi-Cross A, Brown L R, Devi V M, Brawley-Tremblay M and Benner D C 2005 *J. Mol. Spectrosc.* **232** 231–246
[82] Smith M A H, Benner D C, Predoi-Cross A and Malathy Devi V 2009 *J. Quant. Spectrosc. Radiat. Transf.* **111** 1152–1166
[83] Pine A S, Markov V N, Buffa G and Tsrrini O 1993 *J. Quant. Spectrosc. Radiat. Transf.* **50** 337
[84] Hadded S, Arouli H, Orphal J, Bouanich J P and Hartmann J M 2001 *J. Mol. Spectrosc.* **210** 275
[85] Sharp C M and Burrows A 2007 *Astrophys. J. Suppl.* **168** 140
[86] Baldacchini G, D'Amato F, Buffa G, Tarrini O, De Rosa M and Pelagalli F 2001 *J. Quant. Spectrosc. Radiat. Transf.* **68** 625 – 633
[87] Brown L R and Peterson D B 1994 *J. Mol. Spectrosc.* **168** 593–606
[88] Nemtchinov V, Sung K and Varanasi P 2004 *J. Quant. Spectrosc. Radiat. Transf.* **83** 243 – 265
[89] Levy A, Lacome N and Tarrago G 1993 *J. Mol. Spectrosc.* **157** 172 – 181
[90] Sergent-Rozey M, van Thanh N, Rossi I, Lacome N and Levy A 1988 *J. Mol. Spectrosc.* **131** 66 – 76
[91] Salem J, Bouanich J P, Walrand J, Aroui H and Blanquet G 2004 *J. Mol. Spectrosc.* **228** 23 – 30
[92] Pickett H M, Poynter R L and Cohen E A 1981 *J. Quant. Spectrosc. Radiat. Transf.* **26** 197 – 198
[93] Levy A, Lacome N and Tarrago G 1994 *J. Mol. Spectrosc.* **166** 20 – 31
[94] Bouanich J P, Salem J, Aroui H, Walrand J and Blanquet G 2004 *J. Quant. Spectrosc. Radiat. Transf.* **84** 195 – 205
[95] Salem J, Bouanich J P, Walrand J, Aroui H and Blanquet G 2005 *J. Mol. Spectrosc.* **232** 247 – 254
[96] Butler R A H, Sagui L, Kleiner I and Brown L R 2006 *J. Mol. Spectrosc.* **238** 178–192
[97] Kleiner I, Tarrago G, Cottaz C, Sagui L, Brown L R, Poynter R L, Pickett H M, Chen P, Pearson J C, Sams R L, Blake G A, Matsuura S, Nemtchinov V, Varanasi P, Fusina L and Di Lonardo G 2003 *J. Quant. Spectrosc. Radiat. Transf.* **82** 293 – 312
[98] Nguyen-Van-Thanh, Rossi I and Sergent-Rozey M 1989 *J. Mol. Spectrosc.* **135** 410 – 414
[99] Salem J, Aroui H, Bouanich J P, Walrand J and Blanquet G 2004 *J. Mol. Spectrosc.* **225** 174 – 181
[100] Nerf R B J 1975 *J. Mol. Spectrosc.* **58** 451–473
[101] Jacquemart D, Laraia A, Tchana F K, Gamache R, Perrin A and Lacome N 2010 *J. Quant. Spectrosc. Radiat. Transf.* **111** 1209–1222

[102] Cazzoli G and Puzzarini C 2012 *J. Quant. Spectrosc. Radiat. Transf.* **113** 1051–1057
[103] Tasinato N, Charmet A P, Stoppa P, Giorgianni S and Buffa G 2014 *Spectrochimica Acta Part A - Molecular and Biomolecular Spectroscopy* **118** 373–379
[104] Yang C, Buldyreva J, Gordon I E, Rohart F, Cuisset A, Mouret G, Bocquet R and Hindle F 2008 *J. Quant. Spectrosc. Radiat. Transf.* **109** 2857–2868
[105] Devi V M, Benner D C, Smith M A H, Rinsland C P, Sharpe S and Sams R L 2004 *J. Quant. Spectrosc. Radiat. Transf.* **87** 339–366
[106] Rinsland C P, Devi V M, Benner D C, Smith M A H, Sharpe S and Sams R L 2004 *J. Quant. Spectrosc. Radiat. Transf.* **82** 343–362
[107] Devi V M, Benner D C, Smith M A H, Rinsland C P, Sharpe S and Sams R L 2003 *J. Quant. Spectrosc. Radiat. Transf.* **82** 319–342
[108] Mehrotra S C, Mader H, de Vreede J P M and Dijkerman H A 1985 *Chem. Phys.* **93** 115 – 125
[109] Hurtmans D, Henry A, Valentin A and Boulet C 2009 *J. Mol. Spectrosc.* **254** 126–136
[110] Pine A S and Fried A 1985 *J. Mol. Spectrosc.* **114** 148–162
[111] Toth R A, Hunt R H and Plyler E K 1970 *J. Chem. Phys.* **53** 4304–4307
[112] Rank D H, Eastman D P, Rao B S and Wiggins T A 1963 *J. Mol. Spectrosc.* **10** 34–50
[113] Blanquet G, Walrand J and Bouanich J P 1999 *J. Mol. Spectrosc.* **198** 408–415
[114] Misago F, Lepére M and Bouanich J P 2009 *J. Mol. Spectrosc.* **254** 16–19
[115] May R D and Webster C 1989 *J. Mol. Spectrosc.* **138** 383–397
[116] Tennyson J, Bernath P F, Campargue A, Császár A G, Daumont L, Gamache R R, Hodges J T, Lisak D, Naumenko O V, Rothman L S, Tran H, Zobov N F, Buldyreva J, Boone C D, De Vizia M D, Gianfrani L, Hartmann J M, McPheat R, Murray J, Ngo N H, Polyansky O L and Weidmann D 2014 *Pure Appl. Chem.* **86** 1931–1943
[117] Humlicek J 1979 *J. Quant. Spectrosc. Radiat. Transf.* **21** 309–313
[118] Johnson S G 2012 *The Faddeeva package* http://ab-initio.mit.edu/wiki/index.php/Faddeeva_Package
[119] Tan X 2013 *J. Quant. Spectrosc. Radiat. Transf.* **129** 101–108
[120] Amundsen D S, Baraffe I, Tremblin P, Manners J, Wolfgang H, Mayne N J and Acreman D M 2014 *Astron. Astrophys.* **564** A59
[121] Wilzewski J S, Gordon I E, Kochanov R V, Hill C and Rothman L S 2016 *J. Quant. Spectrosc. Radiat. Transf.* **168** 193 – 206
[122] Madhusudhan N 2012 *Astrophys. J.* **758** 36
[123] Tsiaras A, Rocchetto M, Waldmann I P, Tinetti G, Varley R, Morello G, Barton E J, Yurchenko S N and Tennyson J 2016 *Astrophys. J.* **820** 99
[124] Schreier F 2011 *J. Quant. Spectrosc. Radiat. Transf.* **112** 1010 – 1025
[125] Evans T M, Sing D K, Wakeford H R, Nikolov N, Ballester G E, Drummond B, Kataria T, Gibson N P, Amundsen D S and Spake J 2016 *Astrophys. J.* **822** L4
[126] Barton E J, Hill C, Yurchenko S N, Tennyson J, Dudaryonok A and Lavrentieva N N 2016 *J. Quant. Spectrosc. Radiat. Transf.* (in press)
[127] Polyansky O L, Kyuberis A A, Lodi L, Tennyson J, Ovsyannikov R I, Zobov N and Yurchenko S N 2016 *Mon. Not. R. Astron. Soc.*
[128] Polyansky O L, Kyuberis A A, Lodi L, Tennyson J, Ovsyannikov R I and Zobov N 2016 *Mon. Not. R. Astron. Soc.*
[129] Waldmann I P, Rocchetto M, Tinetti G, Barton E J, Yurchenko S N and Tennyson J 2015 *Astrophys. J.* **813** 13
[130] Irwin P G J, Teanby N A, de Kok R, Fletcher L N, Howett C J A, Tsang C C C, Wilson C F, Calcutt S B, Nixon C A and Parrish P D 2008 *J. Quant. Spectrosc. Radiat. Transf.* **109** 1136–1150
[131] Hargreaves R J, Bernath P F, Bailey J and Dulick M 2015 *Astrophys. J.* **813** 12 URL `http://stacks.iop.org/0004-637X/813/i=1/a=12`
[132] Yurchenko S N, Tennyson J, Amundsen D S and Baraffe I 2016 *Astron. Astrophys.*
[133] Barton E J, Hill C, Czurylo M, Li H Y, Hyslop A, Yurchenko S N and Tennyson J 2016 *J. Quant. Spectrosc. Radiat. Transf.*

Line shape models for magnetized hydrogen plasmas

J Rosato, I Hannachi, Y Marandet and R Stamm

Aix-Marseille Université, CNRS, PIIM UMR 7345, F-13397 Marseille Cedex 20, France

joel.rosato@univ-amu.fr

Abstract. We report on Stark-Zeeman line shape models for the diagnostic of magnetic fusion plasmas. Computer simulations, which serve as a reference for an evaluation of the ion dynamics effects, are usually CPU time consuming, so that they cannot be used in a real-time diagnostic. In this framework, a database that allows a fast evaluation of Stark-Zeeman line shapes is currently under preparation. We present preliminary results. Analytical models based on ion collision operators and suitable for strong (near-impact) ion dynamics regimes are also discussed.

1. Introduction

ITER is an international project aimed at demonstrating the feasibility of fusion as a source of energy. This machine, presently under construction in Cadarache (France), is a large scale tokamak designed so as to maintain fusion for long periods of time. It will be the first fusion device to test the integrated technologies, materials, and physics regimes necessary for the commercial production of fusion-based electricity (see the website: www.iter.org). The first plasma is planned for 2025. A "Demonstration Fusion Power Reactor" (DEMO) is next planned to follow ITER.

In accompaniment to ITER, research campaigns are ongoing on present tokamaks. We have taken part to a project devoted to the improvement of passive spectroscopy diagnostics involving hydrogen line shapes for a better characterization of divertor plasmas (MST1, "Medium-Size Tokamak Task Force"). The strong density dependence of Stark broadening renders this broadening mechanism suitable for an investigation of the divertor plasma properties in the so-called "detached regime" where a large amount of neutrals are present. This regime is of interest for the characterization of the plasma-wall interactions. The Stark broadening is important on lines with a high principal quantum number n, and it can be significant as well on lines with a small n like Balmer α (with respect to the Doppler broadening) at the highest density regimes ($N_e \sim 10^{15}$ cm^{-3}); for observations, e.g. see [1,2]. In general, the Stark broadening is affected by the ion dynamics, from the near-static limit (especially on high-n lines) to the near-impact limit (low-n lines at a low density); this effect has to be retained in diagnostic models. The presence of a strong magnetic field (of several teslas) in tokamak plasmas yields an additional splitting due to the Zeeman effect, which has also to be retained. We report here on a database which is currently under preparation. A computer simulation method is used. Preliminary results for Dγ (deuterium Balmer γ transition, $n = 5 \rightarrow 2$) have been obtained last year [3] for an improvement of line shape-based diagnostics of the ASDEX-Upgrade tokamak divertor plasma [4]. We discuss the dependence on the magnetic field and the temperature. Analytical methods based on ion collision operator models and suitable for near-impact ion dynamics regimes are discussed.

2. Computer simulations for Stark broadening with account of Zeeman effect

Computer simulations for the investigation of ion dynamics were initiated in the late seventies and they are now widely used for Stark broadening calculations, e.g. [5–8]. Like the model microfield method (MMM) [9,10], this method can be used in the elaboration of tables, e.g. [6]. The purpose of simulations is to numerically reproduce the motion of the charged particles in the plasma so as to obtain the time dependent electric microfield $\mathbf{E}(t)$ at the location of an emitter. While being CPU intensive, this method has the advantage of providing reference line shapes. We only give here a summary of the technique (it was discussed many times at earlier sessions of the ICSLS conference). Essentially, a numerical simulation consists of: (i) the calculation of a set of realizations for the electric field; (ii) the numerical integration of the Schrödinger equation for each realization, which provides an expression for the atomic evolution operator $U(t)$; (iii) an evaluation of the dipole autocorrelation function by ensemble averaging and a Fourier transform that provides the spectrum:

$$I(\omega) \propto \frac{1}{\pi} \mathrm{Re} \int_0^\infty dt \left\langle \mathbf{d}_\perp \cdot U^+(t) \mathbf{d}_\perp U(t) \right\rangle e^{i\omega t} . \tag{1}$$

The $\perp$ symbol used here indicates that the dipole is projected onto the polarization plane; this specification is required because of the anisotropy induced by the magnetic field (Zeeman effect). In the simulations performed for this work, we have used a code [11] developed according to the method reported in early papers [5]. We consider that the ions move along straight line trajectories with constant velocities, sampled among the particles according to an equilibrium Maxwell distribution function. The electrons are not simulated here, but are described with an impact collision operator. The treatment of the correlations between ions and electrons is retained by using Debye screened fields. A cubic cell with periodic boundary conditions is considered. For each history of the electric field, the code solves the time dependent Schrödinger equation for the evolution operator $U(t)$ according to the algorithm $U(t + \Delta t) = U(t + \Delta t, t)U(t)$, with $U(t + \Delta t, t)$ being the infinitesimal evolution operator between times t and $t + \Delta t$. The latter operator is approximated as $\exp\{-i[H_0 + V(t)]\, \Delta t/\hbar\}$ and the exponential is evaluated using the scaling and squaring method [12]. Note that the Doppler broadening is not retained in Eq. (1); it can be accounted for by convoluting Eq. (1) with the atomic velocity distribution function (except in specific conditions, viz. if the velocity $\mathbf{v}$ is correlated with the electric field $\mathbf{E}(t)$ or if the Lorentz field $\mathbf{v}\times\mathbf{B}$ is important, e.g. [13]). An example of simulated spectrum is shown in Fig. 1. The spectral profile of the D-α (deuterium Balmer α) line in conditions relevant to high density divertor plasmas is shown. As can be seen in the figure, the Stark broadening is comparable to the Doppler broadening. The magnetic field yields a Zeeman-Lorentz triplet structure, which is visible on the spectrum. This structure tends to disappear as the principal quantum number increases, in accordance with the Stark effect trend (Fig. 2).

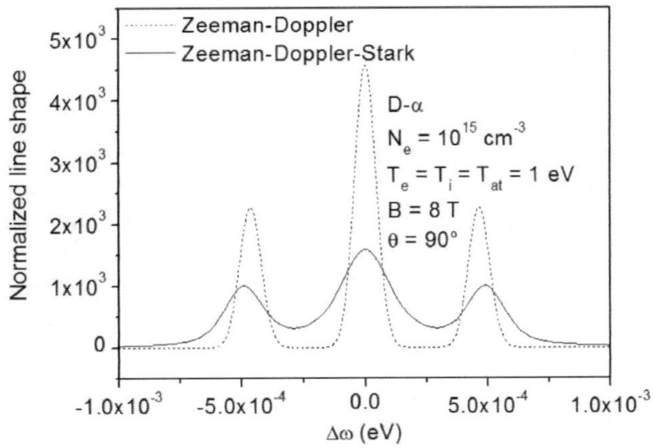

Figure 1. Plot of the D-α line shape obtained from a computer simulation, at conditions relevant to tokamak divertor plasmas at high density regimes. The Stark broadening is comparable to the Doppler broadening. The triplet structure stems from the Zeeman effect.

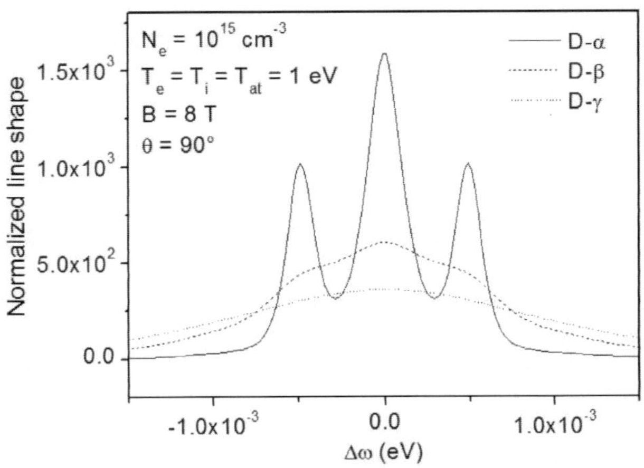

Figure 2. The Stark broadening increases with the principal quantum number and tends to mask the Zeeman triplet structure. Here, the profiles of D-α, D-β, and D-γ have been simulated assuming the same plasma conditions as in Fig. 1.

A line shape database is currently under preparation for the diagnostic of divertor plasmas in detached regime. Spectral profiles of the first Balmer lines (up to $n = 7$) of deuterium have been calculated for a set of densities, temperatures, and magnetic field strengths relevant to magnetic fusion edge plasmas, in such a way to provide a table that can be used in an analysis of experimental spectra [3]. The table provides Stark-Zeeman line shapes for equal ion and electron temperatures with values comprised between 0.316 and 31.6 eV, a plasma density ranging from 10^{13} to 10^{16} cm^{-3}, and a magnetic field up to 5 T. The Doppler broadening is not retained in the table because the atomic velocity distribution can be strongly dependent of the tokamak plasma discharge under consideration (e.g., it can take the form of a superposition of Maxwellian functions with different temperatures owing to various nonthermalized populations [14,15]). Figure 3 shows an example of result: the half-width at half-maximum of the D-α line is plotted against the density for the two temperatures 1 eV and 10 eV, in the absence of a magnetic field (hence no Zeeman effect is present here). As can be seen, the width roughly follows a power law in terms of the density, and it is weakly sensitive to the temperature. Note the change in trend: according to the density value, a higher temperature can result either in a higher or a smaller line width. The former effect is expected at high densities as a result of ion dynamics broadening, whereas the line width reduction with an increasing temperature at low densities can be explained within a collisional picture (note the $T^{-1/2}$ trend of the ion collision operator). An adjustment of the plot using the empirical formula HWHM = $C \times N_e^{\alpha}$, with C and α being constants, is currently ongoing, without and with the Zeeman effect. Preliminary results have indicated a power close to unity.

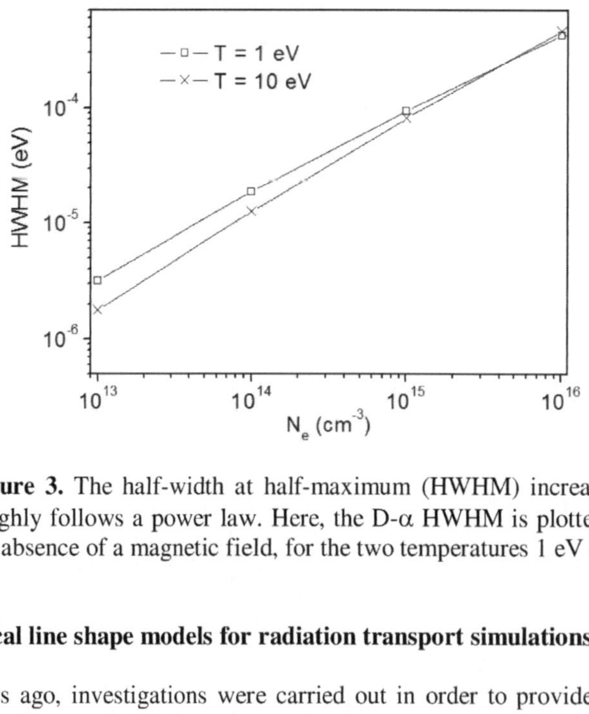

Figure 3. The half-width at half-maximum (HWHM) increases with the density. It roughly follows a power law. Here, the D-α HWHM is plotted against the density in the absence of a magnetic field, for the two temperatures 1 eV and 10 eV.

3. Analytical line shape models for radiation transport simulations

A few years ago, investigations were carried out in order to provide analytical formulas for Stark-Zeeman line shapes suitable for a fast evaluation in radiation transport simulations. This concerns the Lyman α and Lyman β lines in particular, given the short absorption length they provide (the corresponding photons can be absorbed after traveling over less than 1 mm; e.g. [16] for a mean free path estimate; see also [17,18] for experimental observations). The corresponding atom excitation can play an important role on the ionization-recombination balance. In this framework, simulations of the coupled plasma-atom-radiation transport are presently ongoing [19]. The EIRENE code utilized in the magnetic fusion community (see www.eirene.de) employs a kinetic Monte Carlo method for a description of the Lyman line radiation. Photons are described as point-like particles performing random walks and interacting with the atoms according to prescribed statistical properties. The spectral line shape function presented in Eq. (1), when normalized to unity, provides a key quantity for the generation of the photon frequency (in an emission event) or its path length (when dealing with absorption). The broadening of the Lyman lines with a low principal quantum number is usually governed by the Doppler effect. The Zeeman effect provides a correction either visible in the Lorentz triplet form or as an additional broadening if the magnetic field is not sufficiently strong. The Stark effect, which is usually visible in the wings, provides a significant additional broadening on the whole spectrum at the highest density conditions ($N_e > 10^{15}$ cm^{-3}, e.g. [20]). Several approaches were used to model the Stark-Zeeman spectral profile in the atomic frame of reference (see [20] and Refs. therein). Because of the low principal quantum number, the ions are in a strong dynamics regime, which renders their description tricky. An example of approach that can used for the lowest densities is the impact approximation adapted to ions. The ratio b_W/r_0 between the ion Weisskopf radius and the mean interparticle distance is smaller than 10% at $N_e = 10^{13}$ cm^{-3} and $T_{e,i} = 1$ eV, suggesting that the binary assumption can be used as a reasonable approximation. We have adapted the early approaches to impact broadening developed by Griem et al. to ionic perturbers [11]. The original treatment of hydrogen line broadening (Griem-Kolb-Shen 1959 model [21]) cannot be applied in a strictly sense due to the Zeeman energy degeneracy removal, but it can be readily adapted following the same method as for helium lines (Griem-Baranger-Kolb-Oertel 1962 model [22]). Figure 4 shows an example of application to Lyman α. The computer simulation result (where many perturbers are retained) is also shown. The collision operator model is in a good agreement with the simulation.

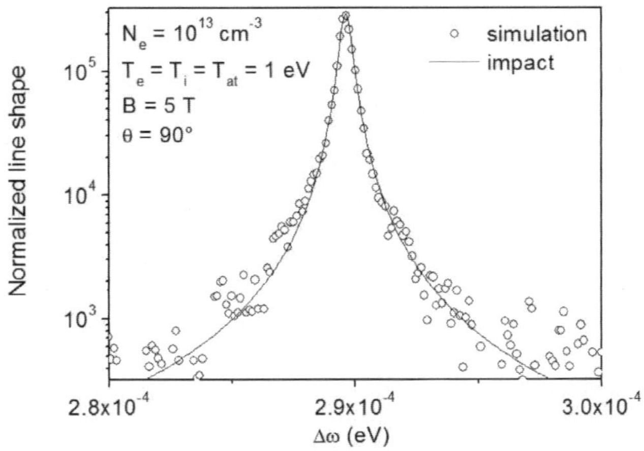

Figure 4. Plot of the Lyman α blue lateral component calculated using an impact model for ions, at conditions such that the binary assumption can be used as a reasonable approximation. The Doppler broadening is not retained here. The Zeeman energy degeneracy removal is accounted for through the GBKO method (see text). As can be seen, the model is in a good agreement with the simulation.

E. Oks made criticisms during the ICSLS conference regarding both the impact model we used and computer simulations [23], with the claim that his models "GT" and "AGT" are more accurate. The "GT" and "AGT" are nothing but a reformulation of the impact theory, which involves a binary approximation like the older models by Griem et al. The subtlety mentioned by Oks concerns the interaction picture used in his theories. In a previous study [24], we pointed out an inconsistency in Oks' theories, which can be avoided readily (in technical points: this concerns the description of the strong collisions). A cut-off procedure in the evaluation of integrals, as done in the models by Griem et al., would be appropriate.

4. Conclusion

The operation of magnetic fusion devices requires passive spectroscopy diagnostics in the edge plasma region. In this framework, we are developing a database for the interpretation of the hydrogen Balmer lines in conditions of recombining divertor plasmas. This database accounts for the simultaneous action of the Zeeman and Stark effects, and it can be used for a density diagnostic routine involving Stark broadening. In the framework of radiation transport simulations, analytical formulas for Stark-Zeeman line shapes in the atomic frame of reference are suitable. An example of method concerns the impact theory adapted to ions. Comparisons to computer simulations have indicated that the use of a collision operator model following early (Griem et al.) approaches is appropriate when the emitter-perturber interactions are binary. Extensions to non-binary interactions can be performed using kinetic theory methods [25–27].

References

[1] Welch B L, Griem H R, Terry J, Kurz C, LaBombard B, Lipschultz B, Marmar E and McCraken G 1995 *Phys. Plasmas* **2** 4246–51
[2] Lipschultz B, Terry J L, Boswell C, Hubbard A, LaBombard B and Pappas DA 1998 *Phys. Rev. Lett.* **81** 1007–10
[3] Rosato J, Bufferand H, Koubiti M, Marandet Y and Stamm R 2015 *J. Quant. Spectrosc. Radiat. Transfer* **165** 102–7
[4] Potzel S, Dux R, Müller H W, Scarabosio A, Wischmeier M and ASDEX Upgrade team 2014 *Plasma Phys. Control. Fusion* **56** 025010-1–9
[5] Stamm R, Smith E W and Talin B 1984 *Phys. Rev. A* **30** 2039–46
[6] Gigosos M A, Gonzalez M A and Cardenoso V 2003 Spectrochim. Acta B **58** 1489–504
[7] Halenka J and Olchawa W 1996 *J. Quant. Spectrosc. Radiat. Transfer* **56** 17–28
[8] Stambulchik E, Alexiou S, Griem H R and Kepple P C 2007 *Phys. Rev. E* **75** 016401-1–9
[9] Brissaud A and Frisch U 1971 *J. Quant. Spectrosc. Radiat. Transfer* 1767–83
[10] Stehlé C 1994 *Astron. Astrophys. Suppl. Ser.* **104** 509–27
[11] Rosato J, Marandet Y, Capes H, Ferri S, Mossé C, Godbert-Mouret L, Koubiti K and Stamm R 2009 *Phys. Rev. E* **79** 046408-1–7
[12] Moler C and Van Loan C 2003 *SIAM Rev.* **45** 3–49
[13] Rosato J, Marandet Y and Stamm R 2014 *J. Phys. B: At. Mol. Opt. Phys.* **47** 105702-1–8
[14] Marandet Y, Genesio P, Koubiti M, Godbert-Mouret L, Felts B, Stamm R, Capes H and Guirlet R 2004 *Nucl. Fusion* **44** S118–22
[15] Hey J D, Chu C C, Mertens Ph, Brezinsek S and Unterberg B 2004 *J. Phys. B: At. Mol. Opt. Phys.* **37** 2543–67
[16] Reiter D, Wiesen S and Born M 2003 J. Nucl. Mater. **313–6** 845–51
[17] Goto M and Morita S 2010 *Plasma and Fusion Research* **5** S2089-1–4
[18] Goto M, Sawada K, Oishi T and Morita S, these proceedings
[19] Rosato J, Marandet Y, Bufferand H, Reiter D and Stamm R *Contrib. Plasma Phys.* in press
[20] Rosato J, Reiter D, Kotov V, Marandet Y, Capes H, Godbert-Mouret L, Koubiti M and Stamm R 2010 *Contrib. Plasma Phys.* **50** 398–403
[21] Griem H R, Kolb AC and Shen K Y 1959 *Phys. Rev.* **116** 4–16
[22] Griem H R, Baranger M, Kolb AC and Oertel G 1962 *Phys. Rev.* **125** 177–95
[23] Oks E, these proceedings
[24] Rosato J, Capes H, Godbert-Mouret L, Koubiti M, Marandet Y and R Stamm 2012 *J. Phys. B: At. Mol. Opt. Phys.* **45** 165701-1–7
[25] Omar B, Günter S, Wierling A and Röpke G 2006 *Phys. Rev. E* **73** 056405-1–13
[26] Lin Ch, Reinholz H and Röpke G, these proceedings
[27] Rosato J, Capes H and Stamm R 2012 *Phys. Rev. E* **86** 046407-1–8

IOP Conf. Series: Journal of Physics: Conf. Series **810** (2017) 012012 doi:10.1088/1742-6596/810/1/012012

Self-absorption characteristics of measured laser-induced plasma line shapes

C G Parigger, D M Surmick and G Gautam

University of Tennessee Space Institute, Center for Laser Applications, 411 B.H. Goethert Parkway, Tullahoma, TN 37388, USA

E-mail: `cparigge@tennessee.edu`

Abstract. The determination of electron density and temperature is reported from line-of-sight measurements of laser-induced plasma. Experiments are conducted in standard ambient temperature and pressure air and in a cell containing ultra-high-pure hydrogen slightly above atmospheric pressure. Spectra of the hydrogen Balmer series lines can be measured in laboratory air due to residual moisture following optical breakdown generated with 13 to 14 nanosecond, pulsed Nd:YAG laser radiation. Comparisons with spectra obtained in hydrogen gas yields Abel-inverted line shape appearances that indicate occurrence of self-absorption. The electron density and temperature distributions along the line of sight show near-spherical rings, expanding at or near the speed of sound in the hydrogen gas experiments. The temperatures in the hydrogen studies are obtained using Balmer series alpha, beta, gamma profiles. Over and above the application of empirical formulae to derive the electron density from hydrogen alpha width and shift, and from hydrogen beta width and peak-separation, so-called escape factors and the use of a doubling mirror are discussed.

1. Introduction

The study of line shapes is essential for remote diagnosis of radiation emitted from a source. Plasma spectroscopy often relies on capturing emission characteristics for the purpose of determining physical parameters including assessments of dynamics [1, 2]. Equally, combustion physics addresses inferences from emitted atomic and molecular radiation. In this work, analysis of laser-induced plasma generated by focused, pulsed laser radiation is of interest, especially in view of self-absorption phenomena. A review is presented of general, expected appearances of emitted atomic line profiles that indicate re-absorption or self-absorption [3–8]. Standard techniques of Abel inversion are applied to gain information of the radiation source that appears very close to symmetric and optically thin for time delays of the order of 0.15 to 10 μs delay for ultra-high-pure (UHP) hydrogen gas and 0.4 to 10 μs for standard ambient temperature and pressure (SATP) laboratory air. In the context of the treatment of self-absorption and line-reversal, the escape factor approach is discussed [9–11]. The experiments also utilize a so-called doubling mirror for the evaluation of the degree of self-absorption [12, 13]. The measurements primarily concentrate on the hydrogen Balmer series lines with some references to the recorded ionized nitrogen lines in the SATP moisture-containing laboratory air [14, 15]. The experimental arrangement includes the use of Q-switched, 1064-nm Nd:YAG laser radiation of 13 to 14 ns duration with pulse energies of 170 mJ. Laser-induced plasma is generated by focusing the laser beam from the top, aligned parallel to the the slit of a Czerny-Turner spectrometer equipped with a 1200 grooves/mm holographic grating and an intensified charge coupled detector (ICCD) to record temporally and spatially resolved spectra with a 10-ns gate-width that are wavelength and sensitivity corrected. Further details regarding the focusing and collection optics, laser device, cell for the UHP hydrogen and SATP air experiments are reported elsewhere including video of typical procedures [16].

2. Escape factor and self-absorption

The escape factor describes the amount of radiation that emanates from a radiation source, and it can be utilized to determine the amount of self-absorption. Three separate integrations are usually required to evaluate this factor [12]. First, an integration over the frequency interval of the emitting line profile while assuming complete redistribution of radiation, second, a surface integral to find the amount of radiation that escapes, and third, a volume integral over the extent of the source. For a point-like source that is usually described by a Dirac delta distribution, both surface and volume integrals are easily evaluated leading to the expected transmission factor, $T(\tau_0)$, as function of the optical thickness, τ_0,

$$T(\tau_0) = \int L(\nu) \exp\{-\tau_0 L(\nu)/L_0\} d\nu, \quad (1)$$

with the line profile, $L(\nu)$, normalized by the maximum, L_0, at line center. The optical thickness is related to the opacity, K_0, and absorption length, ℓ, as usual, $\tau_0 = K_0 \ell$. For the line profile of the laser-induced plasma source, a Lorentzian and a double-peaked profile are typically used to describe the central regions of Stark-broadened hydrogen Balmer alpha and beta lines, respectively [17–19]. For a delta-distributed line profile [12], Equation (1) yields the expected result for a transmission factor of $\exp\{-\tau_0\}$. Figure 1 illustrates the integrand of the transmission factor for increasing values of τ_0.

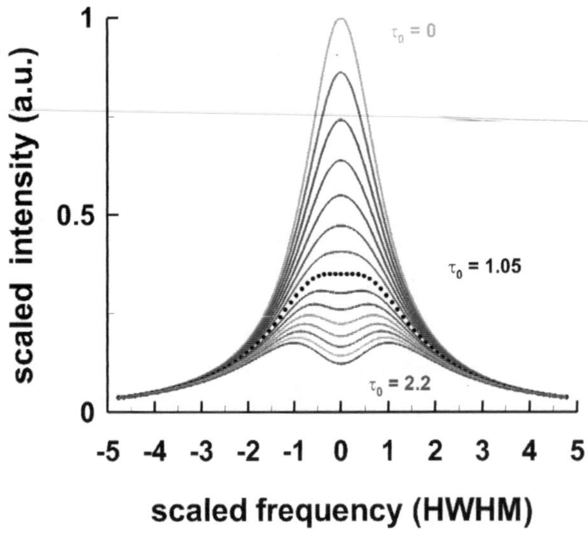

Figure 1. Integrand of the transmission factor for optical thickness values of τ_0 = 0 to 2.20 for a Lorentzian vs. frequency, scaled with respect to the half-width-at-half-maximum (HWHM). For $\tau_0 = 1.05$, the central region shows a reduction by a factor of three from the optical thin line limit $\tau_0 \to 0$ and an increase of the HWHM by a factor of 2. For $\tau_0 = 2.20$, line reversal is clearly developed.

Computations of the escape factors for constant, spherically symmetric distributions of radiators [10] show diminished self-absorption effects and absence of line-reversal in the 0 to 2.20 range for the optical thickness, τ_0.

3. Doubling mirror method

Assessment of the level of self-absorption can be accomplished by including a so-called signal doubling mirror in the experimental arrangement [12]. In a practical realization, also a second lens is utilized for the purpose of matching the radiation source volume that is imaged onto the slit of a spectrometer [13]. The signal should be doubled for an ideal lens and mirror. Consideration of optically thin conditions, i.e., for $\tau_0 \to 0$ or when $\exp\{-\tau_0\} \approx 1 - \tau_0 + \tau_0^2/2$, leads to the correction factor, $K_{\lambda,\mathrm{corr}}$, that can be applied to measured line-of-sight spectra to adjust for self-absorption. The self-absorption corrections are based on the ratios of the wavelength dependent signals, R_λ, and the continuum contribution, R_C,

$$K_{\lambda,\mathrm{corr}} = \frac{\ln\{y_\lambda\}}{y_\lambda - 1} \qquad \text{with} \qquad \mathrm{y}_\lambda = \frac{\mathrm{R}_\lambda - 1}{\mathrm{R_C} - 1}. \tag{2}$$

The ratios R_λ and R_C are found by comparing recorded data with and without the doubling mirror. As the correction factors are applied, not only values for the spectral peaks but rather the entire profile is corrected for distortions including slight shifts of the line's center wavelength. The inverse of the average of the correction factor, $1/\langle K_{\lambda,\mathrm{corr}}\rangle$, corresponds to the self-absorption factor, SA.

A method for determination of self-absorption due to free-free or inverse Bremsstrahlung would be direct measurement of the optical thickness from the plasma emission coefficient and the plasma radius [20]. The value of the self-absorption, SA, at line center is evaluated using

$$SA = \frac{1 - \exp\{-\tau_0\}}{\tau_0} \qquad \text{and} \qquad \mathrm{SA} \approx 1 - \tau_0/2 \qquad \text{for} \qquad \tau_0 \ll 1. \tag{3}$$

The determination of the value of the average self-absorption using this method, however, in absence of spectral lines and with the continuum radiation dominating the emission spectrum, would require tedious absolute calibration of the detector. In comparison, the doubling mirror method yields wavelength-dependent factors for correction of the self-absorption distortions.

4. Abel inversion of line-of-sight data

Spatial information about the radiation source can be extracted from laterally recorded data. The initially expanding laser induced plasma reveals asymmetries associated with focusing and induced including plasma dynamics. However, from recorded shawdowgraphs of the plasma dynamics, largely spherically symmetric expansion appears to occur for time delays of 0.1 to to 10 μs after initiation of optical breakdown. Given a close to symmetric plasma kernel, standard Abel inversion techniques are applied in the study of the expanding plasma [21].

In the experiments, lateral spectral data are recorded with an intensified charge coupled device (ICCD), the horizontal data corresponds to spectral information, the vertical data correspond to the lateral spatial information. For time delays of 0.15 and 0.40 μs for the UHP hydrogen gas in a cell, and for the time delay of 0.40 μs in SATP air, Abel inversion is performed using an adapted Abel-inversion method [22].

In a variety of laser-induced breakdown spectrosocpy (LIBS) arrangements, an optical fiber is utilized to couple the plasma radiation to a multi-order echelle-type spectrograph for highly resolved spectra across a substantial coverage from typically 200 to 900 nm. However, when employing an ICCD with a reasonable two-dimensional coverage of 1024×1024 pixels, a narrower spectral range is measured horizontally but with spatial resolution along the vertical axis. Stark-broadened hydrogen profiles are adequately covered, e.g., over a 24 nm wavelength range for the hydrogen beta, H_β, line when using a 1200 grooves/mm grating in a 0.64 m Czerny-Turner type spectrometer. The spatial resolution along the vertical direction amounts to about 13.8 mm that is sufficient to capture the expanding plasma when using 1:1 imaging.

The line-of-sight measurements portray a sum of contributions along the radiation source,

$$I(z, \lambda) = 2\int_z^R I(r, \lambda)\frac{r}{\sqrt{r^2 - z^2}}dr. \tag{4}$$

Here, $I(z,\lambda)$ and $I(r,\lambda)$ denote the measured lateral spectra (along the slit height, z) and radial wavelength-dependent intensity distributions, respectively. The dependence on wavelength, λ, of the measured and radial distributions is explicitly included due to the use of the dispersive, holographic grating. The upper integration limit, R, is usually much larger than the object dimension, δ, or in other words $R \gg \delta$. The factor of 2 indicates top/bottom symmetry.

In this work, the Abel inverse transform is computed without the need for pre-treating the recorded data [22, 23], leading to inverted intensity values, $I(r,\lambda)$, with errors of only a few percent for selected test cases. Moreover, the strict requirement for axial symmetry, specifically spherical symmetry, can be slightly relaxed following procedures that were utilized originally in the analysis of inductively coupled plasma sources but have been adapted to the analysis of expanding laser-induced plasma [24].

In the analysis of the UHP hydrogen experiments recorded for the time delays of 0.15 and 0.40 μs, the spectra for hydrogen alpha, beta and gamma lines are each Abel inverted, and subsequently fitted to known line profiles. For H_α and H_γ Lorentzians are used to extract background, shift and full-widths-at-half-maximum, and for H_β asymmetric line profiles are directly fitted to find the electron density, N_e. Boltzmann plots are generated to evaluate the electron excitation temperature. The SATP laboratory data at a time delay of 0.4 μs show the hydrogen alpha, H_α, line together with two ionized nitrogen lines, the H_β line is not visible due to the relatively large free electron background contribution caused by the presence of mainly nitrogen and oxygen. The electron densities are evaluated from H_α widths and shifts using recently communicated formulae [18].

5. Electron density results

The electron density in the range of 0.5 to 50×10^{17} cm^{-3} is evaluated from the hydrogen alpha, H_α, Balmer series line. For electron densities of 0.5 to 5×10^{17} cm^{-3}, the hydrogen beta, H_β, line is preferred. Due to the asymmetric appearance, direct fitting is accomplished using predicted H_β line profiles including comparisons with results from width and peak-separation predictions [25–30].

The recorded H_α spectra are fit to Lorentzians to extract the FWHM, Δw_{H_α}, and red-shifts, $\Delta\delta_{H_\alpha}$, and then the empirical formulae [18],

$$\Delta w_{H_\alpha}[nm] = 1.31\left(\frac{\mathrm{N_e[cm^{-3}]}}{10^{17}}\right)^{0.64\,\pm\,0.03}, \tag{5}$$

$$\Delta\delta_{H_\alpha}[nm] = 0.055\left(\frac{\mathrm{N_e[cm^{-3}]}}{10^{17}}\right)^{0.97\,\pm\,0.03}, \tag{6}$$

yield measures for the electron densities. The red shift shows an almost linear dependence on electron density but would show relatively large error bars for electron densities of the order of 1×10^{17} cm^{-3} due to the spectral resolution of $\delta_R = 0.1$ nm. The width however yields results with an error of the order of 25%. Fitting with Lorentzians is quite acceptable considering the minute effect of Gaussian instrument broadening that would in principle result in Voigt profiles.

The H_β empirical formulae can be applied as well for electron densities up to $\simeq 5\times10^{17}$ cm^{-3} for pure hydrogen optical breakdown experiments [18],

$$\Delta w_{H_\beta}[nm] = 4.50\left(\frac{\mathrm{N_e[cm^{-3}]}}{10^{17}}\right)^{0.71\,\pm\,0.03}, \tag{7}$$

$$\Delta\lambda_{ps}[nm] = 1.32\left(\frac{\mathrm{N_e[cm^{-3}]}}{10^{17}}\right)^{0.61\,\pm\,0.03}. \tag{8}$$

The peak separation, $\Delta\lambda_{ps}$, has been investigated in view of assessing self-absorption effects. For electron densities of $\simeq 1\times10^{17}$ cm^{-3}, the H_β peak separations yield results consistent with the ones obtained from the widths. Measurements of the widths are preferred to determine electron densities of the order of $\simeq 5\times10^{17}$ cm^{-3} rather than finding the values for the peak separation [29, 30].

5.1. Air plasma results

Application of the Abel inversion and reconstruction of the spectra along the line of sight yield spatially resolved electron densities along the line of sight. Figure 2 and 3 display the recorded line profile corresponding to the edge of the plasma and the radial variation, respectively.

Figure 2 also shows the Savitzky-Golay filtered data and the results of fitting the data including provision for the self-absorption but excluding the regions for the singly ionized nitrogen lines. For a density of $N_e = 20\times10^{17}$ cm^{-3}, self-absorption effects are illustrated in Fig. 2 by the apparent line reversal in the dip near line center of the fitted Lorentzians. The H_α line shape reveals distortions with contributions to line-widths primarily due to Stark broadening. Evidence of self-absorption is also confirmed by the indicated red shift and by comparison of the electron densities with results obtained from the N^+ lines [14]. Figures 4 and 5 illustrate a 3-dimensional cross-sections of N_e obtained from fitting widths and shifts, respectively.

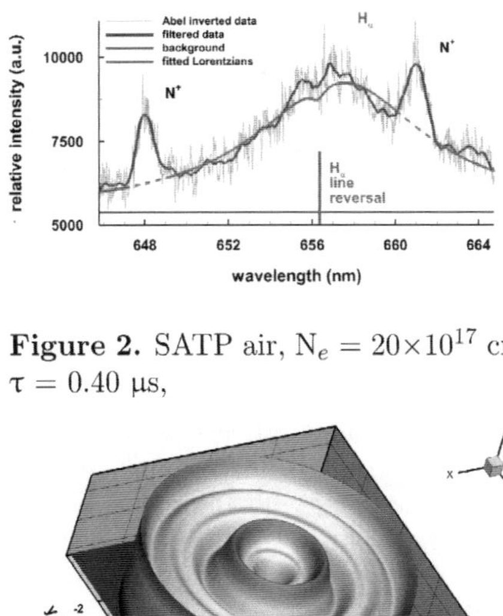

Figure 2. SATP air, $N_e = 20\times10^{17}$ cm^{-3}, $\tau = 0.40$ μs,

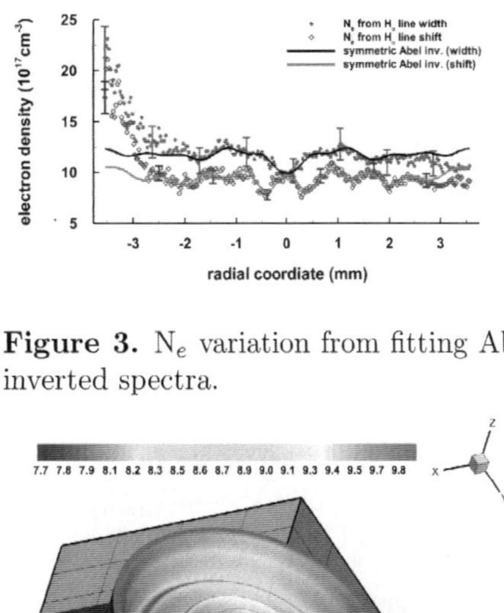

Figure 3. N_e variation from fitting Abel-inverted spectra.

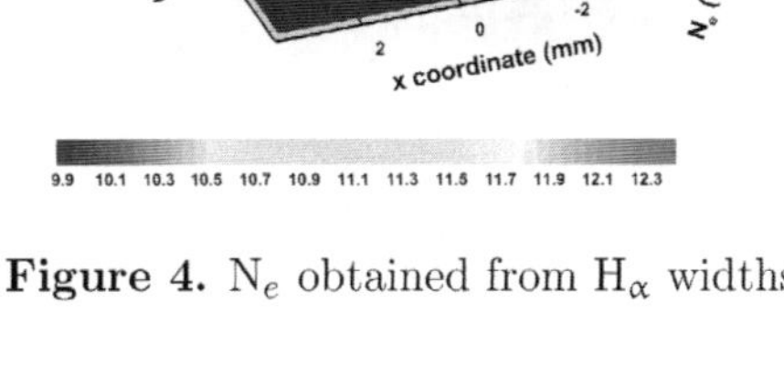

Figure 4. N_e obtained from H_α widths.

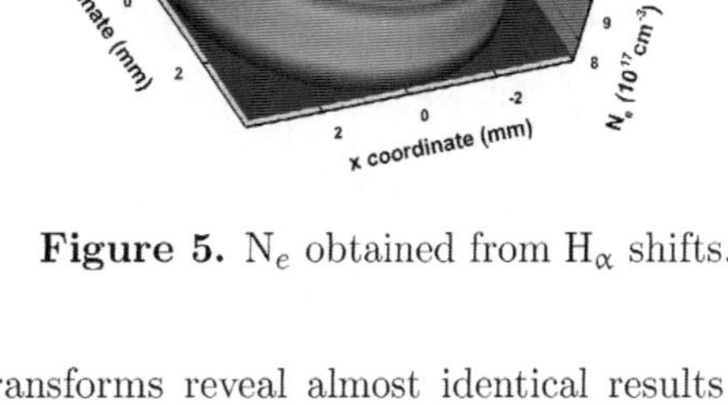

Figure 5. N_e obtained from H_α shifts.

Standard and extended, asymmetric Abel transforms reveal almost identical results in the central 5 mm range. This confirms the hypothesis of spherically symmetric expansion suggested from shadowgraphs.

For a time delay of 0.4 μs, self-absorption effects have been reported based on the doubling mirror method for recorded line-of-sight H_α data [13]. Considering the corrections and the usual, absolute error bars associated with electron density determinations from the hydrogen alpha line, the self-absorption effects caused changes of the electron density within the error bars.

5.2. Hydrogen plasma results

In this section, electron density results are presented of laser-induced plasma in ultra-high-purity (UHP) hydrogen gas contained in a cell. Figure 6 portrays Abel inverted spectra, Savitzky-Golay filtered results, and the fitted Lorentzian. The line reversal manifests self-absorption.

Comparisons of the electron densities inferred from H_α and H_β are expected to yield results within the associated error bars. Figure 7 indicates larger values for N_e obtained from the H_α line near the plasma edge indicating self-absorption. In the 2 mm central region, the electron density results appear to be within the error bars except for the very center. Also, the central region indicates close to spherical symmetric plasma. For a later time delay of 0.4 μs, analysis of the hydrogen beta data shows almost spherically symmetric results for the electron density within the 2 mm central core.

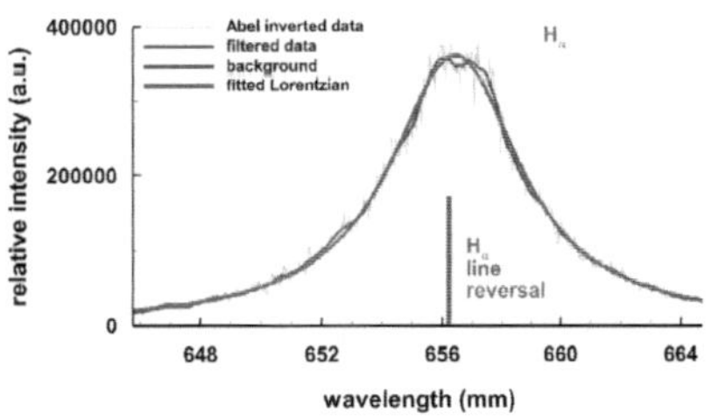

Figure 6. $N_e = 8.4\times10^{17}$ cm^{-3}, UHP hydrogen gas (810 Torr), $\tau = 0.15$ μs.

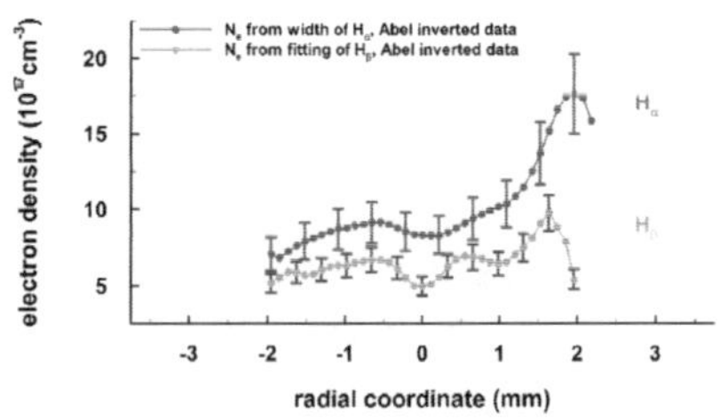

Figure 7. N_e from fitting Abel-inverted H_α and H_β line profiles.

For the recorded data at the time delay of $\tau = 0.40$ μs, Boltzmann plots are constructed from the H_α, H_β and H_γ lines. Integrated line shapes are utilized to provide a measure for the temperature distribution. Figure 8 reveals a cooler central region and a relatively hot ring of the order of 100,000 K (8.6 eV) and . Noteworthy, the kernel expands at or near the speed of sound in hydrogen gas. From Fig. 8, the two temperature peaks near ± 0.5 mm at a time delay of 0.4 μs would indicate an expansion speed of 1.25 km/s. The average temperature along the line of site is consistent with previous experimental investigations. Fig. 9 shows corresponding electron densities. The error bars are in part due to the lower fidelity in the Abel inversion for a time delay of 0.40 μs.

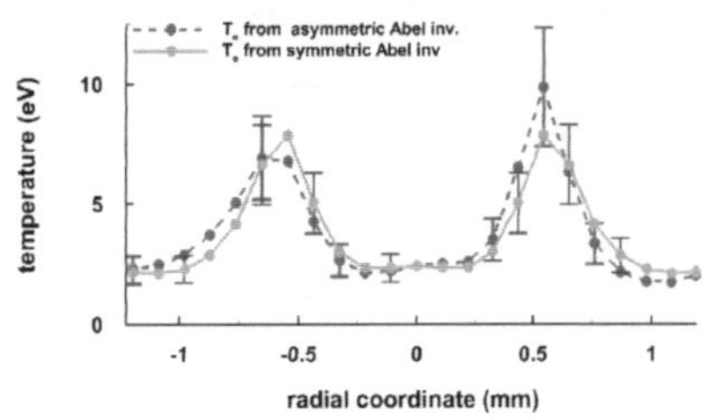

Figure 8. T_e determined from Boltzmann plots, $\tau = 0.40$ μs.

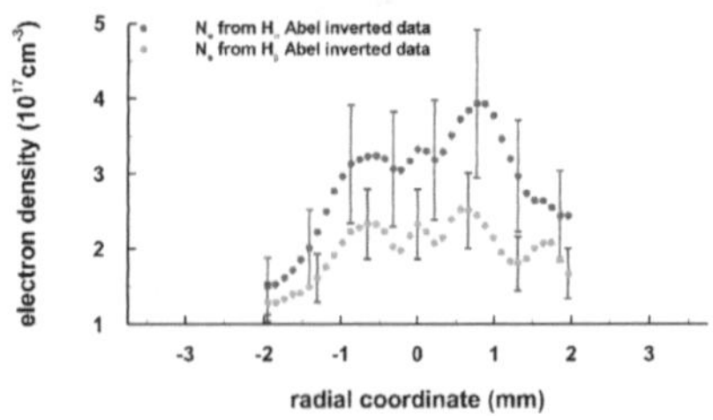

Figure 9. N_e is almost symmetric in the -1 to +1 mm range, $\tau = 0.40$ μs.

6. Conclusions

Analyses of the laser-induced plasma using standard Abel transforms reveal that the central core is close to spherically symmetric at time delays of 0.15 and 0.40 μs for UHP hydrogen gas, and at 0.40 μs for SATP laboratory air. Abel inverted spectra confirm occurrence of self-absorption. The self-absorption phenomena cause line shape distortions and line-reversals. These appear to be insignificant in the electron density determination for $N_e \leq 20\times10^{17}$ cm^{-3} in the presented time-resolved experimental results of laser-induced plasma in UHP hydrogen gas and SATP laboratory air. The temperature distributions in the hydrogen plasma experiments reveal as well a ring structure indicating lower T_e in the central core region, and this central region expands at or near the speed of sound in hydrogen gas.

Acknowledgments

The authors acknowledge A. M. and Th. M. EL Sherbini for their continued interest in this work, and thank for support in part by the Center of Excellence, Center for Laser Applications (CLA) and the University of Tennessee Space Institute.

References

[1] Kunze H-J 2009 *Introduction to Plasma Spectroscopy* (Springer-Verlag, Berlin)
[2] Mulser P and Bauer, D 2010 *High Power Laser-Matter Interaction* (Springer-Verlag, Berlin)
[3] Holstein T 1947 *Phys. Rev.* **72** 1212
[4] Holstein T 1951 *Phys. Rev.* **83** 1159
[5] Cowan R D and Dieke G H 1948 *Rev. Mod. Phys.* **20** 418
[6] Pardini L, Legnaioli S, Lorenzetti G, Palleschi V, Gaudiuso R, De Giacomo A, Diaz Pace D M, Anabitarte Garcia F, de Holanda Cavalcanti G and Parigger C 2013 *Spectrochim. Acta Part B* **88** 98
[7] Aragón C and Aguilera J A 2010 *Spectrochim. Acta Part B* **65** 395
[8] Kielkopf J F and Allard N F 2014 *J. Phys B: At. Mol. Opt. Phys.* **47** 155701
[9] Irons F E 1979 *J. Quant.Spectrosc. Radiat. Transfer* **22** 1
[10] Mancini R C, Joyce R F and Hooper Jr C F 1987 *Phys B: At. Mol. Opt. Phys.* **20** 2975
[11] Fill E E 1988 *J. Quant.Spectrosc. Radiat. Transfer* **39** 489
[12] Moon H-Y, Herrera K K, Omenetto N, Smith B W and Winefordner J D 2009 *Spectrochim. Acta Part B* **64** 702
[13] Parigger C G, Gautam G, Woods A C, Surmick D M and Hornkohl J O 2014 *Trends Appl. Spectrosc.* **11** 1
[14] Parigger C G, Swafford L D, Surmick D M, Witte M J, Woods A C and Gautam G 2014 *J. Phys.: Conf. Ser.* **548** 012043
[15] Swafford L D, Surmick D M, Witte M J, Woods A C, Gautam G and Parigger C G 2014 *J. Phys.: Conf. Ser.* **548** 012049
[16] Parigger C G, Woods A C, Witte M J, Swafford L D and Surmick D M 2014 *J. Vis. Exp.* **84** e51250
[17] Gautam G and Parigger C G 2014 *Int. Rev. At. Mol. Phys.* **5** 115
[18] Surmick D M and Parigger C G 2014 *Int. Rev. At. Mol. Phys.* **5** 73
[19] EL Sherbini A M, Surmick D M, Gautam G and Parigger C G 2014 *Int. Rev. At. Mol. Phys.* **5** 23
[20] Parigger C G, Surmick D M, Gautam G and EL Sherbini A M 2015 *Opt. Lett.* **40** 3436
[21] Parigger C G, Surmick D M and Gautam G 2015 *Int. Rev. At. Mol. Phys.* **6** 41
[22] Pretzler G 1991 *Z. Naturforsch.* **46a** 639
[23] Killer C 2014 http://www.mathworks.com/matlabcentral/fileexchange/43639-abel-inversion-algorithm [last accessed Nov 15 2015]
[24] Blades M W 1983 *Appl. Spectrosc.* **37** 371
[25] Djurović S, Ćirišan M, Demura A V, Demchenko G V, Nikolić D, Gigosos M A and Gonzáles M Á 2009 *Phys. Rev. E.* **79** 046402
[26] Palomares J M, Torres J, Gigosos M A, van der Mullen J J A M, Gamero A and Sola A 2009 *Appl. Spectrosc.* **63** 1023
[27] Palomares J M, Torres J, Gigosos M A, van der Mullen J J A M, Gamero A and Sola A 2010 *J.Phys.: Conf. Ser.* **207** 012013
[28] Ivković M, Konjević N and Pavlović Z 2015 *J. Quant. Spect. Radiat. Transfer* **154** 1
[29] Gautam G, Surmick D M and Parigger C G 2015 *J. Quant. Spect. Radiat. Transfer* **160** 19
[30] Gautam G, Parigger C G, Surmick D M and EL Sherbini A M 2016 *J. Quant. Spect. Radiat. Transfer* **170** 189

IOP Conf. Series: Journal of Physics: Conf. Series **810** (2017) 012013 doi:10.1088/1742-6596/810/1/012013

Superheating effects in line broadening of dense alkali vapors

G. Pichler[1], Y. Makdisi[1], J. Kokaj[1], J. Mathew[1], M. Rakić[2], R. Beuc[2]

[1] Department of Physics, University of Kuwait, Kuwait
[2] Institute of Physics, Zagreb, Croatia

Email: pichler@ifs.hr

Abstract. We used the superheating of the dense alkali vapor in order to separate atomic from the molecular photoionization spectrum. Homonuclear and heteronuclear alkali molecules are essentially two electron systems and therefore simultaneous excitation of two atoms above the ionization limit will form autoionizing potential curves which will bring about new bands in the photoionization spectrum. We present a satisfactory comparison between experimental and theoretical emission coefficient function of diffuse bands (cesium dimer excimer) at temperatures up to 1000 °C, where the metal vapor was heavily superheated. New results for Rb_2 photoionization process reveal similar structured photoionization continuum but with reduced number of photoionization bands.

1. Introduction

Diatomic alkali molecules became of high interest because of the laser cooling effects which bring them into the most interesting quantum mechanical regime with numerous important applications. However, the process of the ionization of alkali molecules did not captured enough attention although it might also have various applications, especially toward efficient energy conversion processes.

The superheating of the dense alkali vapor enables clear separation of atomic from the molecular photoionization spectrum [1]. The molecular photoionization continuum structure is about two orders of magnitude stronger than the atomic photoionization continuum [2].

We performed numerous experiments with different pure and mixed alkali vapors with the main goal to investigate above threshold ionization process of the diatomic alkali molecules. Beside the method of superheating [3] the time-of-flight method [4] was also used in order to distinct the molecular from atomic photoionization process. As the source of electrons in possible novel plasma devices this molecular contribution becomes highly important [5, 6]. This is because of much greater cross-sections than their atomic cross-sections. Thus the strong influence of molecular component in the photoionization process opens up a new vista in the energy conversion research.

2. Experiment

Experiments with sapphire cells are extremely simple. The continuum background source is a xenon high pressure light bulb or a Laser Driven Light Source (LDLS). The absorption or emission cells are either longitudinal all-sapphire cells or T-type sapphire cells with special glass extension as an alkali container. After the light transmission through the cell it was collected on the entrance slit of the OceanOptics digital spectrometer. In order to obtain the absorption coefficient curve we used a Beer-Lambert law using Origin 8.5 software.

IOP Conf. Series: Journal of Physics: Conf. Series **810** (2017) 012013 doi:10.1088/1742-6596/810/1/012013

3. Results

The result of the pure cesium vapor absorption experiment is shown in Fig. 1 for the temperatures of 350, 420, 502 and 900 °C. For temperatures of 350 and 420 °C there was no superheating yet, but above 420 °C, for temperatures of 502 and 900 °C superheating is observable bringing the spectrum at 900 °C close to pure atomic photoionization spectrum. Rydberg spectral lines from the principal series are much better visible approaching the atomic ionization limit at 318.3 nm. Molecular destruction effect is also nicely visible for bands peaks at 352 and 395 nm.

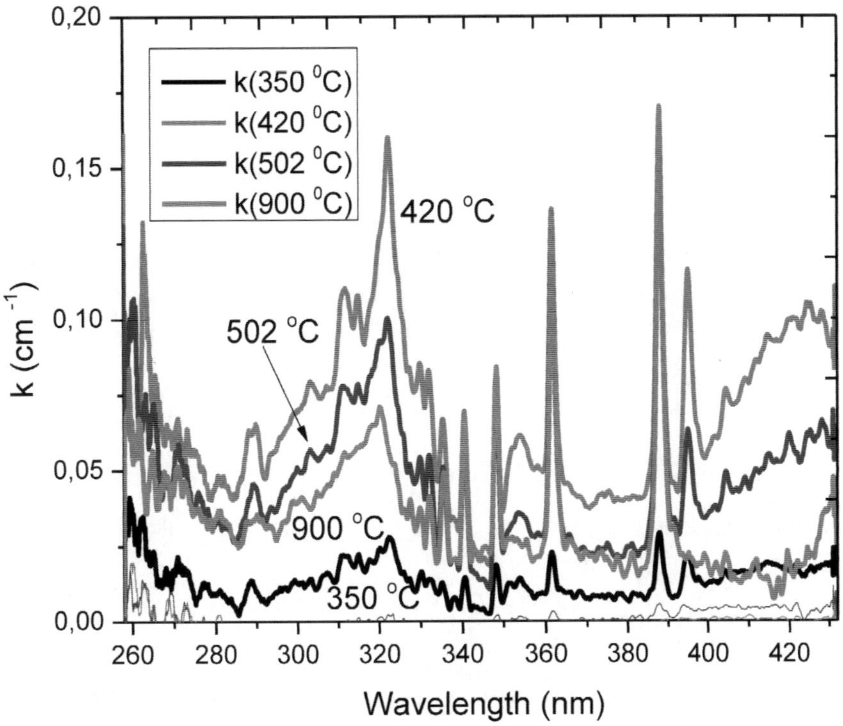

Figure 1. Absorption spectrum for cesium vapor in equilibrium with the liquid droplet (350 °C and 420 °C). Spectrum of superheated cesium vapor (no cesium droplet at 502 °C and 900 °C). A much more detailed description of the absorption photoionization measurements with superheated cesium vapor is given in ref. [1].

Thermal emission from the superheated cesium vapor was additionally studied to very high temperatures from 700 °C to 1000 °C. We observed a number of atomic and molecular spectral features simultaneously in emission and absorption, especially peculiar thermal emission of cesium dimer diffuse bands ($2\ ^3\Pi_g \rightarrow a\ ^3\Sigma_u^+$ transitions) around 710 nm coexisting with absorption bands around first resonance lines at 852 nm and 894 nm. The main result is the comparison between the absorption and emission profiles of the above mentioned diffuse bands. For the emission experiment at high temperatures we obtained the following expression:

$$I(\nu,T) = A\frac{2h\nu^3}{c^2}\frac{(1+B) - e^{-LK(\nu,T)[1-\exp(-h\nu/k_BT)]}}{e^{h\nu/k_BT}-1}$$

There are two distinct contributions in the equation above. The first one is the hot superheated cesium vapor, and the second is the blackbody of the oven. Parameters A and B are optimized in order to obtain best fit to the observed emission profile. K is the absorption coefficient measured in the spectral interval of 680-740 nm. Much more detailed description of the self emission measurements with superheated cesium vapor is given in ref. [3].

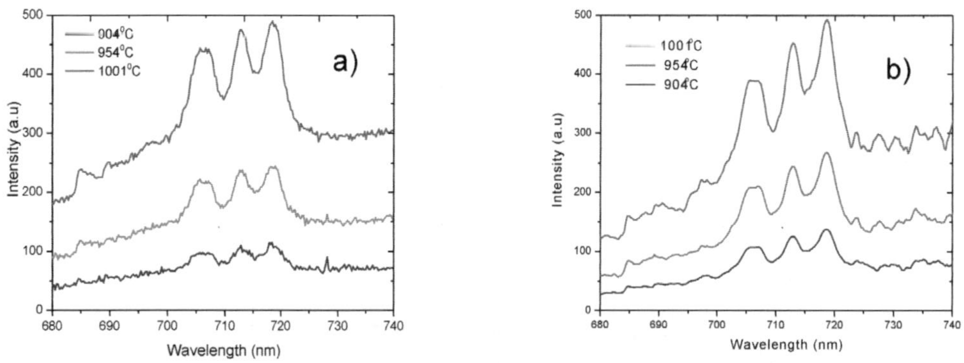

Figure 2. a) Observed emission spectrum of the diffuse bands, b) theoretical calculations of the emission spectrum with absorption coefficient taken into account.

Preliminary rubidium case was studied without efficient superheating of the T-type sapphire cell (L=0.5 cm). In Fig. 3 we present absorption coefficient function plotted from 233 up to 325 nm for temperatures from 550 to 640 °C. The above threshold molecular bands at 245 nm and 268 nm are clearly visible, as well as the Rydberg spectral lines up to the main quantum number n = 17. Previous TOF work [4] already firmly established molecular origin of 245 and 270 nm photoionization bands. This experiment did not enable superheating separation process within present rubidium vapor conditions as previously performed in the case of cesium vapor [1]. However, since the ratio of $[Rb_2]/[Rb]$ densities increases with increasing temperature it is clear that at elevated temperatures photoionization bands of Rb_2 become better visible in the spectrum. Details of the corresponding analysis can be found in ref. [2]. Previous measurements were also done in absorption for Cs, Rb and some heavy alkali mixtures [7].

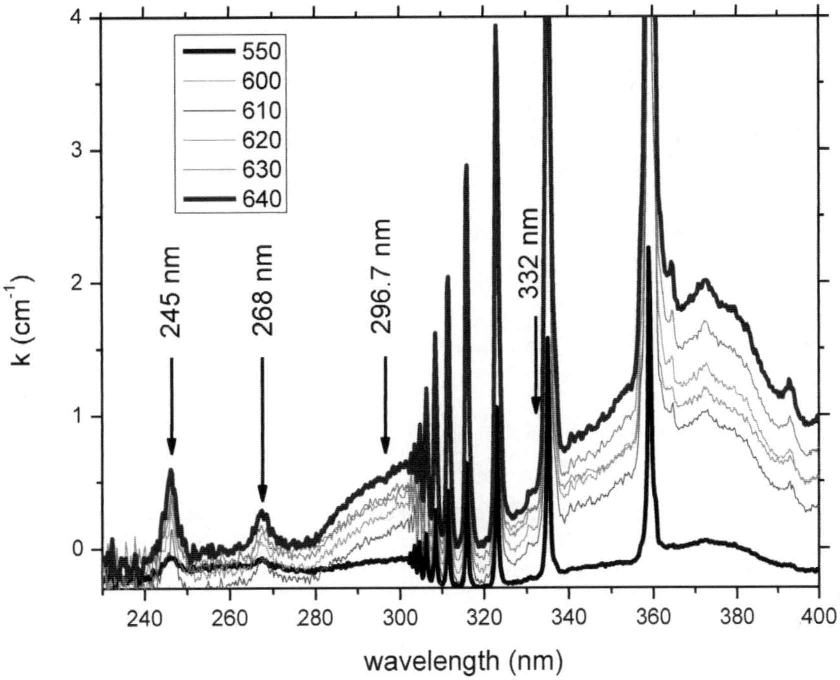

Figure 3. Absorption spectra for temperatures from 550 °C to 640 °C.

Numerous improvements are necessary in order to illuminate the detailed structure and the origin of the photoionization bands of alkali molecules. Accurate molecular calculations are needed in order to explain the differences in the appearance of the heavy alkali cases. This will include the autoionization process connected with the doubly excited molecular states.

4. Discussion

There are many spectral phenomena connected with the ionization of the cesium dimers. Since the coexistence of cesium atoms and molecules is precluding easy separation of the two we decided to study absorption spectrum of dense cesium vapor in an all-sapphire cell employing superheating of the vapor, with a special emphasis on the highly structured photoionization continuum. This continuum appears to be composed of atomic and molecular contributions which can be separated by means of additional superheating of cesium vapor in the sapphire cell. This was possible due to the small amount of cesium filling which completely evaporated at temperature of about 420 °C. This enabled the overheating of cesium dimers which greatly reduced its concentration at a temperature of 900 °C leaving almost pure atomic Cs vapor. The analysis of the thermal destruction indicated that the highly structured molecular component of the photoionization continuum can be attributed to cesium dimers. We discuss the possible origin of the structured photoionization continuum as stemming from the absorption process from the ground state of the Cs_2 molecule to the doubly excited Cs_2^{**} molecule above Cs_2^+ limit. The corresponding potential curves are presently unknown, and their calculations will illuminate this interesting phenomenon. At the moment we assume that the relevant potential curves are subjected simultaneously to mutual interactions and autoionization.

5. Conclusion

Much more relevant data on photoionization structures for heavier alkali molecules are needed. These involve homonuclear and heteronuclear alkali systems. The low resolution measurements could be extended to high resolution experiments at ultracold conditions, which may reveal yet unexpected structures. A possible application of the observed phenomena will be discussed, in the near future, in terms of the solar energy conversion using dense cesium vapor. Another possibility is to make laser action at the excimer transition of diffuse bands if the superheated vapor is located within an appropriate laser cavity. At elevated temperatures in addition to monomers and dimers the coexistence of trimers and tetramers cannot be excluded. Their ionization behavior is not well investigated, but can definitely contribute to better source of free electrons [8].

Acknowledgments

We thankfully acknowledge the Kuwait Foundation for the Advancement of Sciences (KFAS) for funding this work through Research project No: 2012151301. The support by General Facility projects GS03/01 and GS01/08 of Kuwait University Research sector is also gratefully acknowledged. Theoretical and rubidium part of the present work was supported by the Croatian Science Foundation (HRZZ) under the project numbers 7515 and 2753.

References

[1] Pichler G, Makdisi Y, Kokaj J, Thomas N and Mathew J 2015 J. Phys. B: At. Mol. Opt. Phys. **48** 165002-8
[2] Creek D M and G V Marr 1968 J. Quant. Spetrosc. Radiat. Transfer. **8** 1431-6
[3] Pichler G, Makdisi Y, Kokaj J, Thomas N, Mathew J and Beuc R 2016 J. Phys. B: At. Mol. Opt. Phys. **49** 145101-7
[4] Suemitsu H and Samson J A R 1983 Phys. Rev. A **28** 2752-8
[5] Ogino A, Muramatsu T, and Kando M, 2004 Jpn. J. Appl. Phys, **43** 309-14.
[6] Inaguma T, Tsuda N and Yamada 2007 Elect. Eng. Jpn. **158** 14-21.
[7] Hotop R and Niemax K 1980 J. Phys. B: Atom. Molec. Phys. **13** L93-L99
[8] Chaban V V and Prezhdo O V 2016 J. Phys. Chem. A **120** 4302-6

Probing Non-Newtonian gravity by photoassociation spectroscopy

M Borkowski[1], A A Buchachenko[2], R Ciuryło[1], P S Julienne[3], H Yamada[4], K Yuu[4], K Takahashi[4], Y Takasu[4], and Y Takahashi[4,5]

[1] Institute of Physics, Faculty of Physics, Astronomy and Informatics, Nicolaus Copernicus University, Grudziadzka 5, 87-100 Torun, Poland
[2] Skolkovo Institute of Science and Technology, 100 Novaya Street, Skolkovo, Moscow Region, 143025, Russia and Institute of Problems of Chemical Physics RAS, Chernogolovka, Moscow Region 142432, Russia
[3] Joint Quantum Institute, National Institute of Standards and Technology and the University of Maryland, 100 Bureau Drive, Stop 8423, Gaithersburg, Maryland 20899-8423, USA
[4] Department of Physics, Graduate School of Science, Kyoto University, Kyoto 606-8502, Japan
[5] CREST, Japan Science and Technology Agency, Chiyoda-ku, Tokyo 102-0075, Japan

E-mail: `mateusz@fizyka.umk.pl`

Abstract. State of the art photoassociative measurements of bound state energies in the ground state Yb_2 molecule are used to establish limits on non-Newtonian gravity at Yukawa ranges of nanometers.

1. Introduction

Several cosmological theories [1–3] predict deviations from the usual $1/R$ behavior of the gravitational interaction. Those supposed corrections to the gravitational interaction between masses m_1 and m_2 at a distance R are usually expressed in terms of a Yukawa-type potential $V_{\mathrm{corr}}(R)$:

$$V_{\mathrm{Newton}}(R) + V_{\mathrm{corr}} = -G\frac{m_1 m_2}{R}\left(1 + \alpha\exp(-R/\lambda)\right). \tag{1}$$

Since so far no experimental evidence of such deviations has been found, experimental efforts have concentrated on giving limits on the magnitude α of this 'correction' as a function of the Yukawa range λ. In this work we place similar limits in the range of $\lambda = 0.1$ nm to $\lambda = 1000$ nm by measuring bound state energies of Yb_2 and modelling the interactions between Yb atoms using mass-independent Born-Oppenheimer potentials. This way we can place limits on the *mass-dependent* part of the interaction and, by proxy, an upper bound on non-Newtonian gravity.

2. Measurements

Several bound state energies of the ground state Yb_2 have been measured using two-color photoassociation spectroscopy [4–6] in the Raman configuration, with ${}^1S_0 + {}^3P_1\ 0_u^+$ intermediate states. To avoid thermal broadening [7], the measurements were performed in Bose-Einstein condenstates of ytterbium atoms. Techniques for forming such samples have already been

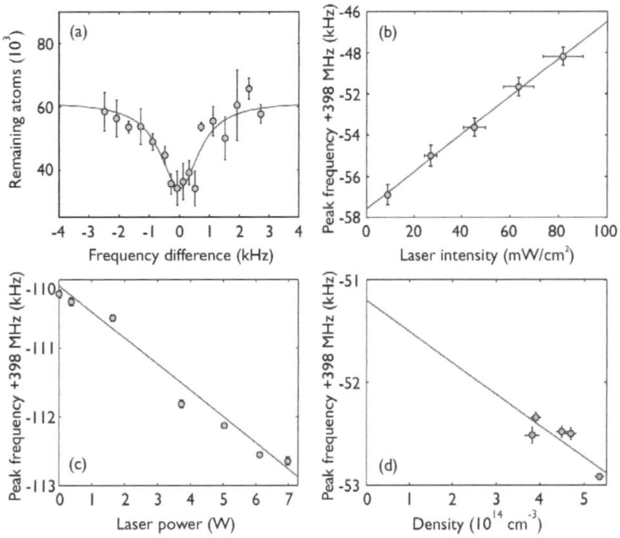

Figure 1. Compensation of systematic errors present in the experiment as shown for the ^{170}Yb $\nu = 2, J = 2$ line position. (a) an example photoassociation spectrum fitted with a Lorentzian lineshape of an FWHM of about 1 kHz, (b) shift due to both photoassociation lasers (molecular optical shift), (c) optical shift due to the far-off-resonant trap (FORT), and (d) the density shift due to the presence of other atoms in the BEC sample.

described elsewhere [8–10]. The photoassociation lines had a Lorentzian FWHM of about 1 kHz, as seen in Figure 1(a). The following systematic shifts were taken into account: shift due to the photoassociation lasers, the trapping laser and the density of atoms. The systematic shifts were eliminated by linear extrapolation, as seen in figs. 1(b)-(d). A total of 13 bound rovibrational state energies with rotational quantum numbers $J = 0$ or 2 has been measured with error bars ≈ 500 Hz: two for ^{168}Yb, six for ^{170}Yb and five for ^{174}Yb and will be published elsewhere [11].

3. Interactions

Thanks to the single spinless $^1S_0$ ground state of the Yb atom, the interactions in the ytterbium dimer are described by a single Born-Oppenheimer potential. We have developed four such potential curves. The short range part of the potential could either be a Lennard-Jones type model potential [5], or an *ab initio* based curve [14]. Our short range curves are shown in the upper panel of Figure 2. The long range part could be the standard R^{-6} van der Waals interaction or one that accounted for the Casimir-Polder effect, as tabulated for Yb_2 in [15].

The potentials were fitted to experimental data using the least-squares method which minimized the difference between experimentally obtained bound state energies and theoretical energies calculated through solving the usual radial Schrödinger equation. Details of the procedure of finding a mass scaled model can be found elsewhere [5, 12, 13]. We find that we are able to reproduce the bound state energies to within a few tens of kHz on average (at $\chi^2 = 307242$ for the best model – *ab initio* short range with a van der Waals long range). While the fact that the *ab initio* short range gives better results than a simple model potential is hardly surprising, it is interesting to note that the Casimir-Polder long range always yielded worse fit evidenced by a larger χ^2.

4. Limits

The final stage of the data analysis was to augment the interaction potentials with the gravitational interaction as shown in the lower panel of Figure 2. The potentials were fitted repeatedly as α was ramped to make it possible for other potential parameters to compensate for the obvious shift in the quantum defect caused by the added interaction. This is important, because we are interested in the change in the mass scaling behavior rather than a simple shift.

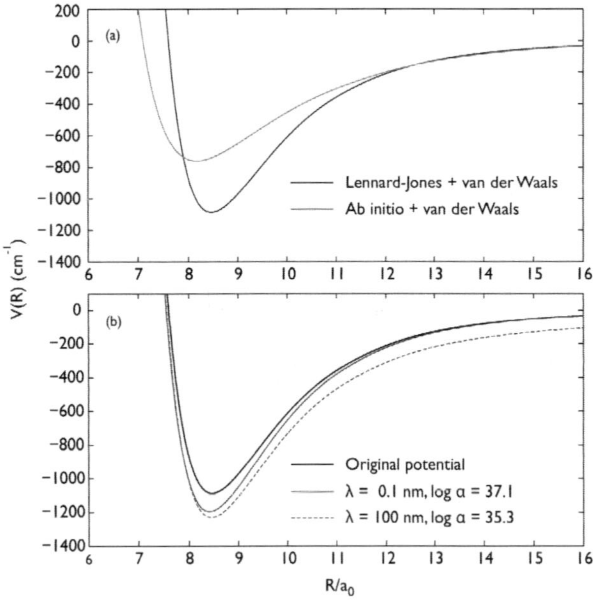

Figure 2. Upper panel: a comparison between the Lennard-Jones (black line) and *ab initio* (red) short range potentials. Despite the different shapes and depths both potentials support the same number of bound states, have similar long range C_6 coefficients and can be used as a basis for a mass-scaled potential model [12, 13]. Lower panel: the impact of the Yukawa potential on the shape of the interaction potential, with the Lennard-Jones potential as an example. For a short $\lambda = 0.1$ nm only the short range part is affected, but for larger λ, the long range part is also changed. For even larger λ no further change at internuclear distances is seen.

Otherwise $\alpha_{\max}$ would be underestimated, because uncompensated shift from the additional interaction would cause χ^2 to grow too quickly. As shown in Figure 3 the behavior was for χ^2 to experience a minimum first, and then quickly grow to infinity. This counter-intuitive phenomenon can be explained by the fact that our models take no account of any (real) mass-dependent effects in the molecule, like the beyond-Born-Oppenheimer effects [16] and the R-dependent isotopic shift [17]. For small α the non-Newtonian interaction may artificially improve the mass scaling of theoretical bound state energies and therefore decrease χ^2. With this in mind, in order to avoid accidental underestimation of the limits on non-Newtonian gravity we pick the

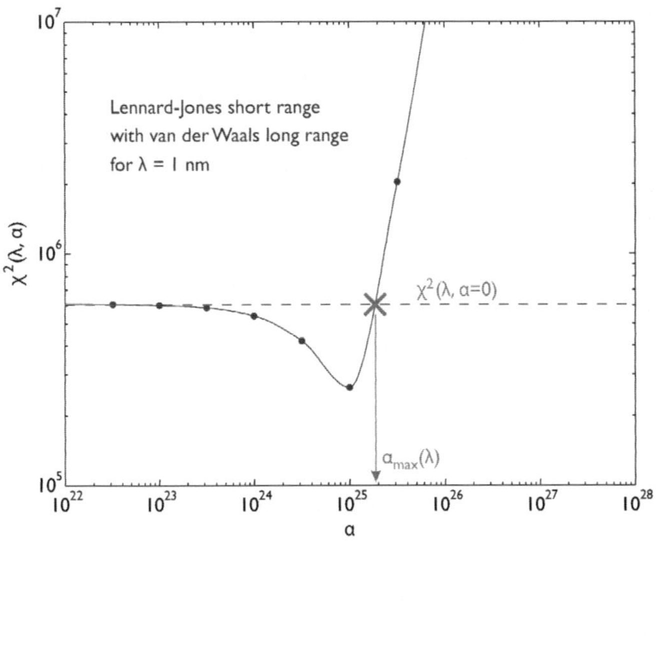

Figure 3. Evolution of the quality of the fit χ^2 as the Yukawa coupling α is ramped. At first the fit is 'improved', due to the compensation of mass-dependent effects not included in the theoretical models, but after a certain point χ^2 quickly rises to infinity. We choose the limit $\alpha_{\max}$ so that $\chi^2(\alpha_{\max}) = \chi^2(\alpha = 0)$, as at this point the non-Newtonian contribution clearly starts to make the fit worse.

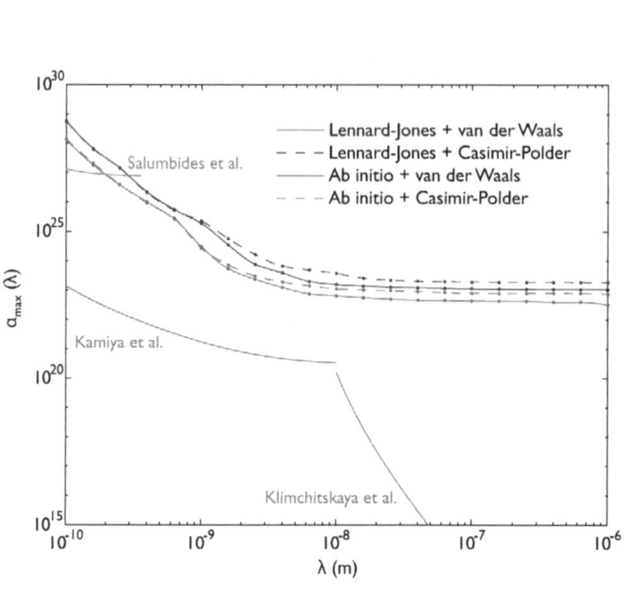

Figure 4. Limits on non-Newtonian Yukawa interaction derived from our interaction models (black and red lines depending on the short range, and dashed or solid for the long range part). The *ab initio* + van der Waals combination provides the most stringent limits. Limits obtained using other methods – neutron scattering (Kamiya *et al.*[18]), Casimir forces between plates (Klimchitskaya *et al.* [20]) and classical spectroscopy of HD^+ (Salumbides *et al.* [21]).

point $\alpha_{\max}$ where $\chi^2(\alpha_{\max})$ becomes larger than $\chi^2(\alpha = 0)$, that is, where the theoretical description of the atomic interactions clearly becomes worse.

Figure 4 shows limits on non-Newtonian gravity derived from the four models. The general behavior of $\alpha_{\max}$ is to decrease up to about $\lambda = 2$ nm, where the limits become constant. This is easily understandable, as the non-Newtonian part becomes $V_{\text{corr}} = -\alpha G \frac{m_1 m_2}{R}$ as λ becomes much larger than the internuclear distances R in the molecule. The differences between models are slight. For small λ the different long range parts have no impact on the limits while for larger λ both the short range and long range parts of the potential matter, as expected.

Our limits can be compared to examples of recent works that derive similar limits using other experimental methods. This is shown as green lines in Figure 4. Both limits from neutron scattering by Kamiya *et al.* [18] and measurements of Casimir forces between plates by Klimchitskaya *et al.* are currently more stringent than ours in their respective ranges of λ. On the other hand, our work is on par with the results of classical molecular spectroscopy of HD^+ by Salumbides *et al.*. It is worth noting, that the probable weakest link in our approach is the quality of the interaction potentials: while the experimental error bars are on the order of 500 Hz, the theory matches experiment to about 50 kHz at best. On the other hand all four models are pure Born-Oppenheimer potentials, with no account taken for known mass-dependent effects, namely the R-dependent isotopic shift [17] and higher orders of the Born-Oppenheimer approximation [16]. Calculation of these effects for heavy systems like Yb is prohibitively difficult and was only recently undertaken [19]. The inclusion of these effects may pay off handsomely, as the limits are rapidly becoming more stringent with decreasing model χ^2. For instance, at $\lambda = 10$ nm, the Lennard-Jones + van der Waals model yields $\log \alpha_{\max} = 23.2$, as opposed to $\log \alpha_{\max} = 22.7$ for the *ab initio* + van der Waals model. The χ^2 of the former is twice as large as the latter. By extrapolation: if χ^2 could be reduced by a factor of about 100 (which would require the models to be ten times more accurate than now), $\log \alpha_{\max}$ could be reduced by about 3, putting photoassociation on par with the best current methods.

5. Conclusion

We have used state of the art photoassociation spectroscopy in a Bose-Einstein condensate of ytterbium atoms to measure the binding energies of several isotopomers of the Yb_2 molecule to an accuracy of ≈ 500 Hz. Four interaction models were fitted to the experimental data in an attempt to reproduce them theoretically. Both *ab initio* and model short range potentials were implemented, either with a standard van der Waals or the quantum-electrodynamic Casimir-Polder long range interaction. The impact of non-Newtonian gravitylike interactions on the mass scaling behavior of bound state energies was assessed to obtain the limits on the Yukawa coupling constant α at distances $\lambda \sim 1\,\mathrm{nm}$. Our best limits are only three orders of magnitude worse than the current best from long standing experimental approaches: neutron scattering and measurements of Casimir-Polder interactions. The presented methodology shows great promise for future research as small improvements in the theoretical description of the atomic interactions result in the rapid tightening of the limits on non-Newtonian gravity.

Acknowledgments
We acknowledge K. Enomoto, M. Ando, J. Hutson, and E. Tiemann for their useful comments. This work has been partially supported by the Grant-in-Aid for Scientific Research of JSPS (25220711, 26247064) and Impulsing Paradigm Changing Through Disruptive Technologies (ImPACT) program. It has also been partially supported by the TEAM Programme of the Foundation for Polish Science, co-financed by the EU European Regional Development Fund and by the COST Action CM1405 MOLIM and also partially supported by the National Science Centre grant no. UMO-2014/13/N/ST2/02591. It is part of an ongoing research program of the National Laboratory FAMO in Toruń, Poland. Calculations took place in Wroclaw Centre for Networking and Supercomputing grant no. 353.

References

[1] Adelberger E, Heckel B and Nelson A 2003 *Annu. Rev. Nucl. Sci.* **53** 77–121
[2] Damour T 2003 *Astrophys. Space. Sci.* **283** 445–456
[3] Su Y, Heckel B R, Adelberger E G, Gundlach J H, Harris M, Smith G L and Swanson H E 1994 *Phys. Rev. D* **50**(6) 3614–3636
[4] Bohn J L and Julienne P S 1999 *Phys. Rev. A* **60** 414–425
[5] Kitagawa M, Enomoto K, Kasa K, Takahashi Y, Ciuryło R, Naidon P and Julienne P S 2008 *Phys. Rev. A* **77** 012719
[6] Jones K M, Tiesinga E, Lett P D and Julienne P S 2006 *Rev. Mod. Phys.* **78** 483–535
[7] Jones K M, Lett P D, Tiesinga E and Julienne P S 2000 *Phys. Rev. A* **61** 012501
[8] Takasu Y, Maki K, Komori K, Takano T, Honda K, Kumakura M, Yabuzaki T and Takahashi Y 2003 *Phys. Rev. Lett.* **91** 040404
[9] Fukuhara T, Sugawa S and Takahashi Y 2007 *Phys. Rev. A* **76** 051604
[10] Sugawa S, Yamazaki R, Taie S and Takahashi Y 2011 *Phys. Rev. A* **84** 011610
[11] Borkowski M, Buchachenko A A, Ciuryło R, Julienne P S, Yamada H, Yuu K, Takahashi K, Takasu Y and Takahashi Y (to be published)
[12] Gribakin G F and Flambaum V V 1993 *Phys. Rev. A* **48** 546–553
[13] Borkowski M, Żuchowski P S, Ciuryło R, Julienne P S, Kedziera D, Mentel L, Tecmer P, Münchow F, Bruni C and Görlitz A 2013 *Phys. Rev. A* **88**(5) 052708
[14] Buchachenko A A, Chałasiński G and Szczęśniak M M 2007 *Eur. Phys. J. D* **45** 147–153
[15] Zhang P and Dalgarno A 2008 *Mol. Phys.* **106** 1525
[16] Przybytek M and Jeziorski B 2012 *Chem. Phys.* **401** 170 – 179
[17] Schlembach J and Tiemann E 1982 *Chem. Phys.* **68** 21 – 28
[18] Kamiya Y, Itagaki K, Tani M, Kim G N and Komamiya S 2015 *Phys. Rev. Lett.* **114** 161101
[19] Lutz J J and Hutson J M 2016 *J. Mol. Spectr.* (in press)
[20] Klimchitskaya G L, Mohideen U and Mostepanenko V M 2013 *Phys. Rev. D* **87** 125031
[21] Salumbides E J, Koelemeij J C J, Komasa J, Pachucki K, Eikema K S E and Ubachs W 2013 *Phys. Rev. D* **87** 112008

Optogalvanic spectroscopy applied to the study of hollow cathode discharges devices

V Gonzalez-Fernandez[1], K Grützmacher[1], L M Fuentes[2], C Pérez[1], M I de la Rosa[1]

[1]Dpto. de Física Teórica, Atómica y Óptica, Universidad de Valladolid, Paseo Belén 7, 47011, Valladolid, Spain
[2]Dpto. de Física Aplicada, Universidad de Valladolid, Paseo Belén 7, 47011, Valladolid, Spain

E-mail: veronica.gonzalez.fernandez@alumnos.uva.es

Abstract. Doppler free two-photon optogalvanic spectroscopy has been applied to measure the electric field strength in the cathode fall region of a hollow cathode discharge, via the Stark splitting of the 2S level of atomic hydrogen. Important improvements in the experimental arrangement and in our understanding of the discharge, have allowed us to measure accurately the dependence of the electric field with different parameters like the cathode material (stainless steel and tungsten), geometry (10 and 15 mm inner diameter) and the buffer gas. The study is done for different pressures and currents. The first results seem to reveal different behaviour in the electric field, depending on the material and geometry of the cathode and the discharge conditions.

1. Introduction

Hydrogen plasmas have been studied for many years due to its multiple applications to industry and research. In particular, the hydrogen plasmas generated in a hollow cathode discharge (HCD) are plasmas widely used in industry, and as they present low pressure, low kinetic temperature and out the local thermodynamic equilibrium, they result very attractive in order to investigate the application of different diagnosis techniques. In hollow cathode devices the electric field (E-field) is the main parameter that controls the behavior of the discharge. In the cathode fall region the E-field shows a very strong variation; this fact makes necessary a non-invasive and high resolution method for its determination.

In this work, we determine the E-field in the cathode region in a low-pressure pure hydrogen plasma, generated in a HCD in glow-discharge regime [1, 2], using optogalvanic spectroscopy and the Stark effect in the 1S-2S hydrogen transition. This technique provides the needed resolution. In fact, its resolution is four times bigger than other kind of spectroscopy commonly use, like laser induced fluorescence (LIF), as it has been proved in [3]. According to our experimental conditions, we use Doppler-free two-photon excitation of the hydrogen 1S–2S transition provided by two counter propagating photons from circularly polarized laser beams of opposite directions at 243 nm, in accordance with the selection rule for the angular momentum $\Delta L = 0$, once the 2S level is reached, it is easy for an atom to catch a third photon and become ionized. This ionization causes a change in the plasma impedance. The optogalvanic spectra are obtained by measuring this change in plasma impedance as a function of the laser frequency.

The Stark effect causes the splitting of the fine structure of the 2S level, theoretical calculations [4] show, Figure 1a, that the intensity of the components depends on the electric field strength, as the E-field increases the intensity of the 2S component decreases, allowing growing the other two components. And the $2P^{½}$ appears much earlier than the $2P^{3/2}$ but soon it disappears. Figure 1b show that the shift of the components as a function of the E-field strength. These two figures will be used to obtain the E-field from the experimental spectra.

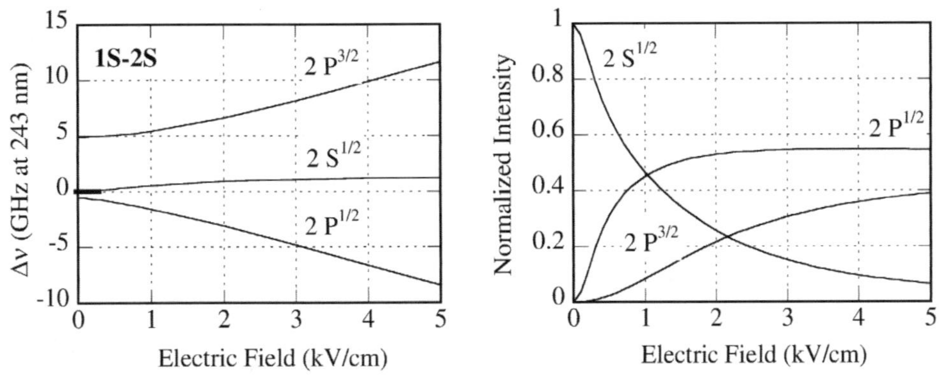

(a) (b)

Figure 1: Stark effect of the 1S-2S two photon transition of hydrogen. (a) Shows the calculated frequency shift versus electric field strength, and the corresponding two-photon absorption normalized lines intensities are given in (b)

2. Experimental arrangement

A comprehensive explanation of the whole experimental arrangement is given elsewhere [4-5]; here we only provide details concerning present work

2.1. The hollow cathode discharge

Picture 1: Peaked anode

In the HCD used in this experiment, the cathode is placed between two peaked anodes, as shown in Picture 1, this is a modified design of the one used before [6]. This new design gives more stability to the discharge. The discharge is operated at a constant gas flow of hydrogen entering at one anode side and leaving at the opposite. The anodes are water-cooled while the cathode is mounted in water-cooled holders made of copper. Thanks to a new system of load resistors, the applied current is equally distributed in both anodes. This device provides symmetric, stationary and stable non-LTE hydrogen plasma with low density and temperature. The power supply allows for discharge currents up to 480 mA, and the pressure can be varied from 0.25 to 1.35 kPa.

2.2. *Experimental set-up*

A schematic view of the whole experimental set-up can be seen in Figure 2. The full potential provided by two-photon spectroscopy is accessible, if a pulsed UV laser radiation of sufficient peak power and SLM spectral quality is available. The system consists of a 10 Hz injection seeded Q-switched Nd:YAG laser (Continuum, Powerlite) and a substantially modified OPO–OPA system (Continuum, Mirage 500). The required 243 nm radiation is obtained by sum-frequency generation (SFG) of 772 nm and the third harmonic radiation in a BBO crystal. This concept provides up to 10 mJ pulse energy in 3 ns and 300 MHz bandwidth that allows resolving the hyperfine structure of 2S level of hydrogen.

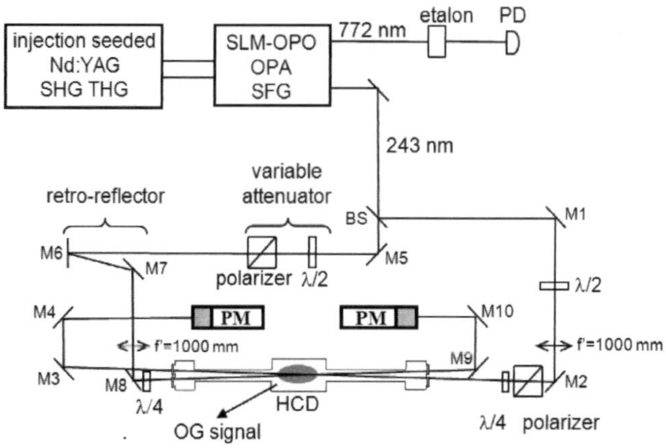

Figure 2: Schematic view of the experimental set-up

The laser beam is divided in two opposite circularly polarized beams. The measurement volume, given by the spatial overlap in focus (f′ = 1 m) of these two beams, is about 30 mm in length and 150 μm in diameter. In order to measure the variation of the electric field strength in the cathode fall, the HCD was mounted on an x–y translator. Displacements of the HCD allow performing measurements in the entire plasma: from the axis up to the cathode-fall region close to the inner surface of the hollow cathode. In order to get the best spatial resolution, the two beams were crossed in a horizontal plane and measurements were performed in the upper central part of the cathode-fall region. The beams direction is counter-propagating, allowing Doppler free measurements. The retroreflector, in Figure 2, assures the spatial stability in the overlap volume by cancelling small displacements of the laser direction. The laser power and the size of the focus (overlap volume) have been chosen according to [3], in order to get the best spectra and spatial resolution.

3. Measurements and first results

Before doing the optogalvanic measurements, it is necessary to make a complete study of the V-I characteristics of the discharge for the four cathodes: stainless steel (10 and 15 mm) and tungsten (10 and 15 mm). Figure 3 is the result of the V-I measurements, this study allow us to verify the high reproducibility of the discharge.

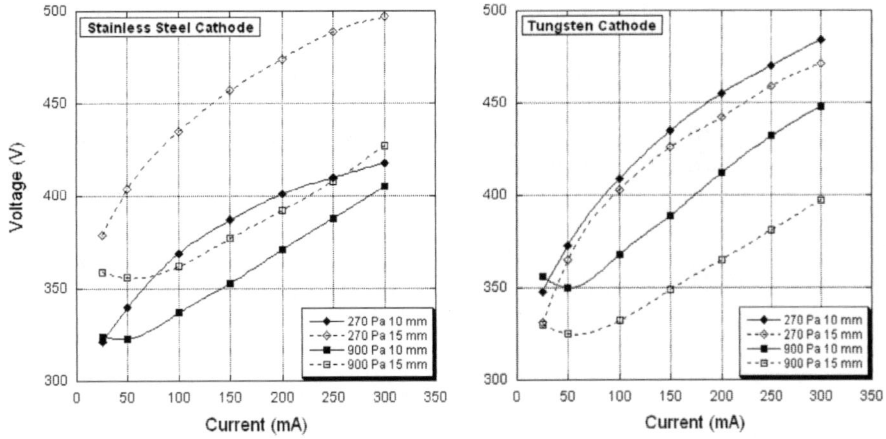

Figure 3: V-I characteristics for stainless steel and tungsten cathodes (10 and 15 mm inner diameter) for different pressures

The main objective of the present work is to measure the variation of E-field in the cathode fall region of a HCD, and to determine its dependence with the cathode geometry and material, with the buffer gas, and for different discharge characteristics: pressures and currents. As it was mentioned before, the E-field in the cathode fall region causes the Stark splitting and shifting of the components. According to Figure 1a, the shifting of the components is directly related to the E-field. Figure 4 shows a typical experimental spectrum (tungsten, 10 mm diameter, 600 Pa and 50 mA and a distance from de cathode of 0.350 mm) without any manipulation acquired shot by shot of the laser. The hyperfine structure of the three components is clearly visible; this means that the laser keeps the SLM mode during a scan. And also confirms the plasma stability.

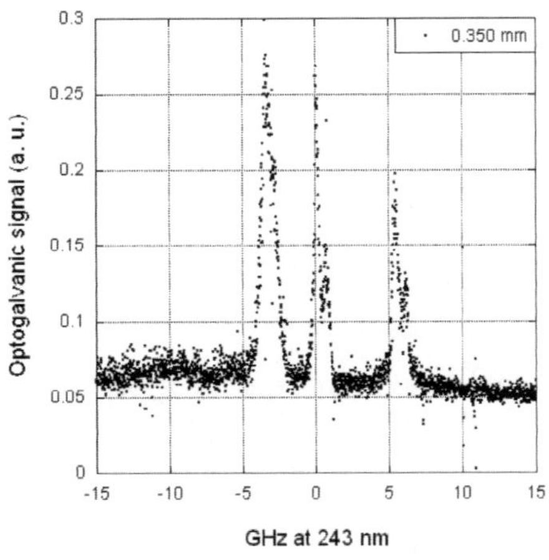

Figure 4: Experimental spectrum. The hyperfine structure of the three components is visible

In figure 5 there are the E-field values obtained for stainless steel for a pressure of 600 Pa and a current of 100 mA vs. the distance from the cathode surface, the parabolic tendency has been obtained before [6]. Nevertheless, the first results seem to reveal different trends of the E-field strength obtained with the two cathode materials for different experimental conditions.

But it is a work still in progress; the immediate project is to complete the measurements in different pressures and currents, in stainless steel and tungsten cathodes with different geometry, in order to compare experimental data with theoretical models and get more accurate conclusions.

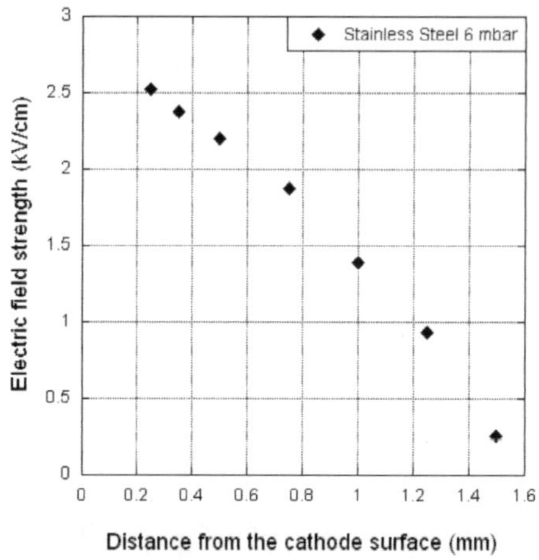

Figure 5: Variation of the E-field in the cathode fall region versus the radial distance from the cathode surface

Acknowledgments
The authors thank DGICYT (Ministerio de Economía y Competitividad) for the project ENE2012-35902, FEDER funds and the grant BES-2013-063248 given to V. González-Fernández.

References

[1] Cobine J D 1957 *Gaseous conductors: Theory and engineering applications* Dover Publications Inc., (New York)
[2] Den Hartog E A, Doughty D A and Lawler J E 1988 *Physical Review A* **38** 2471
[3] García-Lechuga M, Fuentes L M, Grützmacher K, Pérez C and de la Rosa M I 2014 *J. Appl. Phys.* **116** 133103
[4] de la Rosa M I, Pérez C, Grützmacher K, Gonzalo A B and Steiger A 2006 *Plasma Sources Sci. Technol.* **15** 105
[5] de la Rosa M I, Pérez C, Grützmacher K and Fuentes L M 2009 *Plasma Sources Sci. Technol.* **18** 015012
[6] Pérez C, de la Rosa M I and Grützmacher K 2010 *Eur. Phys. J. D* **56** 369-375

Self-reversal in Lyman-α line profile for diagnosis of fusion plasma

M Goto[1,2], K Sawada[3], T Oishi[1,2] and S Morita[1,2]

[1] National Institute for Fusion Science, Toki 509-5292, Japan
[2] SOKENDAI (The Graduate University of Advanced Study), Toki 509-5292, Japan
[3] Department of Applied Physics, Faculty of Engineering, Shinshu University, Nagano 380-8553, Japan

E-mail: goto@nifs.ac.jp

Abstract. We have observed Lyman-α line profile which exhibits the so-called "self reversal" at the line center. A one-dimensional radiation transport equation is numerically solved for simulating the Lyman-α line profile, where the emission and absorption coefficient distributions over the line-of-sight are evaluated based on the Laplace inversion analysis of the Balmer-α line measured simultaneously with the Lyman-α line. While the synthetic profile of the Lyman-α line shows a good agreement with the measured data in the line profile tail, the central part shows a significant discrepancy between the synthetic and measured profiles. It has been known that the neutral density profile in the plasma boundary region derived from the Balmer-α line profile analysis has a relatively large error and the present results shows that the Lyman-α line measurement would play a complementary role to determine an entire neutral density profile from the core to the edge.

1. Introduction

The neutral atom density in the plasma is an important parameter for evaluating the plasma confinement characteristics because the ionization of neutral atoms is essential in the particle source, and neutral atoms in the plasma cause particle and energy loss through the charge exchange process.

Many efforts have been made to determine the hydrogen atom density or the ionization rate in the plasma, most of which are based on spatial inversion of chord-integrated intensities of an emission line such as the Balmer-α line [1] or the Lyman-α line [2]. Nevertheless, this kind of measurement has an essential problem because observed line emissions of neutral atoms are mainly located at the outermost boundary of the plasma and weak line emissions in the core region, if any, are difficult to be detected by a line-integrated measurement with a line-of-sight which passes through both the edge and core regions.

In our previous study, we focused on the tail component of observed Balmer-α line profiles and established a method to deduce a spatial distribution of the hydrogen atom density [3]. This method has a such advantage that the line radiation in the high-temperature core region, which is responsible for the tail component of the line profile, is directly detected without being veiled behind the strong radiation in the low-temperature edge region, which mainly forms the central part of the line profile.

Meanwhile, this method has also a disadvantage in determining the neutral density in the plasma edge region because the central part of the line profile suffers from the Zeeman effect. Moreover, the condition assumed in the analysis that the atom temperature is approximately equal to the proton temperature would be inappropriate in the plasma edge region because atoms should mainly be created by the dissociation processes of molecules and not by the charge exchange process with protons.

On the other hand, we have noticed that the Lyman-α line profile shows a dent near the line center, which is called "self-reversal" and is considered to be due to strong reabsorption effect [4, 5]. Since the large reabsorption effect means the large line-integrated neutral density along the line-of-sight, the degree of the self-reversal should have information on the atom density in the edge region. In this paper, we examine the consistency between the neutral atom density profile derived from the Balmer-α analysis and the Lyman-α line profile.

2. Experiment

For the Lyman-α line observation, a normal incidence VUV spectrometer is used [6]. The spectrometer has a focal length of 3 m and is equipped with a 1200 grooves/mm grating. The Lyman-α line is measured with its second order diffracted light for having a better wavelength resolution. The spectrometer is placed at an outboard side port such that its optical axis overlaps the major radius of the plasma at the toroidal position where the plasma is horizontally elongated. In the present measurement, the field-of-view of the spectrometer is limited to ± 10 mm in the toroidal direction by a vertical aperture placed between the plasma and the spectrometer. The vertical width of the field-of-view is also limited to 20 mm by a horizontal aperture placed between the entrance slit and the grating in the spectrometer.

The measurement was performed for a discharge with $R_{\rm ax} = 3.6$ m and $B_{\rm ax} = 2.75$ T, where $R_{\rm ax}$ and $B_{\rm ax}$ are the major radius of the magnetic axis and the magnetic field strength at the magnetic axis, respectively. The temporal development of the discharge is shown in Fig. 1. The

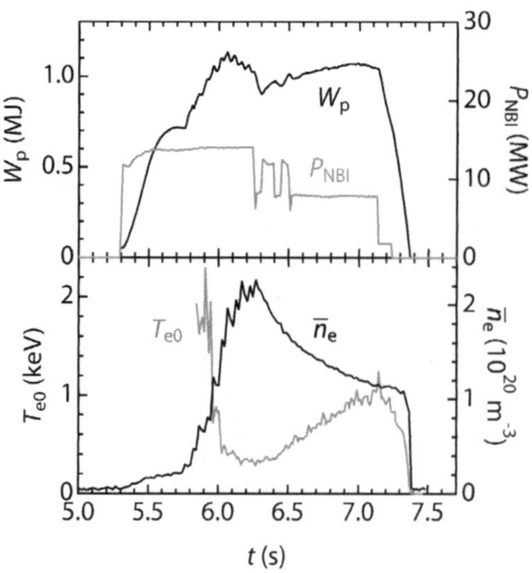

Figure 1. Temporal development of the discharge for the present measurement. The black and gray lines in the upper panel are the stored energy $W_{\rm p}$ and the neutral beam power $P_{\rm NBI}$, respectively. The black and gray lines in the lower panel show the line-averaged electron density $\overline{n}_{\rm e}$ and the central electron temperature $T_{\rm e0}$, respectively.

line-averaged electron density $\overline{n}_e$ is increased by repetitive injection of hydrogen pellets between $t = 5.7$ s and $t = 6.3$ s and gradually decays after termination of the pellet injection. Conversely, the central electron temperature T_{e0} drops by the pellet injection and recovers after $t = 6.3$ s. The spectrum is recorded every 54 ms with an exposure time of 32 ms.

Figure 2 shows the measured profile of the Lyman-α line at around $t = 6.5$ s as indicated by the dotted line in Fig. 1. A salient feature of the spectrum is the dent at the line center, which

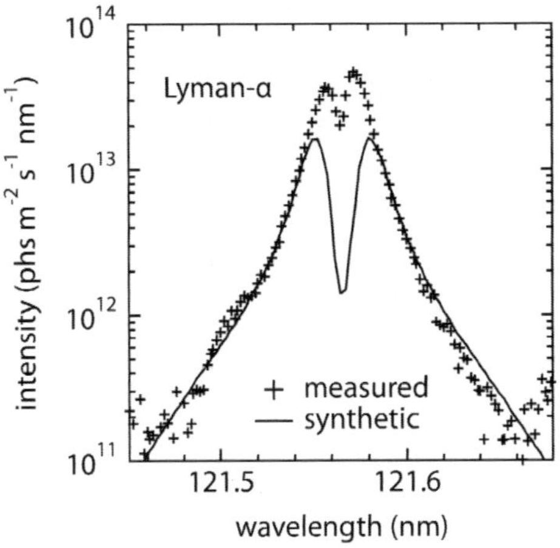

Figure 2. Measured (crosses) and synthetic (solid line) Lyman-α line profiles.

is thought to be due to a strong reabsorption effect by low temperature atoms existing in the plasma boundary region.

3. Results and discussion

For a quantitative evaluation of the line profile with the self-reversal, an equation of radiation transport is solved. We here consider a one-dimensional model which is expressed as

$$\frac{\mathrm{d}}{\mathrm{d}x} I(\lambda, x) = -\kappa(\lambda, x) I(\lambda, x) + \eta(\lambda, x), \tag{1}$$

where the coordinate x is taken along the line-of-sight. The parameters $\eta(\lambda, x)$ and $\kappa(\lambda, x)$ are respectively the emission and absorption coefficients at wavelength λ and at x. They are explicitly expressed as

$$\eta(\lambda, x) = \frac{hc}{4\pi\lambda_0} n_2(x) A_{21} P(\lambda, x), \tag{2}$$

$$\kappa(\lambda, x) = \frac{hc}{4\pi\lambda_0} n_1(x) B_{12} P(\lambda, x), \tag{3}$$

where h and c are the Planck's constant and the speed of light, respectively, λ_0 is the center wavelength, A_{21} and B_{12} are Einstein's coefficients for the spontaneous emission and the absorption, respectively.

The line profile at x, $P(\lambda, x)$, is here assumed to be determined by the Doppler broadening, where the atom temperature is approximated by the proton temperature at the same location because neutral atoms in the confined region are mainly created by the charge exchange process.

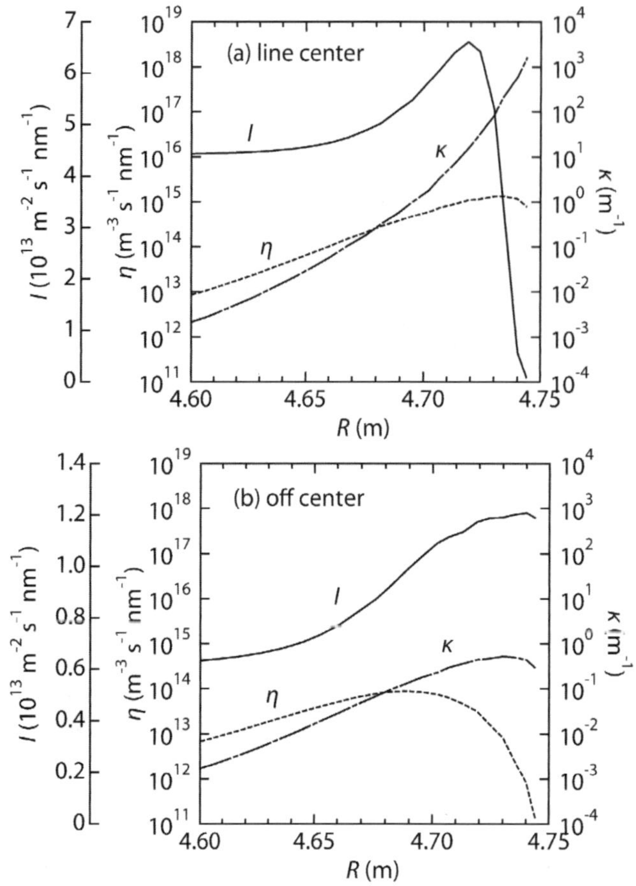

Figure 3. Radial profiles of the emission coefficient η (dashed line), absorption coefficient κ (dot-dashed line), and the resulting intensity (solid line) of the Lyman-α at the line center (a) and at 0.02 nm off the center (b) evaluated from the Balmer-α line profile analysis.

Moreover, the proton temperature is expected to be in equilibrium with the electron temperature T_e because the electron density n_e is high. Therefore, we employ the T_e profile measured by the Thomson scattering system for the atom temperature profile.

The spatial profiles of the ground state density $n_1(x)$ and the $p = 2$ level density $n_2(x)$, where p is the principal quantum number, are derived from the detailed analysis of the Balmer-α line [3] measured simultaneously with the Lyman-α. In this analysis, the Balmer-α line profile is regarded as a superposition of various temperature components and a distribution function against the temperature is derived by the Laplace inversion technique after replacement of variables. The obtained distribution function against the temperature is assigned to actual radial positions in the plasma and thus the entire radial profile of the Balmer-α line emissivity is obtained with the T_e profile measured by the Thomson scattering system, where the atom temperature profile is assumed to be identical to that of T_e as mentioned above. The quantities $n_1(x)$ and $n_2(x)$ are finally derived from the Balmer-α line emissivity, and T_e and n_e values with a help of the collisional-radiative model calculation.

Figure 3 shows the dependence of $\eta(\lambda, x)$ and $\kappa(\lambda, x)$ on the radial position at the line center

(a) and at 0.02 nm off the center (b). It is noted that the coordinate x is here translated into the major radius R. The parameters at the line center and at the off center show larger discrepancy as the position moves outward. This is due to a fact that the line profile in the edge region is narrower than that in the core region. The same figure also shows $I(\lambda, x)$ as a solution of Eq. (1). It is readily noticed that $I(\lambda, x)$ shows a rapid decrease in the edge region only at the line center. This is understandable because $\kappa(\lambda, x)$ at the line center increases rapidly in the edge region.

The complete synthetic line profile is shown in Fig. 2 with the solid line. The absolute intensity of the measured data is normalized to the synthetic profile in the wavelength region $|\lambda - \lambda_0| > 0.04$ nm where an optically thin condition is expected. It is found that the agreement in the tail components of the both profiles is good, which supports our understanding of the penetration of neutral atoms in the plasma core region. However, a significant discrepancy is observed in the central wavelength region of the profile, which indicates the accuracy of the parameters derived in the analysis of the Balmer-α line profile is insufficient especially in the outermost region.

This is understandable because the central region of the Balmer-α line profile suffers from the Zeeman effect and because the assumption of equivalent temperature between atoms and protons would be inappropriate. In other words, the Lyman-α line can be used as a complementary method to the Balmer-α line analysis in determining the neutral density profile.

Acknowledgments

The authors thank all the members of the LHD Experiment Group for their help in conducting valuable discharges and measurements. This research was supported in part by the LHD project (NIFS16ULHH028) and by Grants-in-Aid for Scientific Research (26287148).

References

[1] Colchin R J *et al* 2000 *Nucl. Fusion* **40** 175
[2] Boivin R L *et al* 2001 *Rev. Sci. Instrum.* **72** 961
[3] Goto M *et al* 2011 *Nucl. Fusion* **51** 023005
[4] Katai R *et al* 2007 *Plasma Fusion Research* **2** 014
[5] Goto M *et al* 2010 *Plasma and Fusion Research* **5** S2089
[6] Morita S and Goto M 2003 *Rev. Sci. Instrum.* **74** 2036

Hydrogen dimer features in low temperature collision-induced spectra

M Gustafsson
Applied Physics, Division of Materials Science, Department of Engineering Science and Mathematics, Luleå University of Technology, SE-97187 Luleå, Sweden
E-mail: magnus.gustafsson@ltu.se

Abstract. The absorption of radiation in pure hydrogen (H_2) gas around the S(0) and S(1) rotational transitions is computed at 20 K and compared with laboratory data. All transitions involving free state are included in the calculations of the absolute absorption. These calculations are done with an isotropic approximation for the H_2–H_2 pair potential. Agreement with the experiment is observed around the S(0) transition, while the computational approach appears to be slightly worse around the S(1) transition. The positions for bound-to-bound transitions are computed including the full anisotropic pair potential. The anisotropy seems to be crucial to achieve agreement with the measured bound-to-bound transition frequencies. However, those transitions contribute little to the total absorption. The present computed absolute absorptions will provide improved input for radiative transfer models of planetary atmospheres.

1. Introduction

Despite the weak attraction between hydrogen molecules, if hydrogen gas is sufficiently cool, a significant fraction of the molecules may form weakly bound van der Waals complexes. The importance of these complexes, which are also called dimers, in the absorption spectrum of hydrogen gas has long been recognized. They have been studied both theoretically and experimentally [1, 2, 3, 4], and their influence on Jupiter's and Saturn's spectra has been verified [1]. Recently, it has been discovered that a discrepancy on the order of 10% in the absorption coefficient is significant for the analysis of Jupiter's atmosphere [5]. Therefore, more accurate calculations of absorption data are desired.

The influence of the dimers on the spectrum increases with decreasing temperature. There is a laboratory measurement of the hydrogen spectrum around the rotational transitions S(0) and S(1) at 20 K recorded by McKellar [4] where dimer features are quite pronounced. This temperature is admittedly lower than what is typically considered in modeling of the atmospheres of the giant planets. The measurement provides, however, an excellent testing ground for various computational treatments of the dimer features.

This theoretical study has the purpose to explore the validity of the established treatment [2] of the dimer contributions to the hydrogen spectrum. The contributions are of three different characters.

(i) Free-to-free: This can be computed with proper anisotropic potential accounted for. It has been shown, however, that the influence from the anisotropy is quite small [6, 7]. In this work we have used the isotropic potential approximation (IPA) for this contribution.

(ii) Bound-to-free and free-to-bound: This is only computed in the isotropic potential approximation in this work. Below I will discuss the limitations of this approach and possible ways to include the anisotropy of the pair potential.

(iii) Bound-to-bound: In the laboratory measurement [4] there are one and three bound-to-bound transitions detected in the S(0) and S(1) regions, respectively. For sufficient accuracy the transition frequencies are computed here with the anisotropic potential accounted for.

2. Theory

In this section the method for calculating bound states with a discrete variable representation (DVR) is outlined. It accounts for the complete anisotropic interaction potential of the H_2–H_2 pair. This is the method used for calculations of the bound states in contributions (ii) and (iii) listed above. The calculation of the absorption coefficient is done as described in Ref. [2] where the continuum wave functions are determined with a Numerov integration of the Schrödinger equation. In the present case all bound wave functions and energy levels, whether anisotropic potential is accounted for or not, are computed with a DVR. In section 2.1 the DVR method is described for the case of two distinguishable hydrogen molecules, i.e. one para-H_2 and one ortho-H_2, which have even and odd rotational quantum number, respectively. In section 2.2 the appropriate adjustments for the case of two identical molecules will be described and in section 2.3 the selection rules for dipole transitions for hydrogen dimers are reviewed.

2.1. Bound state calculation with DVR, distinguishable molecules

Consider two interacting hydrogen molecules that remain in their electronic ground states. In a space fixed frame of reference the corresponding Hamiltonian is

$$H(\boldsymbol{r}_1, \boldsymbol{r}_2, \boldsymbol{R}) = H^{\mathrm{H-H}}(\boldsymbol{r}_1) + H^{\mathrm{H-H}}(\boldsymbol{r}_2) - \frac{\hbar}{2m}\nabla^2_{\boldsymbol{R}} + V(\boldsymbol{r}_1, \boldsymbol{r}_2, \boldsymbol{R}) \tag{1}$$

where $\boldsymbol{R}$ is the vector pointing from the center of one hydrogen molecule to the center of the other, and $\boldsymbol{r}_i$ is the vector pointing from one atom to the other in molecule $i = 1$ or 2. $H^{\mathrm{H-H}}$ is the rovibrational Hamiltonian of an isolated diatom, V is the interaction potential, and m is the reduced mass of the two-body system consisting of molecules 1 and 2. In this study the hydrogen molecules will remain in their vibrational ground state and a rigid rotor potential $V(\widehat{\boldsymbol{r}}_1, \widehat{\boldsymbol{r}}_2, \boldsymbol{R})$ is used, Thus the wave function may be expanded in angular basis functions according to

$$\Psi_{JM}(\widehat{\boldsymbol{r}}_1, \widehat{\boldsymbol{r}}_2, \boldsymbol{R}) = \sum_{\beta} \frac{1}{R} F^J_\beta(R)\, Y^{JM}_{j_1 j_2 j l}(\widehat{\boldsymbol{r}}_1, \widehat{\boldsymbol{r}}_2, \widehat{\boldsymbol{R}}) \tag{2}$$

where β is short hand for the quantum numbers $(j_1 j_2 j l)$ and $Y^{JM}_{j_1 j_2 j l}$ is the vector coupling function,

$$Y^{JM}_{j_1 j_2 j l}(\widehat{\boldsymbol{r}}_1, \widehat{\boldsymbol{r}}_2, \widehat{\boldsymbol{R}}) = \sum_{m_1 m_2 m_j m_l} C(j_1, j_2, j; m_1, m_2, m_j)\, C(j, l, J; m_j, m_l, M) \times Y_{j_1 m_1}(\widehat{\boldsymbol{r}}_1)\, Y_{j_2 m_2}(\widehat{\boldsymbol{r}}_2)\, Y_{l m_l}(\widehat{\boldsymbol{R}}) \,, \tag{3}$$

corresponding to $\boldsymbol{J} = \boldsymbol{j} + \boldsymbol{l}$ and $\boldsymbol{j} = \boldsymbol{j}_1 + \boldsymbol{j}_2$; the total angular momentum is the sum of rotational and translational angular momenta. C is a Clebsch–Gordan coefficient. The basis expansion (2) applied to the Hamiltonian (1) yields an R-dependent Hamiltonian matrix

$$H^J_{\beta\beta'}(R) = \left(E_{0j_1} + E_{0j_2} - \frac{\hbar}{2m}\frac{d^2}{dR^2} + \frac{\hbar l(l+1)}{2mR^2} \right) \delta_{\beta\beta'} + V^J_{\beta\beta'}(R) \,. \tag{4}$$

Both J and M are good quantum numbers, and the Hamiltonian is independent of M [8]. For two interacting diatoms it is appropriate to expand the potential energy in angular basis functions according to

$$V(\hat{\boldsymbol{r}}_1, \hat{\boldsymbol{r}}_2, \boldsymbol{R}) = (4\pi)^{3/2} \sum_{\gamma_1\gamma_2\gamma} V_{\gamma_1\gamma_2\gamma}(R) \sum_{m_1 m_2 m} C(\gamma_1\gamma_2\gamma; m_1, m_2 m)\, Y_{\gamma_1 m_1}(\hat{\boldsymbol{r}}_1)\, Y_{\gamma_2 m_2}(\hat{\boldsymbol{r}}_2)\, Y^*_{\gamma m}(\hat{\boldsymbol{R}}) \,. \quad (5)$$

Then the integration over all spatial coordinates except for R, which was carried out in the derivation of Eq. (4), yields the potential energy matrix element,

$$V^J_{\beta\beta'}(R) = \sum_{\gamma_1\gamma_2\gamma} V_{\gamma_1\gamma_2\gamma}(R)\, e_{\gamma_1\gamma_2\gamma}(j_1, j_2, j, l, j'_1, j'_2, j', l'; J) \,, \quad (6)$$

where the coefficient, $e_{\gamma_1\gamma_2\gamma}$, is given in the appendix of Ref. [6].

The radial Schrödinger equation is solved by implementation of a DVR with uniform grid spacing [9] $\Delta R = (R_{\max} - R_{\min})/N$ and corresponding grid points $R_\alpha = R_{\min} + \alpha \cdot \Delta R$ where $\alpha = 1, 2...(N-1)$. This choice of grid applied to the Hamiltonian in Eq. (4) yields

$$H^J_{\alpha\alpha'\beta\beta'} = \left(E_{0j_1} + E_{0j_2} + \frac{\hbar l(l+1)}{2mR^2_\alpha} \right) \delta_{\alpha\alpha'}\delta_{\beta\beta'} + T_{\alpha\alpha'}\delta_{\beta\beta'} + V^J_{\beta\beta'}(R_\alpha)\delta_{\alpha\alpha'} \quad (7)$$

with the kinetic energy matrix

$$T_{\alpha\alpha'} = \frac{\hbar}{2m\,\Delta R^2}(-1)^{\alpha-\alpha'} \begin{cases} \frac{\pi^2}{3} - \frac{1}{2\alpha^2} & \alpha = \alpha' \\ \frac{2}{(\alpha-\alpha')^2} - \frac{2}{(\alpha+\alpha')^2} & \alpha \neq \alpha' \end{cases} \quad (8)$$

and diagonalization of the Hamiltonian (7) yields $(N-1)N_\beta$ eigenstates and eigenenergies, E_k. N_β is the number of angular momentum basis functions in the expansion, Eq. (2).

2.2. Identical molecules

Two homonuclear diatoms of the same species are distinguishable if their individual parities, $(-1)^{j_i}$, are different. This means that para-H_2 and ortho-H_2 are distinguishable molecules. On the other hand, two para-H_2 molecules are identical, as well as two ortho-H_2 molecules, and their wave functions have to be symmetrized correctly for exchange of their nuclear skeletons. Nuclear spin zero for both parts, i.e. the para-H_2–para-H_2 case, is inconsistent with an antisymmetric wave function and thus the para-H_2–para-H_2 pair comes only in symmetric states, while the ortho-H_2–ortho-H_2 pair can be both symmetric (+) and antisymmetric (−).

The angular part of the wave function for the case of a para-H_2 molecule interacting with an ortho-H_2 molecule is the angular momentum coupling function, $Y^{JM}_{j_1 j_2 j l}(\hat{\boldsymbol{r}}_1, \hat{\boldsymbol{r}}_2, \hat{\boldsymbol{R}})$, defined in equation (3). Using this one can construct functions that are symmetric and anti-symmetric under the exchange $(\hat{\boldsymbol{r}}_1, \hat{\boldsymbol{r}}_2, \hat{\boldsymbol{R}}) \rightarrow (-\hat{\boldsymbol{r}}_2, -\hat{\boldsymbol{r}}_1, -\hat{\boldsymbol{R}})$ to be used in calculations involving identical diatoms (see e.g. Ref. [4]). Using those functions potential energy matrix elements, corresponding to those given in Eq. (6) for para-H_2 interacting with ortho-H_2, can be derived for the case of identical molecules. These are given in the appendix of Ref. [6].

In the present calculations interactions that could produce spin flips are not included. Thus the (±) symmetry is conserved in the calculations, much as a para-H_2 molecule cannot become an ortho-H_2.

2.3. Dipole selection rules

Consider first an H_2–H_2 complex which does not interact with radiation. Then total angular momentum J, total parity $(-1)^{j_1+j_2+l}$, diatomic parity $(-1)^{j_i}$ and symmetry, are all conserved. In dipole transitions, however, only the last two of those four constraints still hold. These are the dipole selection rules for H_2–H_2 complexes (unprimed and primed means initial and final, respectively):

Table 1. Bound-to-bound transition frequencies, in units cm^{-1}, around the S(0) and S(1) transitions in hydrogen predicted in the calculations described in section 2.1 together with those found in the laboratory measurement (exp).

j_1	j_2	symmetry	$l, J \rightarrow l', J'$	present calculation, $E_{k'} - E_k$	exp [4]
even	even	+	even, 0 → odd, 1	352.524 − (−2.978) = 355.502	355.425
odd	odd	−	odd, 1 → even, 0	823.826 − 235.625 = 588.201	588.300
				not predicted	588.503
odd	odd	−	odd, 1 → even, 0	823.826 − 234.940 = 588.886	588.803

- Total angular momentum: $J' = J;\ J \pm 1$
- j_i and j'_i are both even or both odd for i=1 or 2
- Parity must change, and including the previous constraint this implies that l changes from odd to even or vice versa
- In the case of identical molecules the (±) symmetry is conserved

2.4. Computational details
For the calculations of the bound-to-bound transition frequencies the DVR Hamiltonian in Eq. (7) is diagonalized with the DSYEV routine from LAPACK [10]. The experimental, rovibrational, hydrogen energy levels, $E_{v_i j_i}$, from Ref. [11] and the H_2–H_2 anisotropic interaction potential by Schäfer and Köhler [12] are implemented. In calculations of absorption coefficients I use the isotropic component of the same potential. For the interaction-induced dipole surface the *ab initio* data from Ref. [13] is used for free-to-free transitions and the data presented in Ref. [2] for free-to-bound and bound-to-free transitions.

3. Results and discussion
Table 1 shows the transition energies between H_2–H_2 bound states. One and three transitions are found in the experiment [4] in the vicinity of the S(0) and S(1) rotational transitions, respectively. One of the three latter transitions is not predicted in the calculation indicating that the potential energy surface may need improvement. A test calculation with a more recent rigid rotor potential by Miliordos and Hunt, which was also used in Ref. [7] yields virtually the same frequency, 355.5 cm^{-1}, for the transition close to S(0). On the other hand, that potential predicts no bound-to-bound transitions close to S(1). A test calculation with the Schäfer and Köhler [12] potential and the anisotropy turned off also fails to predict the transitions close to S(1). The transition energies are clearly quite sensitive to the anisotropy of the potential. Note that in each transition the higher of the two levels is a truly bound state because the basis expansion for the corresponding parity does not contain the $j_1 = j_2 = 0$ and $j_1 = j_2 = 1$ functions for para-para and ortho-ortho, respectively, but at least one of the hydrogens must be rotationally excited.

In Fig. 1 an experimental absorption coefficient for pure para-hydrogen at a temperature of 20 K around the S(0) rotational transition is displayed together with calculations. The bound-to-bound transition frequency is indicated while the corresponding absolute absorption is not computed in this work. From the experimental data it is clear that the corresponding absorption is relatively small, however. Adding the free-to-free absorption to the free-to-bound/bound-to-free appears to give an absorption in agreement with the experiment. Fig. 1 shows the corresponding calculation around the S(1) rotational transition for normal-hydrogen. The agreement between the measured and computed absolute absorption is not as good as in Fig. 1. This may indicate that the anisotropy of the potential plays a bigger role when ortho-H_2 is involved.

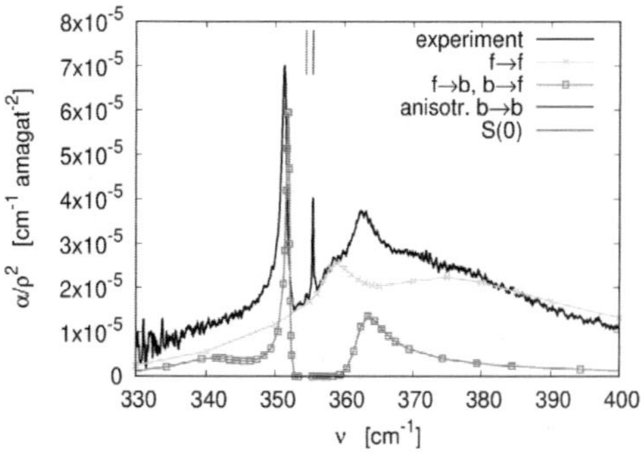

Figure 1. The absorption coefficient, normalized by the square of the density, around the S(0) transition for pure hydrogen gas at a temperature of 20 K. The experimental data is from Ref. [4].

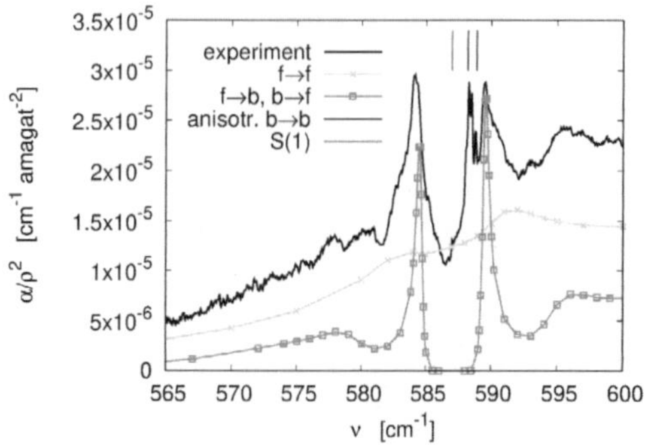

Figure 2. As in Fig. 1 but around the S(1) transition for normal hydrogen gas (25% para-H_2 and 75% ortho-H_2) at a temperature of 20 K.

The anisotropy of the potential could be included in the calculations of the free-to-bound/bound-to-free in order to improve the agreement with experimental data. For example it could be done with a scattering calculation with propagation of the dipole matrix element on-the-fly as in Ref. [14]. The area under the absorption curves in both Figs. 1 and 2 are, however, relatively well reproduced by the calculation. Temperatures below 40 K are normally not involved in the modeling of planetary atmospheres and there is often spectral broadening. I conclude that the present method, with isotropic interaction potential, and including free-to-free, free-to-bound, and bound-to-free transitions in the calculations of absolute absorption coefficients, is sufficient for the purpose of modeling of the atmospheres of the giant planets. The present calculation will be extended to higher temperatures so that it may be added to the presently available free-to-free data table [15, 16].

References

[1] Frommhold L, Samuelson R and Birnbaum G 1984 *Astrophys. J.* **283** L79 – L82
[2] Meyer W, Frommhold L and Birnbaum G 1989 *Phys. Rev. A* **39** 2434 – 2448
[3] Schäfer J and McKellar A R W 1990 *Z. Physik D* **15** 51 – 65 erratum: Z. Physik D 17 (1990) 231
[4] McKellar A R W and Schäfer J 1991 *J. Chem. Phys.* **95** 3081 – 3091
[5] Fletcher L 2016 private communication
[6] Gustafsson M, Frommhold L, Bailly D, Bouanich J P and Brodbeck C 2003 *J. Chem. Phys.* **119** 12264–12270
[7] Karman T, van der Avoird A and Groenenboom G C 2015 *The Journal of Chemical Physics* **142** 084305
[8] Green S 1975 *J. Chem. Phys.* **62** 2271–2277
[9] Colbert D T and Miller W H 1992 *J. Chem. Phys.* **96** 1982 – 1991
[10] Anderson E, Bai Z, Bischof C, Blackford S, Demmel J, Dongarra J, Du Croz J, Greenbaum A, Hammarling S, McKenney A and Sorensen D 1999 *LAPACK Users' Guide* 3rd ed (Philadelphia, PA: Society for Industrial and Applied Mathematics)
[11] Stoicheff B P 1957 *Can. J. Phys.* **35** 730–741
[12] Schäfer J and Köhler W E 1989 *Z. Physik D* **13** 217 – 229
[13] Meyer W, Borysow A and Frommhold L 1989 *Phys. Rev. A* **40** 6931 – 6949
[14] Karman T, Miliordos E, Hunt K L C, Groenenboom G C and van der Avoird A 2015 *J. Chem. Phys.* **142** 084306
[15] Orton G S, Gustafsson M, Burgdorf M and Meadows V 2007 *Icarus* **189** 544 – 549
[16] Richard C, Gordon I, Rothman L, Abel M, Frommhold L, Gustafsson M, Hartmann J M, Hermans C, Lafferty W, Orton G, Smith K and Tran H 2012 *J. Quant. Spectroscopy and Rad. Transfer* **113** 1276 – 1285

IOP Conf. Series: Journal of Physics: Conf. Series 810 (2017) 012018 doi:10.1088/1742-6596/810/1/0120

Interatomic potentials of van der Waals dimers Hg_2 and Cd_2: Probing discrepancies between theory and experiment

T Urbańczyk[1], **M Krośnicki**[2], **M Strojecki**[3], **A Pashov**[4], **A Kędziorski**[5], **P Żuchowski**[5] **and J Koperski**[1]

[1] Smoluchowski Institute of Physics, Jagiellonian University, prof. S. Łojasiewicza 11, 30-348 Kraków, Poland
[2] Institute of Theoretical Physics and Astrophysics, University of Gdansk, Wita Stwosza 57, 80-952 Gdańsk, Poland
[3] Jerzy Haber Institute of Catalysis and Surface Chemistry, Polish Academy of Sciences, Niezapominajek 8, 30-239 Kraków, Poland
[4] Department of Physics, Sofia University, 5 James Bourchier Boulevard, 1164 Sofia, Bulgaria
[5] Institute of Physics, Nicolaus Copernicus University, Grudziądzka 5/7, 87-100 Toruń, Poland

E-mail: ufkopers@cyf-kr.edu.pl

Abstract. Results of new all-electron *ab initio* calculations and revisit of experimental studies of the interatomic potentials of lower-lying *ungerade* excited and ground electronic energy states of the Hg_2 and Cd_2 van der Waals complexes are used as probes of discrepancies between theory and experiment. From simulations of the previously and presently measured LIF excitation and dispersed emission spectra new analytical representations of the excited- and the ground-state interatomic potentials are proposed. An inverted perturbation approach was also used to improve the studied interatomic potentials. The comparison of the new *ab-initio* calculated potentials with the results of the analyses illustrates an improve theory-to-experiment agreement for such a demanding system like Hg_2 or Cd_2.

1. The objective of the study

The main objective of the study presented here was calculation of new all-electron *ab initio* interatomic potentials for the Hg_2 and Cd_2 lower-lying *ungerade* excited and the ground states, and their subsequent assessment against the potentials that are the result of analyses of newly recorded and re-analyses of previously recorded experimental spectra using pulsed and continuous supersonic beams [1–9]. Some of the previous analyses were performed using limited resources (*e.g.*, simulation procedures only for low υ' and υ'', and with only a Morse representation of interatomic potentials in their bound region), and without systematic consideration of the isotopic composition and rotational energy structures that otherwise can completely change results of the simulation (*e.g.*, intensities of the vibrational bands and their shapes). Consequently, in order to achieve the objective, the following agenda was realized:

- New all-electron *ab initio* calculations of the interatomic potentials for the lowest *ungerade* and the ground states of Hg_2 and Cd_2. These was executed to improve the theory-to-experiment agreement;
- Analysis of the newly recorded and revisit of the previously measured experimental spectra in

order to perform simulations of the LIF excitation and LIF dispersed emission spectra using programs and procedures that were originally omitted due to their unavailability *i.e.*, LeRoy's LEVEL [10] and BCONT [11] as well as Western's PGOPHER [12];

• Implementation of analytical representations for the Hg_2 and Cd_2 lowest *ungerade*- and ground-state potentials as well as point-wise representations, obtained using the IPA method [13] for which an analytical representation was found to be impossible;

• Comparison of new *ab-initio* calculated potentials with results of new measurements and new analyses as well as with other *ab initio* calculations and experimental results.

2. Computational results: *ab initio* calculations of Hg_2 and Cd_2 interatomic potentials

The all-electron *ab initio* calculations were performed using the MOLCAS 8.0 package [14]. Details of the calculations for Hg_2 are given in [15].

For Cd_2 Atomic Natural Orbitals (ANO-RCC) basis set was used as designed by Roos *et al.* [16]. The basis set was formed from 21*s*19*p*13*d*6*f*4*g*2*h* Gaussians contracted as 10*s*8*p*8*d*6*f*4*g*2*h* and it allowed for correlation of semi core electrons.

The Hamiltonian used in electron correlation calculations was spin-free second-order Douglas-Kroll-Hess (D-K-H) Hamiltonian [17,18] which accounted for scalar relativistic effects. In order to account for static correlations, complete-active-space self-consistent-field (CASSCF) [19-21] calculations were performed. In the CASSCF calculations molecular orbitals and configuration interaction coefficients of the wave functions were simultaneously optimized. The active space was formed by distributing four active electrons on molecular counterparts of atomic orbitals Cd-5*s*, Cd-5*p* and Cd-6*s*. All closed orbitals were optimized. As starting orbitals, canonical Hartree-Fock orbitals were used. The starting dimension of the small configuration interaction (CI) matrix in the Davidson procedure for the $X^1\Sigma_g^+$ and $(1)^1\Sigma_u^+$ states was set to two; and for the $(1)^3\Sigma_u^+$ state it was set to three [(1) denotes first on the energy scale of the electronic energy state with the given symmetry]. This avoided convergence problems for short interatomic distances. The molecular orbitals and wave functions were optimized in CASSCF calculations separately for each molecular excited electronic state of *ungerade* symmetry that originated from the 5*s*6*p* atomic configuration, with only one exception in case of $(1)\Pi_u$ state which was optimized simultaneously with second state of this symmetry (having ionic character) due to strong configuration interaction.

In the second step, dynamic electron correlations using restricted active space multi-state second-order perturbation theory calculations (MS-CASPT2) [22-25] were included. In CASPT2 calculations, 24 electrons (4 active and 20 closed-shell) were fully correlated. The active space was the same as on the level of CASSCF theory. The 10 molecular counterparts of atomic 5*d* orbitals were correlated through single and double excitations. The rest of core orbitals were not optimized and kept frozen.

In the last step, spin-orbit (S-O) coupling effects were included by means of restricted-active space state-interaction spin-orbit (RASSI-SO) calculations [26]. In this step, many-electron Hamiltonian with the s-o coupling terms of the second-order D-K-H Hamiltonian [17,18] were supplemented and the atomic mean-field approximation (AMFI) for their integrals [27] was adopted. In the RASSI code S-O matrix element were calculated in a wave function basis which is formed from individually optimized CI expansions in the previous CASSCF/CASPT2 calculations. All *ungerade* triplet and singlet states originating from $(5s5p)^3$P Cd and $(5s5p)^1$P Cd configuration, as well as the ground state were included in this step. The transition dipole moments (TDM) were determined within RASSI by calculation the interaction matrix elements between the ground and the excited states.

3. Experimental results: spectroscopy of Hg_2 and Cd_2 in supersonic beams

Laser spectroscopy of 12-group vdW complexes produced and ro-vibrationally cooled in a free-jet expansion beams is one of methods for investigation of molecular energy structure [1]. It has been applied, among others, in experimental studies of the lower-lying $E^31_u(6^3P_2)$-, $F^30_u^+(6^3P_1)$-, $D^31_u(6^3P_1)$- and $G^10_u^+(6^1P_1)$-state interatomic potentials of Hg_2, and $A^10_u^+(5^1P_1)$-, $B^11_u(5^1P_1)$-, $a^31_u(5^3P_1)$-, $b^30_u^+(5^3P_1)$- and $c^31_u(5^3P_2)$-state interatomic potentials of Cd_2 excited, and the $X^10_g^+$ ground electronic-energy states of Hg_2 and Cd_2. Details of the experiments for Hg_2 are given in [15] while for the Cd_2 are reported in [2-9].

4. Results and discussion

As an example of results, analysis of the LIF excitation spectrum recorded using the $A^1 0_u^+ \leftarrow X^1 0_g^+$ transition in Cd_2 is presented. The excitation spectrum consists of the $\upsilon' \leftarrow \upsilon''=0$ progression with υ' spanning the range from $\upsilon'=18$ to 53. As can be seen in Fig. 1a, the approximate locations of the excited υ' are in the lower half of the $A^1 0_u^+$-state potential well *i.e.*, closer to the bottom of the well than to the dissociation limit.

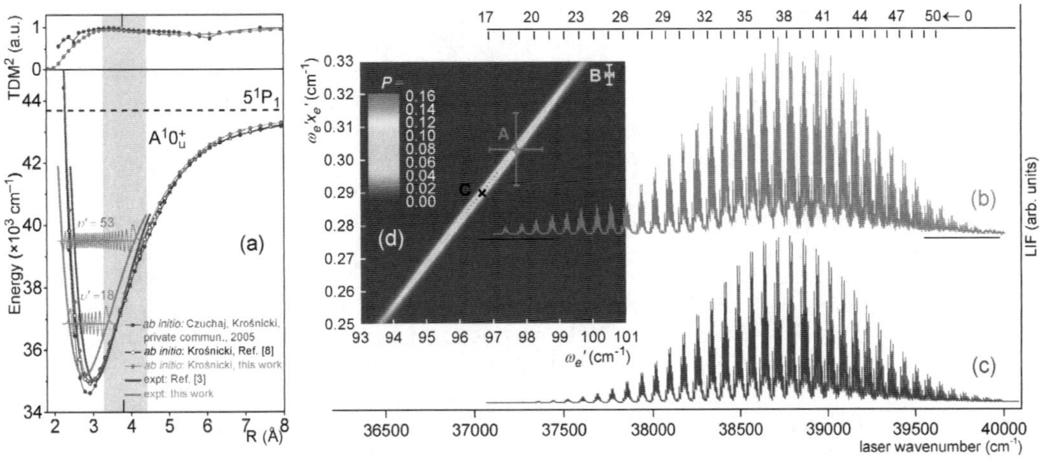

Figure 1. (a) Interatomic potential of the $A^1 0_u^+$ state of Cd_2. The determined TDM^2 (normalized) of Czuchaj and Krośnicki and of this study are shown in the upper. Grey area: Franck-Condon window for excitation from $\upsilon''=0$. Vertical lines: position of the $R_e''=3.76$ Å. (b) LIF excitation spectrum recorded previously [3] using the $A^1 0_u^+ \leftarrow X^1 0_g^+$ transition. (c) Simulation [10,12] performed assuming shifted (-656.9 cm^{-1}) Morse representation of the $A^1 0_u^+$-state potential (D_e'=7882.5 cm^{-1}, β' = 1.0072 Å^{-1}, R_e' = 2.75 Å, see part 1a). For the $X^1 0_g^+$-state potential, a L-J(n–6)-Morse representation (D_e''= 328 cm^{-1}, R_e''= 3.76 Å, n = 3.86) was used; the TDM^2 of this study was included. (d) The ω_e'–$\omega_e' x_e'$ *agreement plot* drawn using vibrational transition energies $E_{\upsilon'}$ of the most abundant $^{114}Cd^{112}Cd$ isotopologue recorded in the spectrum. The simulation-to-experiment agreement [parameter P, see text] in function of the ω_e' and $\omega_e' x_e'$ is presented using a colour scale. Point A corresponds to the ω_e' and $\omega_e' x_e'$ and their uncertainties obtained as a result of this work (ω_e' = 97.74 cm^{-1} and $\omega_e' x_e'$ = 0.303 cm^{-1}). Point B shows the ω_e' and $\omega_e' x_e'$ and their uncertainties reported in [3]. Center of the plot (point C) is the ω_e' and $\omega_e' x_e'$ for which one obtains the best simulation-to-experiment agreement in the analysis of the excitation spectrum.

4.1. *The agreement plot*

Re-analysis of LIF excitation spectrum recorded using the $A^1 0_u^+ \leftarrow X^1 0_g^+$ transition [3] shown in Fig. 1b began with employing of so-called *agreement plot* method *via* examination of the $E_{\upsilon'}^{expt}$ energies of vibrational transition recorded in the spectrum for the most abundant $^{114}Cd^{112}Cd$ isotopologue. The *agreement plot* illustrates a new method of determination of pairs of the ω_e' and $\omega_e' x_e'$ vibrational constants. The method is used under an assumption that in the region of the recorded $E_{\upsilon'}^{expt}$, the $A^1 0_u^+$-state potential is represented by a Morse function. The main advantage of the method [in comparison to well-known Birge–Sponer (B–S) plot] is acknowledging the fact that vibrational constants ω_e' and $\omega_e' x_e'$ are highly correlated. In other words, it can be shown that there are exist many of the (ω_e' , $\omega_e' x_e'$) pairs which can be used to perform simulations that are in satisfactory agreement with the experimental spectrum, especially when the rotational structure of the spectrum is not resolved.

For each combination of the ω_e' and $\omega_e' x_e'$ in the specified ranges of their values (*e.g.*, around their expected values obtained using the B–S plot), the corresponding $A^1 0_u^+$ -state Morse representation was determined. Next, for each Morse representation using the LEVEL, $E_{\upsilon'}{}^{sim}$ energies of the vibrational transitions corresponding to the transitions observed in the experimental spectrum were calculated. Then, the $E_{\upsilon'}{}^{sim}$ were compared with the recorded $E_{\upsilon'}{}^{expt}$. Finally, for each Morse representation, the so-called *agreement coefficient* $P = 1/(0.01+\chi^2)$ was calculated, where $\chi^2 = \Sigma_{\upsilon'}(\Delta E_{\upsilon'}{}^{exp} - \Delta E_{\upsilon'}{}^{sim})^2$ is a sum of squares of the differences between $\Delta E_{\upsilon'}{}^{expt}$ experimental and $\Delta E_{\upsilon'}{}^{sim}$ simulated energy separations, and $\Delta E_{\upsilon'}{}^{sim} = E_{\upsilon'}{}^{sim} - E_{int\,\upsilon'}{}^{sim}$, where $E_{int\,\upsilon'}{}^{sim}$ is an energy of the selected *e.g.*, more intense, $\upsilon' \leftarrow \upsilon''=0$ transition in the experimental spectrum; the $\Delta E_{\upsilon'}{}^{expt}$ were calculated similarly to the $\Delta E_{\upsilon'}{}^{sim}$. The coefficient P describes the agreement between the simulation and experimental results in a quantitative way.

4.2. The spectrum

The *agreement plot i.e.*, a contour plot of the agreement coefficient P in function of the ω_e' and $\omega_e' x_e'$ is shown in Fig. 1d. As one can see, the (ω_e', $\omega_e' x_e'$) pair obtained in [3] depicted with cross B (also representing error bars) is located out of the area where the P gains high values. In the *plot*, the result of this study (ω_e' =97.74 cm^{-1}, $\omega_e' x_e'$ =0.303 cm^{-1}) is depicted with cross A. Centre of the *plot* (see cross C) that gives the best simulation-to-experiment agreement (ω_e' =96.65 cm^{-1}, $\omega_e' x_e'$ =0.29 cm^{-1}) does not match the values accepted in this study. This is caused by the result of simulation of LIF dispersed emission spectrum [3].

We focused our attention at simulation of the LIF excitation spectrum previously recorded using the $A^1 0_u^+ \leftarrow X^1 0_g^+$ transition (Fig. 1b). Due to the high values of υ', the isotopic structure in each of the vibrational components was easily resolved. In previous simulation of the excitation spectrum [3] neither an influence of TDM nor rotational structure of the $\upsilon' \leftarrow \upsilon''=0$ components has been taken into account.

Simulation of the excitation spectrum began with an assumption of a Morse representation for the $A^1 0_u^+$ state and with the (ω_e', $\omega_e' x_e'$) pair as concluded from the *agreement plot* (cross C). However, parallelly performed simulation of LIF dispersed emission spectrum from [3] indicated that, despite the simulation-to-experiment agreement was high, the (ω_e', $\omega_e' x_e'$) pair chosen from the centre of the *agreement plot* is not the best choice as far as both simulations are concerned. Consequently, other (ω_e', $\omega_e' x_e'$) pair was selected (cross A). It is necessary to emphasize that in the (ω_e', $\omega_e' x_e'$) selection process, a trend towards the result of [3] together with high values of P was preferred. In the simulation procedure described in details elsewhere [9], LEVEL 8.0 and PGOPHER 8.0 were employed, taking into account the *full* isotopic composition of Cd_2 as well as rotational energy structure of the $^1\Sigma_u^+ \leftarrow {}^1\Sigma_g^+$, $\Omega'=0 \leftarrow \Omega''=0$ transition. An influence of TDM squared calculated in this work (see upper part in Fig. 1a) was also included in the simulation. Result is shown in Fig. 1c. It can be verified that a correct distribution of intensities of Cd_2 isotopologues has been reconstructed what was not achieved in previous study [3]. Another conclusion that came from the simulation was a lack of strong influence of TDM squared as the TDM has almost constant values for R for which the transition takes place (see Franck-Condon window for excitation in Fig. 1a, grey area).

5. Summary

The objective formulated in Sec. 1 was realized. Specific points are the following.

- New all-electron *ab initio* calculations of the $E^3 1_u$, $F^3 0_u^+$, $D^3 1_u$ and $G^1 0_u^+$-state interatomic potentials of Hg_2, and $A^1 0_u^+$, $B^1 1_u$, $a^3 1_u$, $b^3 0_u^+$ and $c^3 1_u$-state interatomic potentials of Cd_2 were performed in order to improve the theory-to-experiment agreement. Indeed, a significant improvement and increased agreement with the experimental interatomic potentials were achieved as compared to the earlier calculations.
- Presently and previously recorded LIF excitation and LIF dispersed emission spectra were analyzed or revisited, and their simulations were performed employing the LEVEL 8.0, BCONT 2.2 and PGOPHER 8.0 programs. The full isotopic composition of Hg_2 and Cd_2, as well as the rotational energy structure of their transitions were taken into account.

• New analytical representations of the Hg_2 and Cd_2 lowest excited *ungerade*-state potentials, *i.e.* Morse, Morse-vdW or ground-state potentials, *i.e.* Morse-L–J(n–6) or Morse-L–J(n–m) were obtained. For the D^31_u and $G^10_u^+$ electronic states of Hg_2 the new representations were obtained using an inverted perturbation approach (IPA).

• A comparison of the new *ab initio* calculated potentials with results of new analyses as well as other *ab initio* calculations was performed. These comparisons illustrate an improved theoretical and experimental characterization of the lowest excited *ungerade*- and ground-state potentials of Hg_2 and Cd_2.

Acknowledgements
This work was supported by the National Science Centre Poland under grant number UMO-2015/17/B/ST4/04016.

References

[1] Koperski J 2002 *Phys Rep.* **369** 177
[2] KoperskiJ, Łukomski M and Czajkowski M 2002 *Spectrochim. Acta A* **58** 927
[3] Łukomski M, Koperski J, Czuchaj E and Czajkowski M 2003 *Phys. Rev. A* **68** 042508
[4] Łukomski M, Ruszczak M, Czuchaj E and Koperski J 2005 *Spectrochim. Acta A* **61** 1835
[5] Strojecki M, Ruszczak M, Łukomski M and Koperski J 2007 *Chem. Phys.* **340** 171
[6] Ruszczak M, Strojecki M, Łukomski M and Koperski J 2008 *J. Phys. B: At. Mol. Opt. Phys.* **41** 245101
[7] Strojecki M, Krośnicki M, Zgoda P and Koperski J 2010 *Chem. Phys. Lett.* **489** 20
[8] Koperski J, Strojecki M, Krośnicki M and Urbańczyk T 2011 *J. Phys. Chem. A* **115** 6851
[9] Urbańczyk T and Koperski J 2014 *Mol. Phys.* **114** 2486
[10] LeRoy R J 2002 *LEVEL 8.0 A Computer Program for Solving the Radial Schrödinger Equation for Bound and Quasibound Levels* (University of Waterloo Chemical Physics Research Report) CP-655R
[11] LeRoy R J, Kraemer G T 2004 *BCONT 2.2 A Computer Program for calculating Bound→ Continuum Transition Intensities for Diatomic Molecules* (University of Waterloo Chemical Physics Research Report) CP-650R2
[12] Western C M 2013 *PGOPHER 8.0, a Program for Simulating Rotational Structure* (University of Bristol)
[13] Pashov A, Jastrzębski W and Kowalczyk P 2000 *Comput. Phys. Commun.* **128** 622
[14] Karlström G, Lindh R, Malmqvist P-Å, Roos B O, Ryde U, Veryazov V, Widmark P O, Cossi M, Schimmelpfennig B, Neogrady P and Seijo L 2003 *Comput. Mater. Sci.* **28** 222
[15] Krośnicki M, Strojecki M, Urbańczyk T, Pashov A and Koperski J 2015 *Phys. Rep.* **591** 1
[16] Roos B O, Lindh R, Malmqvist P-Å, Veryazov V and Widmark P-O 2005 *J. Phys. Chem. A* **109** 6575
[17] Douglas M and Kroll N M 1974 *Ann. Phys. (N.Y.)* **82** 89
[18] Hess B A 1986 *Phys. Rev. A* **33** 3742
[19] Roos B O, Taylor P R and Siegbahn P E M 1980 *Chem. Phys.* **48** 157
[20] Siegbahn P E M, Heiberg A, Roos B O and Levy B 1980 *Phys. Scr.* **21** 323
[21] Siegbahn P E M, Heiberg A, Almlöf J and Roos B O 1981 *J. Chem. Phys.* **74** 2384
[22] Andersson K, Malmqvist P-Å, Roos B O, Sadlej A J and Wolinski K1990 *J. Phys. Chem.* **94** 5483
[23] Andersson K, Malmqvist P-Å and Roos B O 1992 *J. Chem. Phys.* **96** 1218
[24] Zaitsevskii A and Malrieu J P 1995 *Chem. Phys. Lett.* **233** 597
[25] Finley J, Malmqvist P-Å, Roos B O and Serrano-Andrés L 1998 *Chem. Phys. Lett.* **288** 299
[26] Malmqvist P-Å, Roos B O and Schimmelpfennig B 2002 *Chem. Phys. Lett.* **357** 230
[27] Hess B A, Marian C M, Wahlgren U and Gropen O 1996 *Chem. Phys. Lett.* **251** 365

Helicity-induced shapes of resonant four-wave mixing responses from photofragments

A Kouzov[1], **P Radi**[2], **N Filippov**[1] **and T Sinyakova**[1]
[1]Faculty of Physics, Saint Petersburg State University, Ulyanovskaya str. 3, Saint Petersburg 198504, Russia
[2]Paul Scherrer Insitute, CH-5232 Villigen, Switzerland
E-mail: `alex@ak1197.spb.edu (A Kouzov)`

Abstract. Doppler profiles of the Degenerate Resonant Four-Wave Mixing (DFWM) spectral responses from rapidly recoiling products of photodecomposition by plane-polarized radiation are theoretically studied. Emphasis is placed on the shapes produced by rotational helicity when orientations of the angular momenta and the recoil velocities of nascent particles are correlated. Illustrations are provided for cases when a decay of resonating states tangibly affects the helicity-induced DFWM profiles.

1. Introduction

Reaction pathways are sharply imprinted on characteristics of the nascent particles [1] typically born in non-equilibrium states. To read off these characteristics, linear spectroscopies have proved their unexelled efficiency, especially in tandems with the mass spectrometric and ionization techniques. In particular, the laser-induced fluorescence (LIF) techniques have developed into the most useful means to study quantum properties of photofragments, i.e., in the case when a reaction itself is ignitied by photons. In so doing, many photofragment properties as well as a number of correlations between them [1, 2, 3, 4, 5, 6] have been investigated, both experimentally and theoretically. Subsequently, in a review of the day [7] an advent of nonlinear optical methods to such studies, especially of the resonant four-wave mixing (RFWM) with its uniquely flexibile polarization setup, was prognosticated. Yet, except the recent experiments [8] performed at the Paul Scherrer Institute, the nonlinear optical tools have been never used for that aim despite the fact that RFWM had gained wide recognition as one of the most advanced all-optical means to track the quantum states of molecules at equilibrium [9, 10].

Photofragments produced by polarized laser beams have anisotropic distributions of their angular momenta $\mathbf{j}$ and velocities $\mathbf{v}$. To study the anisotropy of the OH radicals produced by the H_2O_2 photolysis by a plane-polarized laser radiation at 266 nm, the DFWM setup was established [8]. Absorption of photons with the energy of about 37 600 cm^{-1} much exceeding the H_2O_2 dissociation threshold (17 300 cm^{-1}) creates pairs of rapidly recoiling OH radicals which are almost exclusively generated in the substates belonging to the ground vibronic level [11]. Based on this fact, a quick estimation of the Doppler halfwidths of the DFWM frequency scanning the $A^2\Sigma^+$-$X^2\Pi(0,0)$ OH vibronic band around 32 300 cm^{-1} results in ≈ 0.4 cm^{-1}, approximately five times larger than the laser linewidth, thus allowing reliable resolution of the Doppler broadened profiles.

2. Third-order susceptibility of photofragments

Assuming that the laser line width is negligible, all four DFWM beams (three of the input and one of the generated signal) have the same frequency ω tuned close to a certain dipole-allowed transition $f \leftarrow i$. The latter circumstance allows us to consider only the on-resonance part ($\chi^{(3)}_{res}$) of the third-order susceptibility of nascent particles. To derive an expression for the induced polarization, one should average $\chi^{(3)}_{res}$ over the $\mathbf{j}-$ and $\mathbf{v}$-distributions and then calculate $|\chi^{(3)}_{res}|^2$ that defines the useful signal. Taken together, these distributions define the form of the product density matrix ρ, the key consequence of the photolysis path. In so doing, we have extended classical Dixon's expression [12] of ρ to the case of a quantized linear rotator using the classical representation of the translational motion which is well justified for rapidly recoiling fragments. Another property to be introduced to ρ is the (linear) polarization of a dissociation laser, namely, the orientation (Ω_D) of its polarization vector $\mathbf{e}_D$ which is a primary source of anisotropy. Derivation of such semiclassical form of ρ was reported at the previous ICLS conference [13].

Transformations needed to explicitly present the functional dependence of $\chi^{(3)}_{res}$ on $\mathbf{e}_D$, unit polarization vectors $\mathbf{e}_k$ ($k = 1, 2, 3, 4$) of the RFWM beams and on the quantum numbers of the resonating states f and i require complicated algebra and will be published separately [14]. The line-space (Liouville) techniques [15] naturally combined with the angular momenta recoupling machinery [16] allows to decouple the field characteristics (i.e., tensors $G^{(K)}$ build on four unit vectors $\mathbf{e}_k$) from the photofragments properties. In accord with the conditions of measurements [8], one can assume the predominance of the Doppler broadening, or $\varepsilon \equiv c\Gamma_{fi}/\omega_{fi}v_0 \ll 1$, where v_0 is the mean recoil velocity and Γ_{fi} denotes the decay rate which can be caused both by the collisional and natural lifetime effects. At the near-collinear geometry, when all five beams can be considered as copropagating along the common axis $\mathbf{k} \| OZ$, the decoupling procedure results in

$$\chi^{(3)}_{res} = \sum_{KL\lambda} \{F^{(K)}(L, \lambda; f, i) \otimes C^{(\lambda)}(\Omega_D)\}^{(L)}_0 S_L(\Delta\omega) \tag{1}$$

where $\otimes$ symbolizes contraction of two irreducible spherical tensors (IST) into a third one; the zero subscript denotes the OZ component of the resultant IST of rank L; $C^{(\lambda)}(\Omega_D)$ is an IST whose components are the Racah spherical harmonics [16]. In turn, $F^{(K)}$ is a linear combination of the polarization tensors $G^{(K)}$ and depends also on the quantum numbers of the combining states. We do not provide here its explicit form since it is of minor importantance for the RFWM shape analysis.

The allowed values of $\lambda(= 0, 2)$ are defined by the electric-dipole mechanism of photolytic decomposition [12] whereas K runs from 0 up to 4 with the unity step. Finally, the rank L characterizing the recoil anisotropy may vary from $|K - \lambda|$ up to $K + \lambda$.

It is pertinent to note that the odd K and L values are forbidden in the LIF case implying that the correlations between odd powers of $\mathbf{j}$ and $\mathbf{v}$ remain silent in LIF. In contrast, RFWM opens a direct way to study such correlations, the leading one being the photofragment helicity for which $K = L = 1$. At $\lambda = 0$, the helicity suggests that the mean value of $(\mathbf{v}, \mathbf{j})$ is distinct from zero so that a certain part of photofragments has angular moments parallel (or antiparallel) to $\mathbf{v}$ and moves like bullets from a rifle's barrel.

Remarkably, the RFWM polarization setup allows to directly select the signals produced by helicity. This can be achieved by using the properties of the ISTs $G^{(K)}$ from which $F^{(K)}$ are built up. To annul all terms of $\chi^{(3)}_{res}$ with even K, we just set $\mathbf{e}_D = \mathbf{e}_1 = \mathbf{e}_2 = \mathbf{e}_3 = \mathbf{e}_X$ and use the perpendicular polarization ($\mathbf{e}_4 = \mathbf{e}_Y$) in the detection channel [8, 14].

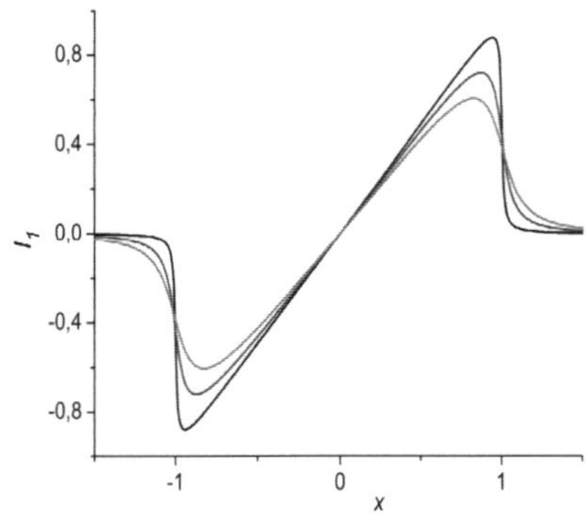

Figure 1. $I_1(x)$ shape transformation with reduced decay rate ε: $\varepsilon = 0.01$ (black line), $\varepsilon = 0.05$ (blue line), $\varepsilon = 0.1$ (red line).

3. Helicity-induced spectral distributions

The key objective of this study is the shape function $S_L(\Delta\omega)$ which corresponds to the translational anisotropy of rank L and depends on the detuning $\Delta\omega$ from the resonance frequency ω_{fi}. For rapidly recoiling fragments, the DFWM shape function can be written as a weighted sum of two terms

$$S_L(\Delta\omega) = \int f(v)v[\int_{-v}^{v}[\frac{a}{(\Delta\omega - kv_Z)^2 + \Gamma^2} + \frac{b}{[\Delta\omega - kv_Z + i\Gamma]^2}]P_L(\kappa)dv_Z dv \quad (2)$$

where a and b are certain real coefficients, $f(v)$ denotes the distribution of the absolute values of $\mathbf{v}$ and $P_L(\kappa)$ is the Legendre polynomial of $\kappa = \cos(\mathbf{v}, \mathbf{k}) = v_Z/v$.

To proceed, we may consider v_Z as a complex variable and close the integration path by adding a semicircle of radius v in the upper half plane. In the fast-recoil limit when $\varepsilon = \Gamma/kv \to 0$ the integral along the semicircle vanishes. By taking the residues inside the thus closed integration loop we immediately conclude that the b-term disappears whereas the first integral over v_Z equals

$$I_1 = \frac{a\pi}{k\Gamma}P_L(\Delta\omega/kv)\theta(1 - (\Delta\omega/kv)^2) \quad (3)$$

where $\theta(x)$ the Heaviside step function. In this limit, the Doppler shape (2) drops to zero when the pole crosses the integration loop at $(\Delta\omega/kv)^2 > 1$ and the residue contribution vanishes. The result (3) exactly corresponds to that of the LIF theory when decay processes are discarded. The only difference is that, to obtain a measurable DFWM profile, we must finally square the spectral function while this operation is unneeded for LIF. Therefore, when the helicity ($L = 1$) is the sole mechanism of Doppler broadening, the DFWM profile $|S_1(\Delta\omega)|^2$ reduces to a parabola (x^2) with the intensity dropping to zero both at the center and at $|\Delta\omega| \geq kv$.

However, for small but finite decay rates the DFWM and LIF shape expressions differ since the first one (2) is complex whereas the LIF shape is a real part of I_1. Here, illustrations are provided for the spectral behaviour of the helicity ($L = 1$) shape at finite ε. Both I_1 and I_2 integrals (the factor standing at b coefficient in Eq. (2), correspondingly) can be expressed via elementary functions. For brevity sake, we do not provide the explicit bulky expressions but plot on Figs. 1 and 2 I_1 and I_2 as functions of dimensionless detuning $x = \Delta\omega/kv$ at different values of ε. The functions are normalized by the factor $\pi/k\Gamma$ (see Eq. (3)) that makes the maximal value of I_1 close to unity. We see that the effect of the relaxation rate increasingly smoothes the sharp edges at $x = \pm 1$ and induces the intensity in the "forbidden" range in which $|\Delta\omega| > kv$. The $I_2(x)$ component is entirely due to the finite decay rate; moreover, its real part behaves as

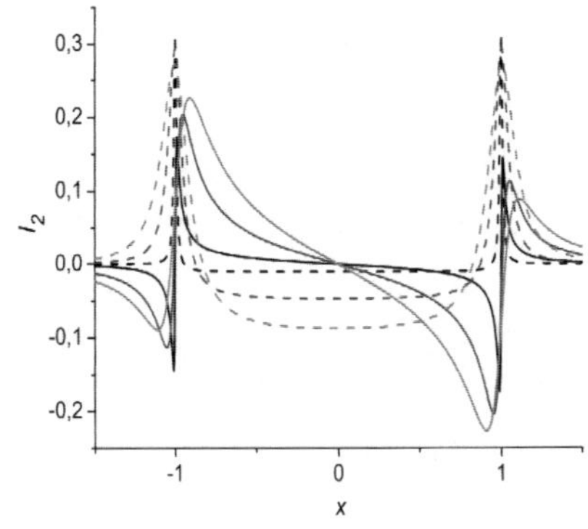

Figure 2. Real (solid lines) and imaginary (dashed lines) components of $I_2(x)$ at different decay rates ε: $\varepsilon = 0.01$ (black line), $\varepsilon = 0.05$ (blue line), $\varepsilon = 0.1$ (red line).

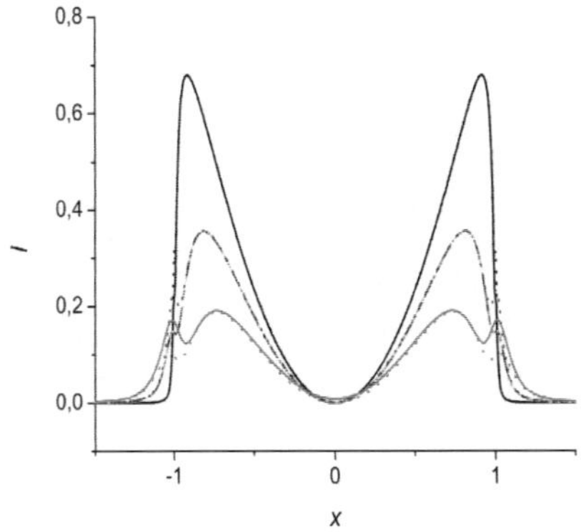

Figure 3. Simulated DFWM helicity-induced profiles at different decay rates ε (line colours correspond to decay rates used for Figs. 1 and 2). Solid and dotted lines represent the total and $|I_1(x)|^2$ profiles, correspondingly.

an odd function of x whereas the imaginary part is even. It means that - since a and b are real multipliers - the interference between the odd and even components is absent and the observed Doppler DFWM helicity profile is therefore an even function of the frequency detuning.
The observable (modulus squared) shapes are plotted on Fig. 3 for $a = b = 1$ and for different values of ε. Because of squaring, the periphery intensity is somewhat dampened but peaks near $x = \pm 1$ survive. The main peak produced by the sum of I_1 and $\Re I_2$ is broadened and shifted towards the line center. As seen from Fig. 2, the positions of the $\Im I_2$ component maxima are weakly affected by ε and remain close to $x = \pm 1$. Finally, because of the increasing role of $|\Im I_2|^2$, new maxima at $x = \pm 1$ appear at $\varepsilon \approx 0.1$ (Fig. 3). However, we should note that in the practically important cases the collisional relaxation is inevitably accompanied by the velocity-changing collisions which also affect the Doppler shapes and therefore the theory should be refined.

4. Synopsis
The DFWM profiles caused by the rotational helicity of photofragments are derived for the case of inhomogenious Doppler broadening accompanied by decay processes of resonating states. The results reveal the difference between the DFWM and LIF spectral distributions which becomes more pronounced at faster decays.

Acknowledgments
This study was supported by the Swiss National Science Foundation (grant 200021 153170/1), the Russian Foundation for Basic Research (grant 15-03-04997) and by the Saint-Petersburg State University (grant 11.38.265.2014).

References
[1] Sato H 2001 *Chem. Rev.* **101** 2687
[2] Houston P L 1987 *J. Phys. Chem.* **91**, 5388
[3] Alexander A J and Zare R N 2000 *Acc. Chem. Res.* **33** 199
[4] Siebbeles L D A, Glass-Maujean M, Vasyutinskii O S, Beswick J A, and Roncero O 1994 *J. Chem. Phys.* **100**, 3610
[5] Rakitzis T P and Zare R N 1999 *J. Chem. Phys.* **110**, 3341
[6] Costen M L, Marinakis S, and McKendrick K G 2008 *Chem. Soc. Rev.* **37** 732
[7] Orr-Ewing A J and Zare R N 1994 *Annu. Rev. Phys. Chem.* **45** 315
[8] Maksyutenko P, Radi P P, Kozlov D N, and Kouzov A P 2013 *J. Raman Spectrosc.* **44** 1349
[9] Eckbreth A C 1996 *Laser Diagnostics for Combustion Temperature and Species* (2nd ed., Amsterdam: Gordon and Breach)
[10] Hemmerling B, Radi P, Stampanoni-Panariello A, Kouzov A, Kozlov D 2001 *C. R. Acad. Sci. Paris*, **2**, Serie IV, 1001
[11] Gericke K-H, Klee S, Comes F, and Dixon R N 1986 *J. Chem. Phys.* **85**, 446
[12] Dixon R N 1986 *J. Chem. Phys.* **85** 1866
[13] Kouzov A, Radi P, Maksyutenko P, and Kozlov D 2014 *J. Physics: Conference Series* **548** 012016
[14] Kouzov A and Radi P 2016 *Chem. Phys. Lett.*, submitted for publication
[15] Kouzov A and Radi P 2014 *J. Chem. Phys.* **140** 194302
[16] Varshalovich D A, Moskalev A N, and Khersonskii V K 1988 *Quantum Theory of Angular Momentum* (Singapore: World Scientific)

IOP Conf. Series: Journal of Physics: Conf. Series **810** (2017) 012020 doi:10.1088/1742-6596/810/1/012020

On the nature of ion dynamics

E Stambulchik[1] **and A V Demura**[2]

[1]Faculty of Physics, Weizmann Institute of Science, Rehovot 7610001, Israel
[2]National Research Centre "Kurchatov institute", Moscow 123182, Russia

E-mail: Evgeny.Stambulchik@weizmann.ac.il, Demura_AV@nrcki.ru

Abstract. A study of physical mechanisms responsible for the ion dynamics effects is presented. It is shown that the effect is primarily caused by the fluctuations of ion microfield direction, which can naturally be represented as rotation of the electric-microfield vector. Various aspects of this problem are discussed.

1. Introduction

Computer simulations (CS) provided a unique tool for calculation and analysis of spectral line broadening when conventional notions of quasistatic or impact broadening were not applicable [1]. While the broadening by electrons in a rather wide range of plasma conditions is within the impact regime ($h \ll 1$), the broadening by heavy perturbers often corresponds to $h \sim 1$, where h is the number of particles in the Weisskopf sphere [2]. This means that neither quasistatic nor impact broadening approximations could be applied for the latter case. So well before appearance of CS, the deviation of broadening by ions from the results of the quasistatic approximation was thought to be attributed to effects of ion dynamics (see [3] and references therein). But although the CS technique has proved its power for variety of cases [4], there had for a long time been no serious attempts to explain its results in terms of physical mechanisms, responsible for broadening under these conditions. The experimentally observed ion dynamics effects on profiles of the hydrogen Balmer-α and Balmer-β lines, although relatively minor, allowed for discovering the dependence of ion dynamics on the temperature T and the reduced mass μ as $\propto \sqrt{T/\mu}$ [5]. The real boom in studying the ion dynamics effects began after the experimental discovery [6] of an anomalous broadening of hydrogen Ly-α, whose width was about 2.5 times larger than the value, obtained within the Standard Theory (ST) model [7, 8] (i.e., quasistatic ions + impact electrons). Seven years later it was shown, also using CS, that even larger deviations from the ST results could be observed for Ly-α profiles of highly charged hydrogen-like emitters in DT plasmas [9]. During the series of Spectral Line Shapes in Plasmas code comparison workshops held recently (2012 [10], 2013 [11], and 2015) the task of physical interpretation of CS results became again a subject of interest, including studies specifically focused on the Ly-α profiles [3, 12, 13] of hydrogen and hydrogen-like species.

2. The role of microfield rotation

Historically, the first concise study of influence of the ion microfield time dependence was limited to changes of its absolute value due to the thermal fluctuations [2]. This, as well as many other studies were done within the notion of the theory of thermal corrections (TC's) to the total profile assuming applicability of the perturbative approach, i.e., that these TC's are small enough with respect to the initial broadening profile. Several years later it was shown [14] that besides the contribution of the fluctuations of the microfield absolute value, there was also an influence of the fluctuation of the microfield *orientation* via two effects: (i) the amplitude modulation, related to changing of dipole projection in the course of of the microfield rotation in time, and (ii) non-adiabatic effects, related to inertia in the evolution of the quantum states with respect to the rotation of the quantization axis along the microfield direction. It was shown that numerically TC's due to the fluctuation of the microfield orientation are many times larger than those related to the fluctuations of the field magnitude. However, as experiments indicated, this influence could be very significant and located in the central part of the line, so that the validity of the perturbative approach would be violated. Recently performed CS, taking into account the time evolution of only microfield magnitude or only its direction, showed explicitly that fluctuations of the microfield orientation play the dominant role in the formation of the central part of the line profile [3, 12], when the ion dynamics effects are important. This was shown for various models for weakly to moderately coupled plasmas using "rotational" and "vibrational" pseudocomponents of the total microfield $\vec{F}(t)$, defined as

$$\vec{F}_{\rm rot}(t) = F_0 \frac{\vec{F}(t)}{F(t)} \tag{1}$$

and

$$\vec{F}_{\rm vib}(t) = \vec{n}F(t), \tag{2}$$

respectively, where F_0 is the Holtsmark field [15] and $\vec{n}$ is a unity vector aligned, e.g., along the z axis.

A comparison of Ly-α partial profiles of neutral hydrogen assuming the ideal one component plasma (OCP) model and the linear Stark effect interaction is presented in Fig. 1 for a few values of the reduced mass. Here and below, results of a variant of CS calculations, described in Ref. [16], are used. It is seen that the "rotational" microfield component has a significantly more pronounced effect on the total line shapes. Most notably, the width of the central component due to the rotational microfield component scales as $\propto 1/\sqrt{\mu}$, similar to the ion-dynamical effects observed experimentally [5], which is a strong indication of the rotational nature of ion dynamics.

Contrary to that, changing only the magnitude of the field while keeping its direction constant (the "vibrational" component) has only a minor effect on the shape of the lateral components (and the central component remains a δ-function, evidently). Moreover, the effect has an opposite sign, i.e., the width of the lateral component *increases* with μ. Recalling that in an ideal OCP varying μ as $s^2\mu_0$ is equivalent to varying T as T_0/s^2, we see that the lateral component becomes slightly *narrower* due to the thermal-motion effect on the magnitude of the microfield. There is also a minor shift of the maxima of the lateral components toward the far wings of the line.

3. Rotational broadening regime

Based on rather general dimensionality considerations, it was shown [17] that in an ideal OCP, the rotational broadening of a spectral line with a central unshifted component of a hydrogen-like species must have the following dependencies:

$$w = \bar{\beta}\left(\frac{T}{\mu}\right)^{1/2} N_p^{1/3}, \tag{3}$$

where N_p is the plasma density, and $\bar{\beta}$ is a *universal* dimensionless constant, found from CS to be ≈ 1.2. In other words, the broadening in this regime depends neither on plasma particle charges Z_p nor on *any* atomic property—including the atomic number Z and the principal quantum numbers of the initial n and final n' levels—of the radiator! The only condition required for entering this paradoxical regime is that the (quasi)static Stark broadening $w_{\rm qs} \sim (d_i - d_j)F_0$ significantly exceeds the typical microfield frequency $w_{\rm dyn} \sim \sqrt{T/\mu}N_p^{1/3}$, or $R \gg 1$, using the dimensionless quantity R [18]

$$R = w_{\rm qs}/w_{\rm dyn} \approx 7\frac{|Z_p|(n^2 - n'^2)\mu^{1/2}N_p^{1/3}}{ZT^{1/2}}. \tag{4}$$

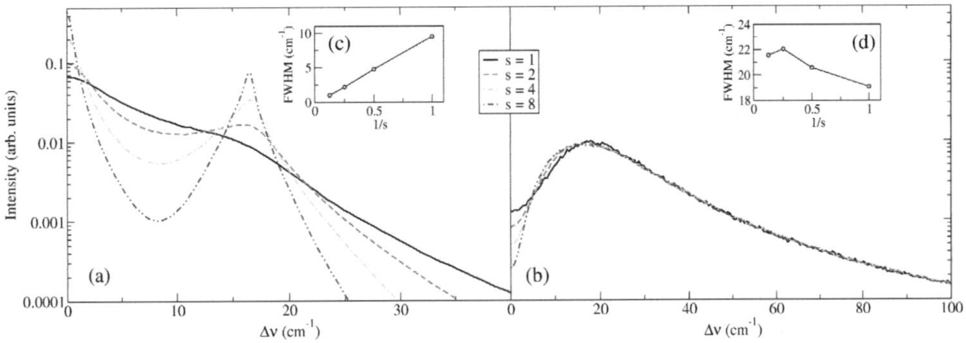

Figure 1. Ly-α profiles, broadened by (a) $\vec{F}_{\rm rot}$ and (b) $\vec{F}_{\rm vib}$ of an OCP, assuming $N = 10^{17}\,{\rm cm}^{-3}$ and $T = 1\,{\rm eV}$. The ion radiator reduced mass $\mu = s^2\mu_0$, where $\mu_0 = 0.5$. The FWHM's of the rotational and vibrational broadenings are shown in the insets (c) and (d), respectively. Note that in the later case, the width of a single lateral component is shown (the central component remains a δ-function, not shown here).

In spite of the counter-intuitive nature of this conclusion, it was fully confirmed using CS results [17]. Furthermore, a simple expression can be used interpolating between the impact and rotational broadening regimes, well suited to Ly-α of any H-like species:

$$w = w_{\rm qs}\left(\frac{\bar{\alpha}}{R} + \frac{R}{\bar{\beta}}\right)^{-1} \equiv w_{\rm dyn}\left(\frac{\bar{\alpha}}{R^2} + 1/\bar{\beta}\right)^{-1}, \qquad (5)$$

where $\bar{\alpha} \approx 2$, similarly to $\bar{\beta}$, is a universal dimensionless constant. In Fig. 2 we demonstrate a good quality of this interpolation. In particular, it is seen that in the rotational regime ($R \gg 1$), the widths of both Ly-α and Balmer-α lines asymptotically approach the same value. We also show there the width of Ly-α broadened by the "rotational" microfield projection (1) of the same OCP. Strikingly, whether the full microfield histories or just their rotational properties are used, the results are practically indistinguishable. This is another proof of the rotational nature of ion dynamics.

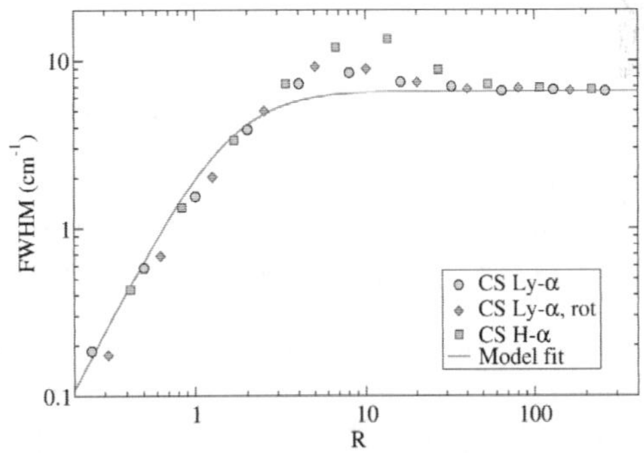

Figure 2. FWHM of Ly-α and Balmer-α in an ideal OCP with $T = 1\,{\rm eV}$ and $N_p = 10^{17}\,{\rm cm}^{-3}$ as a function of R (4). CS results are shown, as well as a model fit according to Eq. (5).

We stress again that Eq. (5), interpolating between the impact and rotational broadening regimes, is universal, i.e., it is applicable to any kind of the plasma perturbers, including electrons. This allows for a straightforward generalization to the case of two-component plasma (TCP) model, assuming, as is often done in plasma line broadening, independent effects of ion and electrons. Since the lineshapes are Lorentzian in both the impact and rotational limits, the total width is approximated by an arithmetic sum of the both contributions, $w_{\rm tot} = w_i + w_e$. As shown in Fig. 3, this simple model gives very accurate ($\sim$15%) results, when compared to rigorous CS calculations. We note that the effective N_p dependence of the Ly-α width in the TCP (close to $N_p{}^{2/3}$ over many orders of magnitude of plasma density) *coincidentally* matches that of the quasi-static approximation, even though each of the ion and electron contributions scales differently; the quasistatic broadening regime itself is never realized in the central part of the Ly-α shape. For similar reasons, there is practically no T dependence of the total linewidth [17]—resembling, again, the quasistatic broadening.

4. Influence of plasma coupling

The Debye screening by electrons leads to suppression of the influence of remote perturbers (i.e., lower frequencies). Therefore, a weak Debye screening (leaving fields of the perturbers at distances $\lesssim N^{-1/3}$ largely unaffected) should result in a minor *increase* of the rotational broadening. Indeed, calculations shown in Fig. 4 confirm this. For a more strongly coupled plasmas, the screening, including that due to ions, will eventually decrease the broadening. However, except for strongly coupled plasmas, these effects are relatively minor, as seen from Fig. 3 where only minor differences between results of TCP without and with the plasma coupling accounted [19] can be observed. It is important to stress that the study of partial profiles similar to presented in Fig. 1 for real plasmas of moderate coupling also demonstrate the predominance of rotational effects [12, 13].

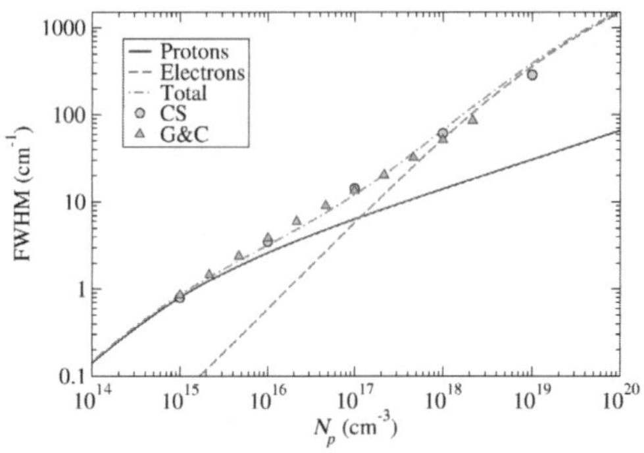

Figure 3. FWHM of Ly-α in an ideal TCP as a function of N_p: comparison of CS results with the model. $N_p = 10^{17}\,\mathrm{cm}^{-3}$ is assumed. Also shown are results for *non-ideal* hydrogen TCP from the reference diagnosis tables [19] (labeled here as "G&C").

From the ratio of the rotational width to the ion plasma frequency ω_{pi}, one can see that it is inversely proportional to the square root of the ion coupling parameter Γ_i or proportional to the cubic root of the number of particles in the ion Debye sphere N_D: $w_{rot}/\omega_{pi} \propto \Gamma_i^{-1/2} \propto N_D{}^{1/3}$. Therefore, it is reasonable to expect that in a strongly coupled plasma, the proportionality coefficient $\bar{\beta}$ in Eq. (3) might deviate significantly from that of the ideal-plasma case. At the same time, one can see that the Debye screening makes all variables interdependent, meaning the existence of some additional imposed constraints. One of these constraints stems from the fact that an infinite time is needed for the Debye screening to be fully established (see [20]).

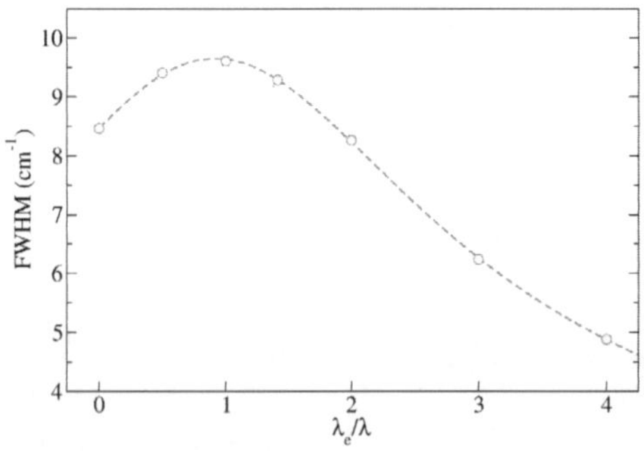

Figure 4. FWHM of Ly-α in a OCP of Debye quasiparticles with a varied screening length. $N_p = 10^{17}\,\mathrm{cm}^{-3}$ and $T = 1\,\mathrm{eV}$.

5. Discussion

The results presented so far were obtained assuming the linear Stark effect, i.e., within the dipole-perturbation approximation, and neglecting interactions between states with different n and the fine-structure effects (the LS coupling). For hydrogen and low-Z species the latter is often justified, however, for heavier elements the LS coupling grows very fast with Z ($\propto Z^4$) and may become important. Notably, even in this case the rotational regime holds, provided the static Stark effect exceeds the fine structure. In Fig. 5 we compare Stark shapes of the Al XIII Ly-α line broadened by Al ions with $N = 10^{22}\,\mathrm{cm}^{-3}$ and $T = 240\,\mathrm{eV}$. It is seen that the strong electric fields decrease the splitting to $2/3$ of its zero-field value, and each of the component acquires width that is nearly equal to that of the central component in the calculations neglecting the LS coupling.

The dipole approximation, used in most Stark broadening studies, usually suffices unless rather fine effects due to the higher members of the multipole expansions need to be described. However, transitions with unshifted Stark components, such as Ly-α, may need to be considered with the quadrupole interaction [21] accounted for. The situation is complicated, however, by the strong dipole effect of the lateral components resulting, due to the rotation of the microfield, in a complex mix of the effects of different multipoles. Therefore, a thorough study of Ly-α broadened by plasma, starting from a OCP model—while accounting for the quadrupole interactions—would be a welcome contribution to the field.

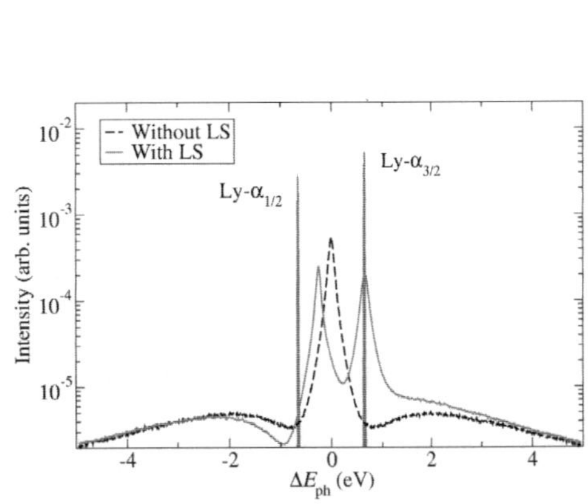

Figure 5. Al XIII Ly-α in a OCP with $N_p = 10^{22}\,\mathrm{cm}^{-3}$, $T = 240\,\mathrm{eV}$, and $\mu = 13.5$. Compared are CS results obtained without and with fine-structure effect taken into account. Also shown are the positions of the unperturbed doublet components.

Finally, we mention that a plain field rotation leaves the central component unshifted; thus, even after averaging over a distribution of the angular frequencies, this component will remain a δ-function. Evidently, a more complicated "motion" of the microfield vector is responsible for the ion dynamical effect in general and the rotational broadening limit in particular. For example, an addition of a constant field perpendicular to the rotation plane (resulting in a cone-rotating microfield vector) removes this degeneracy. Interestingly, the effect of such a field can be described in terms of the Berry phase in an effective non-Abelian gauge potential [22].

Acknowledgments
The work of E. S. was supported in part by the Israel Science Foundation.

References

[1] Stamm R and Voslamber D 1979 *J. Quant. Spectr. Rad. Transfer* **22** 599–609
[2] Kogan V I 1960 Broadening of spectral lines in high temperature plasma *Plasma Physics and the Problems of Controlled Thermonuclear Reactions* vol IV ed Leontovich M A (Oxford: Pergamon) p 305
[3] Demura A V and Stambulchik E 2014 *Atoms* **2** 334–356
[4] Stambulchik E and Maron Y 2010 *High Energy Density Phys.* **6** 9–14
[5] Wiese W L, Kelleher D E and Helbig V 1975 *Phys. Rev. A* **11** 1854–1864
[6] Grutzmacher K and Wende B 1977 *Phys. Rev. A* **16** 243–246
[7] Sobel'man I I 1972 *Introduction to the theory of atomic spectra* (Pergamon Press) ISBN 978-1483127309
[8] Griem H R 1974 *Spectral Line Broadening by Plasmas* (New York: Academic Press) ISBN 0-12-302850-7
[9] Stamm R, Smith E W and Talin B 1984 *Phys. Rev. A* **30** 2039
[10] Stambulchik E 2013 *High Energy Density Phys.* **9** 528–534
[11] Stambulchik E, Calisti A, Chung H K and González M Á 2014 *Atoms* **2** 378–381
[12] Calisti A, Demura A, Gigosos M A, González-Herrero D, Iglesias C A, Lisitsa V S and Stambulchik E 2014 *Atoms* **2** 259–276
[13] Ferri S, Calisti A, Mossé C, Rosato J, Talin B, Alexiou S, Gigosos M A, González M Á, González-Herrero D, Lara N, Gomez T, Iglesias C A, Lorenzen S, Mancini R C and Stambulchik E 2014 *Atoms* **2** 299–318
[14] Demura A V, Lisitsa V S and Sholin G V 1977 *J. Exp. Theor. Phys.* **46** 209–215
[15] Holtsmark J 1919 *Ann. Phys. (Leipzig)* **58** 577–630
[16] Stambulchik E and Maron Y 2006 *J. Quant. Spectr. Rad. Transfer* **99** 730–749
[17] Stambulchik E and Demura A V 2016 *J. Phys. B: At. Mol. Opt. Phys.* **49** 035701
[18] Stambulchik E and Maron Y 2008 *J. Phys. B: At. Mol. Opt. Phys.* **41** 095703
[19] Gigosos M A and Cardeñoso V 1996 *J. Phys. B: At. Mol. Opt. Phys.* **29** 4795–4838
[20] Demura A V 2010 *International Journal of Spectroscopy* **2010** 671073
[21] Demura A V and Sholin G V 1975 *J. Quant. Spectr. Rad. Transfer* **15** 881–899
[22] Matveev V I and Musakhanov M M 1989 *Phys. Lett. A* **135** 209–211

IOP Conf. Series: Journal of Physics: Conf. Series **810** (2017) 012021 doi:10.1088/1742-6596/810/1/012021

On the Stark broadening of Cr VI spectral lines in astrophysical plasma

M S Dimitrijević[1,2], Z Simić[1] and S Sahal-Bréchot[2]

[1]Astronomical Observatory, Volgina 7, 11160 Belgrade 74, Serbia

[2]LERMA, Observatoire de Paris, PSL Research University, CNRS, Sorbonne Universities, UPMC Univ. Paris 06, 5 Place Jules Janssen, 92195 Meudon Cedex, France

E-mail: mdimitrijevic@aob.rs

Abstract. Stark broadening parameters for Cr VI lines have been calculated using semiclassical perturbation method for conditions of interest for stellar plasma. Here are presented, as an example of obtained results, Stark broadening parameters for electron- and proton-impact broadening for Cr VI 4s ^{2}S-4p ^{2}P^o $\lambda = 1430$ Å and Cr VI 4p ^{2}P^o-5s ^{2}S $\lambda = 611.8$ Å multiplets. The obtained results are used to demonstrate the importance of Stark broadening of Cr VI in DO white dwarf atmospheres. Also the obtained results will enter in STARK-B database which is included in Virtual Atomic and Molecula Data Center - VAMDC.

1. Introduction

Rauch et al. [1] have found Cr V, Cr VI and Cr VII lines in the spectrum of LS V +46o21 white dwarf, which is the central star of planetary nebula Sh 2-216. High-resolution, high-signal to noise ratio ultraviolet spectra obtained by FUSE (Far Ultraviolet Spectroscopic Explorer) and HST/STIS (Space Telescope Imaging Spectrograph aboard the Hubble Space Telescope) have been used. Rauch et al. [1] stated that: Theoretical physics should provide data which cover the astrophysical relevant temperature and density space. Better atomic and line-broadening data will then strongly improve future spectral analyses and thus, make determinations of photospheric properties more reliable.

Since the Stark broadening is usually the most important line-broadening mechanism in white dwarf atmospheres (see e.g. [10]), we will calculate, using semiclassical perturbation method [3, 4, 5], Stark broadening parameters of Cr VI spectral lines, due to collisions with electrons, protons and doubly charged helium ions, the main perturbers in white dwarfs, and an example of the obtained results will be presented here. Calculations will be performed in function of perturber density and temperature, for plasma conditions of interest for white dwarfs.

Obtained results will be used to demonstrate the importance of Stark broadening of Cr VI lines for analysis and synthesis of spectra of white dwarfs, using theoretical models of DO white dwarf atmosphere and comparing Stark and Doppler widths in function of optical depth. The obtained Stark broadening parameters of spectral lines, broadened by electron-, proton-, and He III-impacts with Cr VI emitter / absorber will be also implemented in the STARK-B database (http://stark-b.obspm.fr [6]), a part of Virtual Atomic and Molecular Data Center (VAMDC - http://www.vamdc.org [7]).

2. Theory

The theory used here for the calculations of Stark broadening parameters of Cr VI lines is the semiclassical perturbation formalism, which is described in details elsewhere with different inovations and updates [3, 4, 5]. Consequently, only short description, necessary for the understanding of the way of calculations will be presented here.

For isolated lines, the Stark broadened profile $F(\omega)$ is Lorentzian:

$$F(\omega) = \frac{W/2\pi}{(\omega - \omega_{if} - d)^2 + (W/2)^2}. \quad (1)$$

Here,

$$\omega_{if} = \frac{E_i - E_f}{\hbar}$$

where E_i and E_f are energies of the initial and final states, (W) is width (FWHM) in angular frequency units and (d) shift

$$W = N \int v f(v) dv \left(\sum_{i' \neq i} \sigma_{ii'}(v) + \sum_{f' \neq f} \sigma_{ff'}(v) + \sigma_{el} \right)$$

$$d = N \int v f(v) dv \int_{R_3}^{R_D} 2\pi \rho d\rho \sin(2\varphi_p). \quad (2)$$

N is here the electron density, $f(v)$ the Maxwellian velocity distribution function for electrons, ρ the impact parameter of the incoming electron, and with i', f' are denoted the perturbing levels of the initial and final state. The inelastic cross section $\sigma_{jj'}(v)$, $j = i, f$

$$\sum_{i' \neq i} \sigma_{ii'}(v) = \frac{1}{2}\pi R_1^2 + \int_{R_1}^{R_D} 2\pi\rho d\rho \sum_{i' \neq i} P_{ii'}(\rho, v). \quad (3)$$

where $P_{jj'}(\rho, v), j = i, f; j' = i', f'$ is transition probability. The elastic cross section is

$$\sigma_{el} = 2\pi R_2^2 + \int_{R_2}^{R_D} 2\pi\rho d\rho \sin^2\delta + \sigma_r,$$

$$\delta = (\varphi_p^2 + \varphi_q^2)^{\frac{1}{2}}. \quad (4)$$

The phase shifts due to the polarization potential φ_p (r^{-4}) and to the quadrupolar potential φ_q (r^{-3}), are explained in detail in Section 3 of Chapter 2 in [3]. The explanation of cut-offs R_1, R_2, R_3, R_D, where R_D is the Debye radius, can be found in Section 1 of Chapter 3 in [4]. Finally, the contribution of Feshbach resonances, denoted as σ_r, is introduced and explained in detail in [8].

IOP Conf. Series: Journal of Physics: Conf. Series **810** (2017) 012021 doi:10.1088/1742-6596/810/1/012021

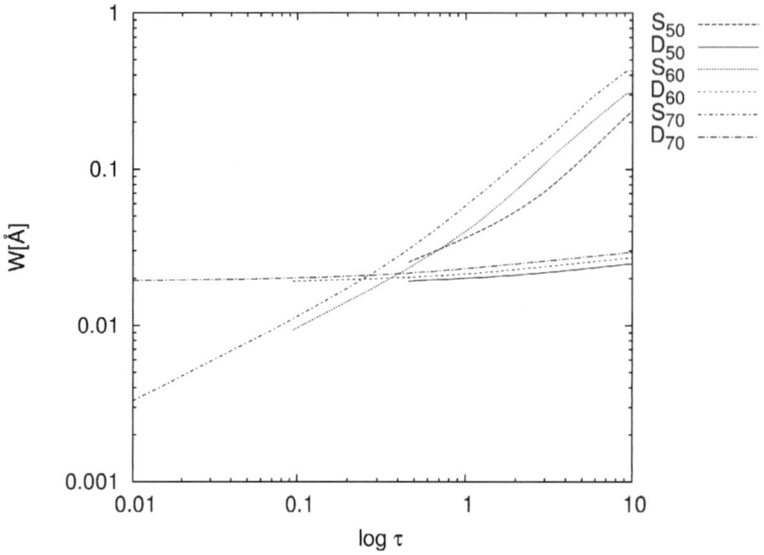

Figure 1. Stark and Doppler widths for Cr VI 4s ^{2}S-4p^{2}P^o (λ = 1430.0 Å) multiplet as a function of logarithm of Rosseland optical depth (log τ). Stark (S) and Doppler (D) widths are shown for three atmospheric models [22] with effective temperatures from T_{eff} = 50 0000 K (S_{50}, D_{50}) to 70 000 K (S_{70}, D_{70}), and log g = 8.

3. Results and discussion

The Stark broadening parameters of Cr VI multiplets have been determined with the code, which is based on semiclassical perturbation theory. The energy levels, needed for present calculations have been taken from [9]. All details of the performed calculations are given e.g. in [10]. The complete obtained results will be published elsewhere. Here, in Table 1, only a sample of the results is presented. The calculations of Stark widths (FWHM) and shifts for electron-, and proton-impact broadening, have been determined for a perturber density of 10^{17} cm^{-3} and for temperatures from 50 000 K to 800 000 K.

In Table 1, a parameter C [11], provides possibility for an estimate of the maximal perturber density for which the line may be treated as isolated, when it is divided by the corresponding full width at half maximum. Denoting with V the collision volume and multiplying it by the perturber density N, for each value presented in Table, $NV < 0.1$. Since this product is much less than one, the impact approximation is valid [3, 4].

Spectral lines of highly charged chromium ions, including Cr VI, have been observed in the spectra of white dwarf atmospheres where, as it has been demonstrated several times [12, 13, 14, 15, 16, 17, 18, 19, 20, 21, 10], Stark broadening is usually dominant line broadening mechanism. With the results obtained here, we also wish to demonstrate the importance of Stark broadening in DO white dwarf atmospheres, which effective temperature is within the range 40 000 K $< T_{eff} <$ 120 000 K.

In Fig. 1 Stark (FWHM) and Doppler widths for Cr VI 4s ^{2}S-4p^{2}P^o (λ = 1430.0 Å) multiplet as a function of logarithm of Rosseland optical depth (log τ) are compared for three atmospheric models of [22], for DO white dwarf atmospheres with effective temperature T_{eff} = 50 000-100 000 K and logarithm of surface gravity log g = 8.

One can notice from Fig. 1, that Stark broadening is more important or comparable to Doppler broadening for atmospheric layers which are significant for radiative transfer calculations

Table 1. This table gives electron-, and proton-impact broadening parameters for Cr VI multiplets, for a perturber density of 10^{17} cm^{-3} and temperatures from 50 000 to 800 000 K. Calculated wavelength of the transitions (in Å) and parameter C are also given. This parameter, when divided with the corresponding Stark width, gives an estimate for the maximal pertuber density for which the line may be treated as isolated. W_e: electron-impact full width at half maximum of intensity, d_e: electron-impact shift, W_p: proton-impact full width at half maximum of intensity, d_p: proton-impact shift. This table is an example of results which, in its entirety, will be published elsewhere.

Transition	T(K)	W_e (Å)	d_e (Å)	W_{H+} (Å)	d_{H+} (Å)
Cr VI 4s ^{2}S-4p^{2}P^o	50000.	0.159E-01	-0.278E-03	0.159E-03	-0.969E-04
1430.0 Å	100000.	0.115E-01	-0.284E-03	0.346E-03	-0.185E-03
C = 0.14E+21	150000.	0.956E-02	-0.255E-03	0.488E-03	-0.249E-03
	200000.	0.847E-02	-0.296E-03	0.575E-03	-0.303E-03
	400000.	0.650E-02	-0.348E-03	0.812E-03	-0.421E-03
	800000.	0.294E-02	-0.188E-03	0.152E-02	-0.870E-03
Cr VI 4p ^{2}P^o-5s^{2}S	50000.	0.494E-02	0.313E-03	0.684E-04	0.155E-03
611.8 Å	100000.	0.363E-02	0.291E-03	0.167E-03	0.240E-03
C = 0.10E+20	150000.	0.309E-02	0.338E-03	0.227E-03	0.292E-03
	200000.	0.277E-02	0.358E-03	0.287E-03	0.331E-03
	400000.	0.220E-02	0.330E-03	0.400E-03	0.397E-03
	800000.	0.984E-03	0.152E-03	0.921E-03	0.722E-03

and modelling of atmospheric plasma, as well as that its importance increases with the increase of the optical depth, as expected. We note also that due to differences between Lorentzian profile for Stark broadening and Gaussian for Doppler broadening, Stark broadening could be important or non negligible in the line wings even when Doppler width is larger than Stark.

It should be noted as well that the atmospheric models of [22] provide LTE models. Obviously that both NLTE and line-blanketing effects should be included to obtain correct values for the line widths, but models of [22] are tabulated with all the needed data for calculations. Looking at Fig.1, one can see that the influence of Stark broadening is so significant that the modifications due to NLTE effects and line-blanketing will not change our conclusions.

The obtained Stark broadening parameters for Cr VI multiplets will be implemented in computer readable form in the STARK-B database (http://stark-b.obspm.fr [6]), suitable especially for the diagnostics, modelling and investigations of stellar atmospheres. STARK-B is also a part of Virtual Atomic and Molecular Data Center - VAMDC (http://www.vamdc.org [7]).

Acknowledgments

This work is a part of the project 176002 "Influence of collisional processes on astrophysical plasma line shapes" supported by the Ministry of Education, Science and Technological Development of Serbia. This work has also been supported by the Paris Observatory, the CNRS and the PNPS (Programme National de Physique Stellaire, INSU-CNRS).

References

[1] Rauch T, Ziegler M, Werner K, Kruk J W, Oliveira C M, Vande Putte D et al. 2007 *A&A* **470** 317
[2] Dimitrijević M S, Simić Z, Kovačević A, Valjarević A and Sahal-Bréchot S 2015 *J. Astrophys. Astron.* **36** 681
[3] Sahal-Bréchot S 1969 *A&A* **1** 91
[4] Sahal-Bréchot S 1969 *A&A* **2** 322

[5] Sahal-Bréchot S, Dimitrijević M S and Ben Nessib N 2014 *Atoms* 225
[6] Sahal-Bréchot S, Dimitrijević M S, Moreau N and Ben Nessib N 2015 *Phys. Scripta* **50** 054008
[7] Dubernet M L, Boudon V, Culhane J L, Dimitrijević M S, Fazliev A Z, Joblin C and Kupka F 2010 *JQSRT* **111** 2151
[8] Fleurier C., Sahal-Bréchot S. and Chapelle J. 1977 JQSRT **17** 595
[9] Sugar J. and Corliss C. 1985 *J. Phys. Chem. Ref. Data* **14** Suppl. 2 1664
[10] Dimitrijević M S, Simić Z, Kovačević A, Valjarević A and Sahal-Bréchot S 2015 *MNRAS* **454** 1736
[11] Dimitrijević M. S. and Sahal-Bréchot S. 1984 *JQSRT* **31** 301
[12] Popović L. Č., Dimitrijević M. S. and Tankosić D. 1999 *A&AS* **139** 617
[13] Tankosić D., Popović L. Č. and Dimitrijević M. S. 2003 *A&A* **399** 795
[14] Milovanović N., Dimitrijević M. S., Popović L. Č. and Simić, Z. 2004 *A&A* **417** 375
[15] Simić Z., Dimitrijević M. S., Popović L. Č. and Dačić M. 2006 *New Astron.* **12** 187
[16] Dimitrijević M. S., Kovačević A., Simić Z. and Sahal-Bréchot S. 2011 *Baltic Astronomy* **20** 580
[17] Dufour P., Ben Nessib N., Sahal-Bréchot S. and Dimitrijević M. S. 2011 *Baltic Astronomy* **20** 511
[18] Larbi-Terzi N., Sahal-Bréchot S., Ben Nessib N. and Dimitrijević M. S. 2012 *MNRAS* **423** 766
[19] Simić Z., Dimitrijević M. S. and Sahal-Bréchot S. 2013 *MNRAS* **432** 2247
[20] Simić Z., Dimitrijević M. S. and Popović L. Č. 2014 *Adv. Space Res.* **54** 1231
[21] Hamdi R., Ben Nessib N., Milovanović N., Popović L. Č., Dimitrijević M. S. and Sahal-Bréchot S. 2008 *MNRAS* **387** 871
[22] Wesemael F. 1981 *ApJS* **45** 177

H-like ions with penetrating collisions

S Alexiou[1] **and A Poquérusse**[2]

[1] University of Crete, TETY, 71409 Heraklion, TK 2208, Greece

[2] LULI, Ecole Polytechnique, CNRS, CEA, UPMC ; 91128 Palaiseau, France

E-mail: moka1@otenet.gr

Abstract. Collisions that penetrate the wavefunction extent are not correctly accounted for by the usual long-range dipole approximation. Since the interaction is softened for penetrating collisions a reduction in line broadening is expected and indeed this is the case with isolated ion lines. Here we report an *enhancement* of broadening for H-like ions due to penetration. This has a simple physical explanation and potentially important applications.

1. Introduction

We consider penetrating electron collisions for the electron impact broadening of H-like ions of spectroscopic charge number Z between upper level a and lower level b. We evaluate the collisional term:

$$\{1 - S_a S_b^\dagger\} = Q_a + Q_b + Q_{ab} \tag{1}$$

with S denoting S-matrix. For instance the upper (Q_a) level contribution is:

$$\langle n_a lm|Q_a|n_a l'm'\rangle = \frac{2}{3}\left(\frac{mv}{\epsilon(Z-1)\hbar}\right)^2 I(n_a, l, n_a, l'')I(n_a, l'', n_a, l') \\ \langle n_a lm|\mathbf{r}|n_a l''m''\rangle\cdot\langle n_a l''m''|\mathbf{r}|n_a l'm\rangle, \tag{2}$$

where the integral I is defined as:

$$I(n, l, n', l') = \frac{1}{2}\int_{-\infty}^{\infty} du C_1(R(t); n, l, n', l')\frac{\epsilon(\epsilon - coshu)}{(\epsilon coshu - 1)^2} \tag{3}$$

and C_1 is a factor, accounting exactly for penetration, defined in Ref.1 and given by Ref.2:

$$C_1(u) = 1 - e^{-aR(t)}P_{2n+1}(aR(t)) \tag{4}$$

with $P_i(x)$ a polynomial of degree i, $aR(t) = \xi(\epsilon coshu - 1)$ and $\xi = \frac{Z(Z-1)e^2}{2\pi\epsilon_0 mna_0 v^2}$, n the principal quantum number of the upper level and a_0 the Bohr radius.

2. Collisional Aspects

The integral I in Eq. (3) essentially includes the atomic collision physics, while the eccentricity ϵ and velocity integrations represent the plasma parameter phase space average. We recover the standard behavior $I = 1$ if $C_1 = 1$, i.e. in the case of no penetration. Penetration is intuitively expected to soften the interaction and hence result in a reduction of broadening. However this

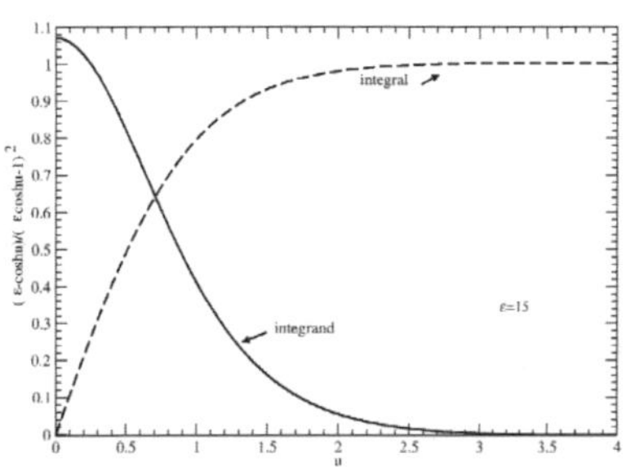

Figure 1. The integrand and integral of I without penetration for a large $\epsilon = 15$ showing a positive contribution of about 1 and a negative of about 0.

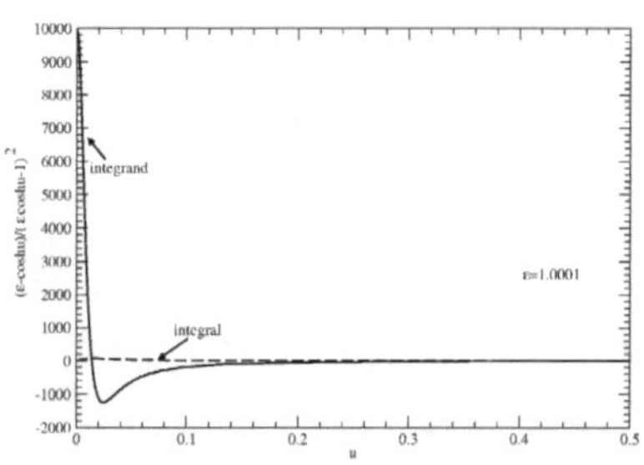

Figure 2. The integrand and integral of I without penetration for a small $\epsilon = 1.0001$ showing a large positive contribution and slightly smaller negative contribution.

is not always the case. As illustrated in Figs. 1-3, the relevant collisional integral involves both positive (small u) and negative (large u) contributions. Without penetration, the positive contributions win and the result is a value of 1.

This result may arise from a positive contribution of about 1 and a negative one of about 0 as in Fig.1, but it can and does for small velocities and ecentricities, also arise from a large positive contribution and a slightly smaller negative contribution as in Fig.2.

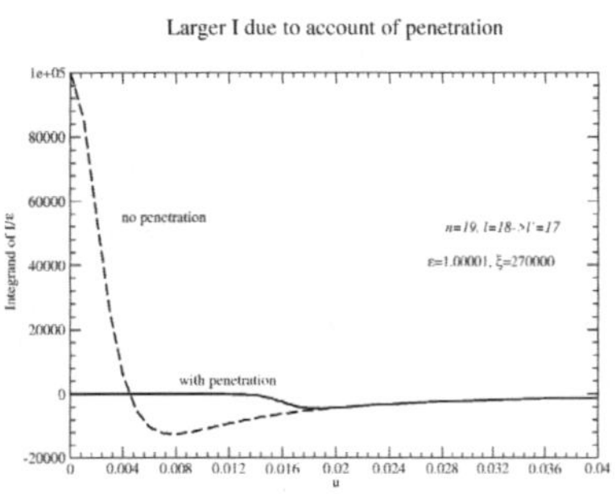

Figure 3. The integrand of I with and without penetration for small $\epsilon = 1.00001$ and large ξ.

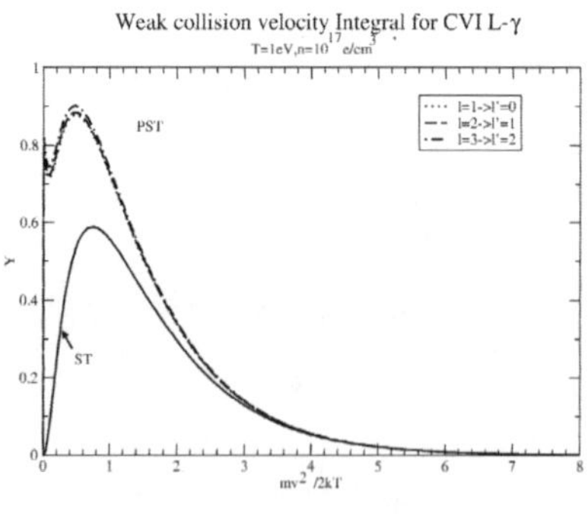

Figure 4. The velocity integrand as a function of $x = mv^2/2kT$ (integrated over eccentricities) without (ST) and with(PST) account for penetration. Notice the very sharp decrease of the PST result to 0 near $x = 0$, which is due to the minimum and maximum eccentricites become equal for very small x.

With penetration, the small u contributions are damped and the negative contributions may win, resulting in possibly large negative values. This is illustrated in Fig.3, where instead of $I = 1$

without account of penetration, one gets a value of $I = -113$ when penetration is accounted for. Since I^2 determines broadening, for the case shown in Fig.3, ignoring penetration underestimates broadening by at least 4 orders of magnitude. If on the other hand the velocity is large, we still get a close to 0 contribution from large u, but the small u contribution which was about 1, becomes smaller and as a result I^2 and consequently broadening is reduced. In contrast, for isolated ion lines, the imaginary exponentials in the integrand of I that vanish for $u = 0$ and oscillate wildly for large u, resulting in a negligible large u contribution. Thus, for isolated ion lines, penetration, which softens the small u contribution can only decrease broadening. Hence the possibility of enhanced broadening over the nonpenetrating case appears unique to hydrogenic ion lines.

When integrated over eccentricities, penetration can give a larger weak collision contribution (and a corresponding smaller strong collision contribution) as illustrated in Fig.4.

3. Computational Aspects

I can be expressed in terms of the modified Bessel functions K_0 and K_1. However, this is numerically very inaccurate and instead a robust method was devised that employs a recursion relation and asymptotics for a related function

$$N_k = \int_0^\infty du(\epsilon coshu - 1)^k e^{\xi(1-\epsilon coshu)} \tag{5}$$

In terms of N_k, I is written as:

$$\begin{aligned} I &= 1 - \xi(N_0 + N_1) - \textstyle\sum_{k=2}^{2n} s_k[\xi^2(\epsilon^2 - 1)\xi^{k-2}N_{k-2} - \xi\xi^{k-1}N_{k-1}] \\ &= 1 - [\xi N_0(1 + s_2\xi(\epsilon^2 - 1)) + \xi N_1(1 - \xi(s_2 - (\epsilon^2 - 1)\xi s_3)) \\ &\quad + \textstyle\sum_{k=2}^{2n-2} \xi^k N_k[\xi^2(\epsilon^2 - 1)s_{k+2} - s_{k+1}] - s_{2n}\xi\xi^{2n-1}N_{2n-1}] \end{aligned} \tag{6}$$

with s_k numerical, plasma-independent coefficients defined for each $l - l'$ channel in Ref. 2. The algorithm based on recursion for small $\xi\epsilon$ and asymptotics for large $\xi\epsilon$ is robust except for parameter combinations where substantial cancellation occurs, in which case integrations were performed. Accounting for penetration decreases the strong collision phase space and hence improves accuracy as shown in Refs.2-3, as the strong collision term is the least certain. More importantly, it can both decrease (in cases similar to Fig.1) and, increase broadening (as illustrated by Figs. 2 and 3). These effects can be substantial and we may expect an increase in broadening for high densities/low temperatures and a corresponding decrease for low densities/high temperatures compared to standard predictions. In turn this can have important applications, for instance in line merging (Ref.4).

[1] Alexiou S and Lee RW 2006 *J. Quant. Spectroscopy Radiative Transfer.* **99** 10
[2] Alexiou S and Poquérusse A 2005 *Phys.Rev.E.* **72** 046404
[3] Poquérusse A and Alexiou S 2006 *J. Quant. Spectroscopy Radiative Transfer* **99** 493
[4] Inglis D.R and Teller E. 1939 *Astrophys. J.* **90** 439

Blue line wings of resonance lines of potassium and sodium perturbed by molecular hydrogen and rare gases

J F Kielkopf[1], **N F Allard**[2,3], **V A Alekseev**[4], **F Spiegelman**[5], **G Guillon**[6] **and H Berriche**[7,8]

[1] University of Louisville, Louisville, Kentucky 40292 USA [2] Observatoire de Paris GEPI, F-75014 Paris, France [3] Institut d'Astrophysique de Paris, Université Pierre et Marie Curie, F-75014 Paris, France [4] St.Petersburg State University, St. Petersburg 199034, Russia [5] Laboratoire de Physique Quantique, Université Paul Sabatier, F-31400 Toulouse, France [6] Université de Bourgogne, 21078 Dijon Cedex, France [7] Faculty of Science, University of Monastir, Monastir, Tunisia [8] American University of Ras Al Khaimah, Ras Al Khaimah, UAE

E-mail: [1] john.kielkopf@louisville.edu [2] nicole.allard@obspm.fr

Abstract. We report on work in progress that compares unified line shape calculations based on *a priori* potentials with experiments to determine the wings of the sodium and potassium resonance lines broadened by H_2, He, and other rare gases.

1. Line profiles in an astrophysical context

The spectra of ultra-cool brown dwarf stars, such as the type T1 ϵ Indi Bab, and inflated hot-Jupiter exoplanets such as HD209458b, show absorption due to neutral Na and K. The strength and shape of these resonance lines is diagnostic of their atmospheric conditions, chemistry, and history. In the coolest and densest of such atmospheres, radiative transfer in the visible and near-IR spectrum may be dominated by far wings of the lines broadened by collisions with neutral H_2 and He. Similarly, DZ class white dwarfs are low temperature Earth-sized degenerate stars of solar mass lacking atomic H lines but showing heavier elements. Detected through recent observations with the Spitzer and Hubble Space Telescopes, these surprising key spectral signatures are attributed to accreted material and inform us about the star's planetary environment when accurate pressure broadened profiles are incorporated into spectral models. The theory of spectral line shapes, especially the unified approach we have developed and refined, makes possible models of stellar spectra that account for the centers of spectral lines and their extreme wings in one consistent treatment. Complete details and the derivation of the theory are given elsewhere. [1]

2. *Ab initio* molecular potentials

Our unified approach to calculating the spectral line profile requires molecular potentials of high accuracy because the shift and shape of the line are very sensitive to the details of the molecular potential curves. New calculations of the potentials of Na/K-H_2 by F. Spiegelman have been reported by Allard et al [2, 3]. We also use here Na-He potentials of Dell'Angelo et al [4], K-He potentials of Santra & Kirby [5], and *ab initio* Na/K-Ar, Ne, Kr, and Xe of V. Alekseev computed using MOLCAS 7.6 [6]. These potential curves are shown in Fig. 1 and 2.

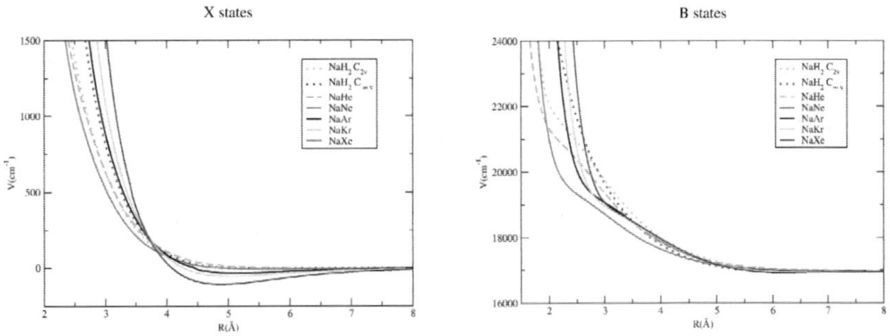

Figure 1. Potential curves for the X and B states of the Na-H_2/rare gas molecules.

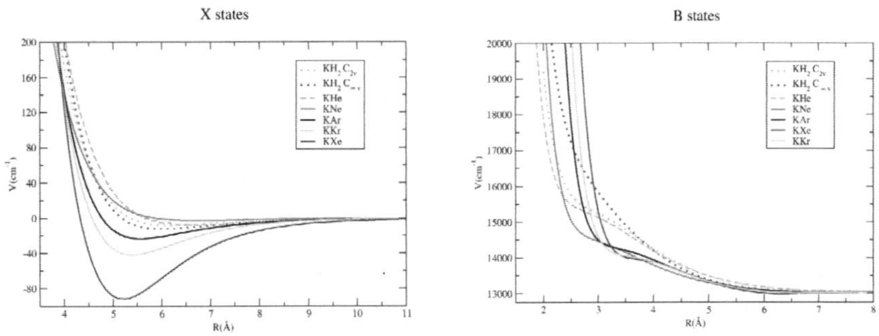

Figure 2. Potential curves for the X and B states of the K-H_2/rare gas molecules.

3. Laboratory spectra and comparisons to line shape theory

The laboratory spectra of both Na and K alkalis with H_2, He, and other rare gases are found to exhibit a systematic pattern of satellites in the blue wings shown in Fig. 3. [7] While Xe and Kr produce the strongest satellites that are closest to the parent line, H_2 and He produce weakly absorbing satellites farthest from the line. In that region the atomic wings of interest overlap alkali dimer absorption. When we use the far wings of Na and K broadened by Kr as a reference, the dimer contribution can be measured in those spectra and removed from the experimental absorption coefficient for the other gases.

Figure 3 shows an exceptionally good match between the K-H_2 experimental spectrum and the model from the unified line shape theory with the new potentials. While laboratory conditions cannot fully duplicate those in brown dwarf or stellar atmospheres, test comparisons such as this confirm the validity of the potentials and the accuracy of the line shape theory. We have compared the line satellite positions for H_2, He and other rare gases in order to assess the accuracy of the potentials and ultimately to guide refinement of their computation. [8, 9, 10] The general properties of the full line profiles predicted by the unified line shape theory are confirmed by these comparisons, even where the specifics of the satellite positions set by the details of the potentials may differ between theory and experiment, especially for perturbers with many electrons.

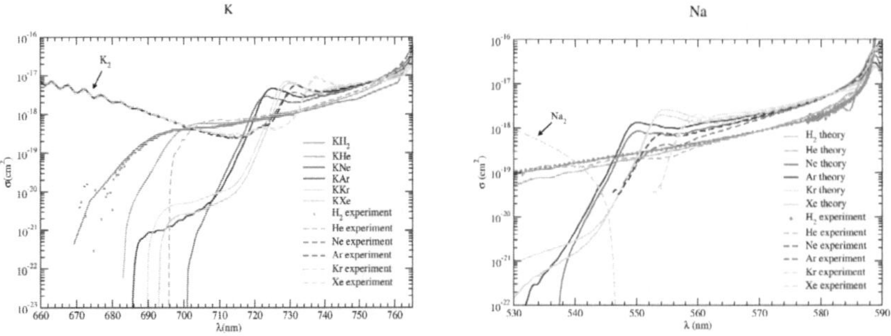

Figure 3. Comparison of the theoretical blue line wings of the resonance line of Na and K (full lines) with the experimental spectra (dashed lines). The density of perturbers is 10^{19} cm^{-3} at 800 K. The experimental spectra have been corrected for dimer absorption.

Pseudopotentials for K-Ar recently computed by H. Berriche improve on the earlier calculations of Li/Na/K-Ar by El Hadj Rhouma *et al* [11], and a comparison with the *ab initio* potentials is shown in Fig. 4. The new K-Ar pseudopotentials lead to a better agreement with the experimental position of the blue satellite.

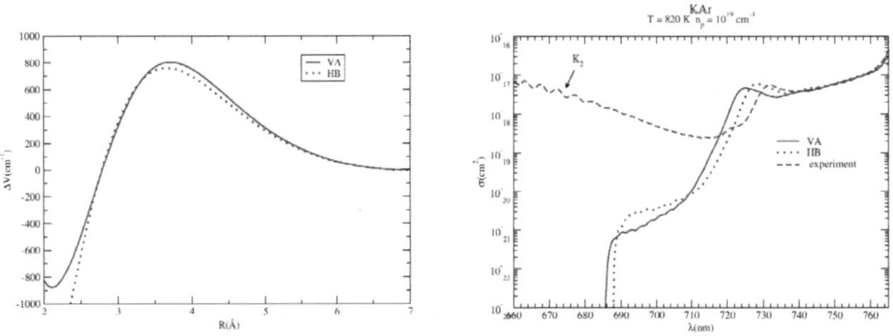

Figure 4. Comparison of difference potentials ΔV for the B-X transitions (left) and the experimental and theoretical blue line wings (right) of the resonance line of K-Ar based on *Ab initio* potentials of V. Alekseev (VA) and pseudopotentials of H. Berriche (HB).

References

[1] N.F. Allard, A. Royer, J.F. Kielkopf, N. Feautrier, Phys. Rev. A. **60**, 1021-1033 (1999)
[2] N. F. Allard, F. Spiegelman, J. F. Kielkopf, G. Tinetti and J. P. Beaulieu, A&A, 543, A159 (2012)
[3] N. F. Allard, F. Spiegelman and J. F. Kielkopf, A&A, 589, A21 (2016)
[4] D. Dell'Angelo, G. Guillon and A. Viel, J. Chem. Phys., 136, 114308 (2012)
[5] R. Santra and K. Kirby, J. Chem. Phys., 123, 214309 (2005)
[6] F. Aquilante, et al., MOLCAS 7: the next generation, Journal of Computational Chemistry, **31**, 224-247, (2010).
[7] J. F. Kielkopf and N. Allard, in preparation (2016).
[8] Allard, N. F., Allard, F., Hauschildt, P. H., Kielkopf, J. F., & Machin, L. 2003, A&A, 411, 473
[9] Allard, N. F., Spiegelman, F., & Kielkopf, J. F., A&A, 465, 1085 (2007)
[10] F. Allard, N. F. Allard, D. Homeier, J. Kielkopf, M. J. McCaughrean, and F. Spiegelman, K-H_2 quasi-molecular absorption detected in the T-dwarf ϵ Indi Ba . A&A 474, L21-L24 (2007)
[11] M.B. El Hadj Rhouma, H. Berriche, Z.B. Lakhdar and F. Spiegelman, J. Chem. Phys., **116**, 1839 (2002)

IOP Conf. Series: Journal of Physics: Conf. Series **810** (2017) 012024 doi:10.1088/1742-6596/810/1/012024

Measurement of oxygen B–band line center frequency in reference to strontium atomic optical clock

K Bielska*, S Wójtewicz, P Morzyński, P Ablewski, A Cygan, M Bober, M Zawada, R Ciuryło, P Masłowski, D Lisak

Institute of Physics, Faculty of Physics, Astronomy and Informatics, Nicolaus Copernicus University in Torun, Grudziadzka 5, 87-100 Torun, Poland

E-mail: *kasia@fizyka.umk.pl

Abstract. We present a proof–of–principle experiment demonstrating the use of atomic optical clocks as a frequency reference in Doppler–limited molecular spectroscopy. We report the determination of an unperturbed line position with a relative uncertainty of 2×10^{-11}.

1. Introduction

We report on a molecular line position determination in the Doppler–limited regime with respect to the frequency reference provided by a strontium atomic optical clock. We apply spectra averaging which significantly reduces the uncertainty of the determined line position. It is a step towards ultra–precise spectroscopy, capable of searching for variations of physical constants in molecular systems, verifying quantum electrodynamics in molecules as well as searching for fundamental effects in atomic interactions [1].

To demonstrate the technique we chose the P7P7 transition from molecular oxygen B–band around 690 nm. New data for this and other B–band lines will enable the improvement of the global fit of the O_2 spectrum model [2]. Its accuracy in the B–band range was mostly determined by uncertainties of line positions in [3] thus we push accuracy limits in this range.

2. Experimental apparatus and measurements

The spectra were collected with the Pound–Drever–Hall locked, frequency stabilised cavity ring–down spectrometer (PDH–locked FS–CRDS) linked to an Er:fibre optical frequency comb (OFC), which was frequency shifted to 1390 nm range and subsequently frequency doubled. For details of spectrometer's operation see [4] and references therein. For the purpose of this work modifications in previously described system considering frequency measurement were applied. Previously, a heterodyne beat–note signal between the probe beam and OFC was employed. At each point of spectrum, the OFC's repetition frequency f_{rep} was changed so that beat–note's frequency remained in the required RF–filter's range during spectra acquisition.

In this work, we used the ^{88}Sr $^1S_0 -^3 P_0$ transition of the strontium atomic optical clock as a frequency reference [5, 6]. The spectrometer's probe laser frequency, which was varying from 434 779 to 434 788 GHz and the clock frequency, which is 429 228.066 418 012(5) GHz, were linked by the OFC. A schematic diagram of the experimental setup is shown in Fig. 1. Two heterodyne beat–note signals were measured: the first one between the clock laser and the OFC, and the second one between the spectrometer's probe beam and the OFC. In order to simultaneously measure both beat–note frequencies, $f_{\mathtt{rep}}$ remains unchanged during entire spectrum acquisition, whereas the probe beam's frequency is additionally shifted with a double–passed acousto–optic modulator so that the beat–note frequency remains in the measurement range. In such case the spectrometer's frequency f can be calculated as:

$$f = f_c + (n_p - n_c)\, f_{\mathtt{rep}} + f_{\mathtt{Bc}} - f_{\mathtt{Bp}} - f_s. \tag{1}$$

Here f_c is the clock laser frequency, n_p and n_c are comb tooth numbers for heterodyne beat–note with the spectrometer's laser and clock's laser beams, respectively, $f_{\mathtt{Bp}}$ and $f_{\mathtt{Bc}}$ are the frequencies of those beat–notes and f_s is sum of all additional frequency shifts in both beams. During the measurements all frequency counters and signal generators were referenced to the same, stable 10 MHz signal [7] from the hydrogen maser [8] transferred from Borowiec near Poznań to our Institute by a 330 km fibre link [6, 9].

With the system described above, we performed measurements of P7P7 self–perturbed transition in the oxygen B–band. Data was acquired at four sample pressures in the range from 1.5 to 6 Torr. At each pressure between 38 and 113 spectra were collected and averaged. Measurements at three pressures were performed within one week, whereas the fourth one was made four months later. First three datasets were referenced to the strontium optical clock, whereas the last one was referenced to the frequency standard from the hydrogen maser.

Spectra were fitted with the speed–dependent Galatry profile (SDGP) [10] with speed dependence described by hypergeometric functions [11]. To test the influence of line shape model on the line position determination, we have also performed fits with Voigt profile (VP). We found that due to the lack of line asymmetry, line position obtained from the VP agrees within fit uncertainty with value resulting from the SDGP fit. Averaged spectrum measured at the highest sample pressure, which was 6 Torr, as well as fit residuals, are shown in Fig. 2.

Referencing measurements to the frequency of the atomic optical clock enabled us to avoid systematic uncertainties caused by frequency standard, whereas spectra averaging led to significant reduction of statistical uncertainty. Other uncertainty sources are associated with

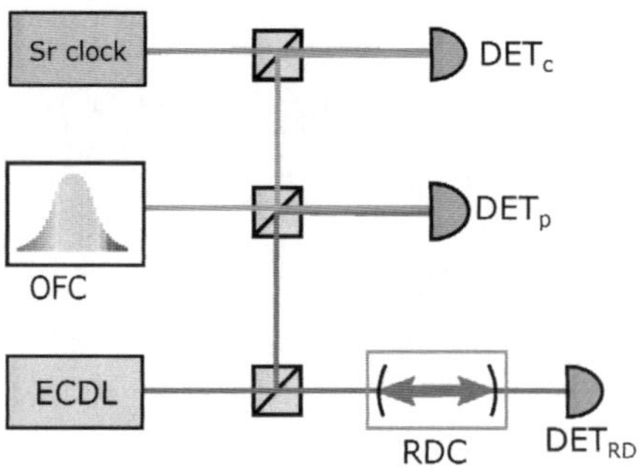

Figure 1. Schematic diagram of optical connections between main parts of the experimental setup. Sr clock is strontium atomic optical clock, ECDL – spectrometer probe laser, RDC – spectrometer's ring–down cavity, DET$_c$, DET$_p$ and DET$_{\mathtt{RD}}$ are detectors for heterodyne beat–notes with clock, probe beam and detector for acquisition or ring–down decays, respectively.

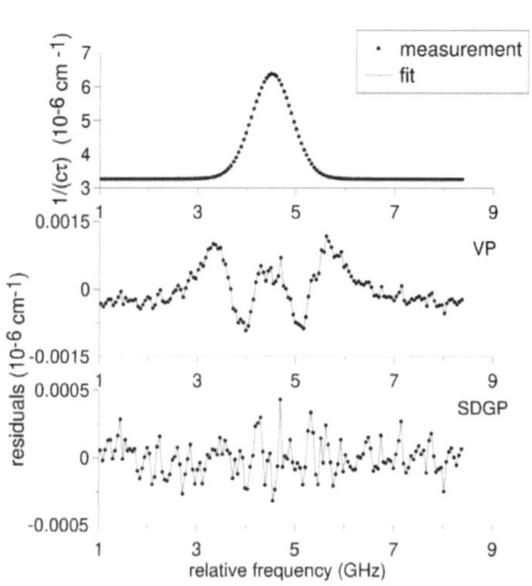

Figure 2. Top panel: Line P7P7 obtained at pressure of 6 Torr with averaging of 38 scans. Central and bottom panels: fit residuals obtained for VP and SDGP, respectively.

Table 1. Uncertainty budget for the determination of unperturbed line position ν_0 and collisional shift coefficient Δ/N. Systematic fit offset indicates differences in line positions obtained while performing fits pressure–by–pressure and multispectrum fit approach.

Uncertainty source	$u(\nu_0)$ (kHz)	$u(\Delta/N)$ (%)
Fit uncertainty	8.1	0.65
Systematic fit offset	2.1	0.12
Pressure reading & calibration (max 5.6 mTorr)	0.60	0.065
Temperature: variations, gradient, reading & calibration (max 240 mK)	0.86	0.094
Total	8.5	0.7

pressure and temperature measurement and calibration, as well as choice of spectra fitting technique. Their estimated influence on the uncertainty of unperturbed transition frequency and collisional shift coefficient is shown in Table 1. We have determined the unperturbed frequency of the P7P7 transition in the oxygen B-band equal to 434 783.508 485 7(85) GHz, which corresponds to relative uncertainty of 2×10^{-11}, and its collisional shift coefficient equal to $-0.09381(66)\times10^{-19}$ GHz/(molecule/cm^3)) which corresponds to relative uncertainty of 7×10^{-3}.

3. Conclusions

We performed a proof–of–principle experiment showing application of atomic optical clocks as frequency reference in molecular spectroscopy. This experimental scheme can be applied for precise and accurate determination of hydrogen line positions and other of lines that cannot be easily accessed by sub–Doppler spectroscopy techniques and are important for fundamental studies.

Acknowledgments

The research is part of the program of the National Laboratory FAMO in Toruń, Poland, and is supported by the National Science Centre, Poland projects no. 2014/15/D/ST2/05281, DEC-2012/05/D/ST2/01914, DEC-2013/11/D/ST2/02663 and 2015/17/B/ST2/02115 as well as by the Foundation for Polish Science TEAM Project co–financed by the EU European Regional Development Fund and by MOLIM COST action.

References

[1] Salumbides E J *et al* 2013 *Phys. Rev.* D **87** 11200
[2] Yu S *et al* 2012 *J. Chem. Phys.* **137** 024304
[3] Domysławska J *et al* 2012 *J. Chem. Phys.* **136** 024201
[4] Domysławska J *et al* 2016 *J. Quant. Spectrosc. Radiat. Transf.* **169** 111
[5] Bober M *et al* 2015 *Meas. Sci. Technol.* **26** 0075201
[6] Morzyński P *et al* 2015 *Sci. Rep.* **5** 17495
[7] Cygan A *et al* 2016 *J. Chem. Phys.* **144** 214202
[8] Jiang Z *et al* 2015 *Metrologia* **52** 384
[9] Krehlik P *et al* 2015 *Metrologia* **52** 82
[10] Ciuryło R and Szudy J 1997 *J. Quant. Spectrosc. Radiat. Transf.* **57** 411
[11] Ward J, Cooper J, Smith E W 1974 *J. Quant. Spectrosc. Radiat. Transf.* **14** 555

Multi-spectrum fitting software for advanced spectral line shapes analysis

A Cygan, D Lisak

Institute of Physics, Faculty of Physics, Astronomy and Informatics, Nicolaus Copernicus University in Toruń, Grudziadzka 5, 87-100 Toruń, Poland

E-mail: agata@fizyka.umk.pl, dlisak@fizyka.umk.pl

Abstract. We present a multi-spectrum fitting software that enables advanced analysis of high-resolution atomic and molecular spectra in a large range of pressures. Individual lines are modeled with theoretical profiles that take into account the non-Voigt line shape effects, such as the speed-dependence of collisional broadening and shift, velocity-changing collisions and their correlations with phase/state changing collisions, first order line mixing and collision time asymmetry.

1. Introduction

Increasing requirements for high-accuracy spectroscopic data has led to a transition from the commonly used Voigt profile to more advanced and physically justified line-shape models in spectroscopic databases, reference laboratories and high-precision remote sensing applications. As shown in many papers, e.g. [1, 2, 3], a sub-percent agreement of experimental data with the fitted line shape in typical atmospheric conditions requires taking into account such physical effects as: speed-dependence of collisional broadening and shift [5], velocity-changing collisions, which leads to Dicke-narrowing [4] and their correlations with phase/state changing collisions [6]. For non-isolated lines, line mixing [7] may also play an important role. In semiclassical line shape theory these effects may be described using the correlated asymmetric speed-dependent Rautian-Sobelman profile (CSDARSP) [8, 9] and its simplifications, such as the Hartmann-Tran profile (HTP), recently recommended as a standard line-shape model to replace the Voigt profile in spectroscopic databases [2, 10], e.g in HITRAN [11, 12]. In real experimental conditions, where the spectra have finite signal-to-noise ratio, numerical correlations between some line-shape parameters make data fitting difficult. In such cases a multi-spectrum fitting approach [13] is used. In this approach shared line parameters are fitted to a set of spectra measured at different gas pressures, which makes it possible to obtain a set of line parameters consistent in a wide range of gas pressures.

2. Software capabilities

Our spectra fitting software consists of two main components. A Fortran-based library contains an implementation of the Levenberg-Marquardt nonlinear least-squares fitting algorithm [14] and algorithm that defines a theoretical model describing a set of fitted spectra. A LabView-based interface enables convenient reading of experimental spectra, setting the line-shape model and initial parameters of spectral lines, provides a tool for graphical representation of the data and fit, and finally saving the fit results.

The Fortran-based library, compiled to a DLL file, provides access to functions which fit the user provided multi-spectrum data with a chosen model. The individual line shapes can be modeled with simplifications of CSDARSP. Speed-dependent effects can either be modeled with the quadratic [15] or hypergeometric approximation [5], while velocity-changing collisions are modeled within the soft- or hard-collision approximation. The correlation between velocity-changing and phase/state changing collisions can be modeled by one of two common approaches, i.e. the speed-independent complex parameter of frequency of optical collisions [9] or the speed-dependent correlation parameter, typically denoted as η [1], used also in HTP. The first order line mixing effects [16] can be modeled by additional pressure-dependent parameters, which may also describe collision-time asymmetry [17]. The relative accuracy of numerical calculations of the line shape functions, mostly limited by the numerical accuracy of the integration methods used, was tested to be at least 10^{-6}. The experimental spectra can be provided in a form of an absorption coefficient, a cavity ring-down time constant or a transmission through absorption cell or optical cavity, according to Lambert-Beer law. Instrumental features of a spectrum, such as a linear background and the etalon fringes, can be fitted together with line shapes.

The LabView-based program, partially shown in Fig. 1, provides a convenient user interface for spectra fitting. Each of the initial parameters of the single- or multi-spectrum fitting model can be set as individual for each pressure or a shared parameter for all pressures, with either a

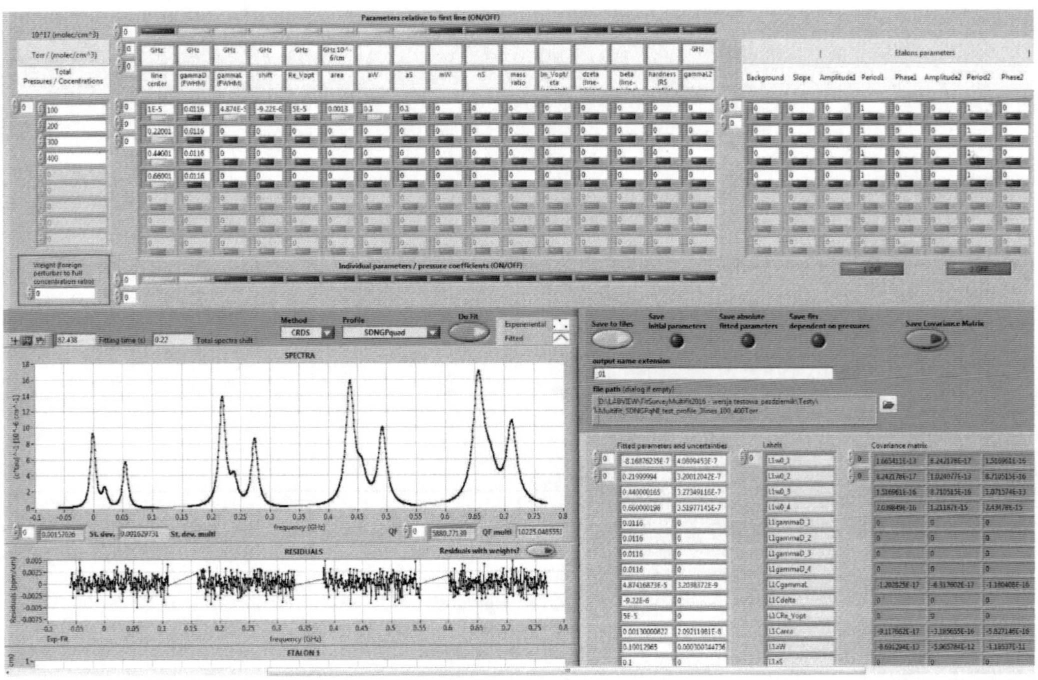

Figure 1. The main part of the LabView-based program interface. The upper part enables settings of gas pressures and initial parameters of the model. Graphs in the lower part represent a multi-spectrum fit of example spectrum of 3 lines, measured at 4 pressures. The lower graph represents fit residuals. In the lower right part a list of fitted parameters with uncertainties and covariance matrix are displayed.

linear pressure dependence (e.g. collisional broadening) or pressure independent (e.g. Doppler broadening). For convenience, in the case of multi-line spectra the parameters can be optionally set as relative to the corresponding parameters of a reference line (e.g. to the strongest line). The output files contain the experimental and fitted spectra and their differences, the fitted parameters of the model with standard uncertainties, the quality of the fit (QF) parameter [18, 19] and optionally the covariance matrix of the fitted parameters.

3. Conclusions

Our multi-spectrum line-shape fitting software enables analysis of high-precision spectra with advanced semi-classical models. So far this software, or its earlier versions, were used to analyze cavity ring-down spectra of O_2 [20, 21], CO [22, 23], CO_2 [19, 24] and H_2O [3, 25]. It was also used in analysis of the O_2 [26, 27], CO [28] and D_2 [29] data, presented as posters at the ICSLS 23 conference.

In the future we plan to extend the program functionality by adding temperature dependencies of line parameters in the multi-spectrum fitting to enable analysis of pressure and temperature dependencies of spectra and fitting temperature dependencies of line-shape parameters. We also plan to extend a list of available line profiles by adding a combination of hard- and soft-collision for velocity-changing collisions.

Acknowledgments

The research is part of the program of the National Laboratory FAMO in Toruń, Poland and is supported by the National Science Centre, Poland, Project Nos. DEC-2013/11/D/ST2/02663 and 2015/18/E/ST2/00585.

References

[1] Pine A S 1999 *J. Quant. Spectrosc. Radiat. Tr.* **62** 397
[2] Ngo N H *et al* 2013 *J. Quant. Spectrosc. Radiat. Tr.* **129** 89
[3] Lisak D *et al* *J. Quant. Spectrosc. Radiat. Tr.* **164** 221
[4] Dicke R H 1953 *Phys. Rev.* **89** 472
[5] Berman P R 1972 *J. Quant. Spectrosc. Radiat. Tr.* **12** 1331
[6] Rautian S G, Sobelman I I 1967 Sov. Phys. Usp. **9** 701
[7] Baranger M 1958 *Phys. Rev.* **111**, 481
[8] Lance B, Robert D, 1999 J. Chem. Phys. **111** 789
[9] Ciuryło R, Pine A S, Szudy J 2001 *J. Quant. Spectrosc. Radiat. Tr.* **68** 257
[10] Tennyson J *et al* 2014 Pure Appl. Chem. **86** 931
[11] Kochanov R V *et al* 2016 *J. Quant. Spectrosc. Radiat. Tr.* **177** 15
[12] Wcisło P *et al* 2016 *J. Quant. Spectrosc. Radiat. Tr.* **177** 75
[13] Chris Benner D *et al* 1995 *J. Quant. Spectrosc. Radiat. Tr.* **53** 705
[14] Press W H *et al* Numerical Recipes in Fortran, Cambridge University Press, 1992, pp 678
[15] Rohart F *et al* 1994 *J. Chem. Phys.* **101** 6475
[16] Rosenkranz P 1975 IEEE Trans. Antennas. Propag. **23** 498
[17] Szudy J, Baylis W E, 1996 *Phys. Rep.* **266** 127
[18] Cygan A *et al* 2012 *Phys. Rev. A* **85** 022508
[19] Bui T Q *et al* 2014 *J. Chem. Phys.* **141** 174301
[20] Wójtewicz S *et al* 2015 *J. Quant. Spectrosc. Radiat. Tr.* **165** 68
[21] Domysławska J *et al* 2016 *J. Quant. Spectrosc. Radiat. Tr.* **169** 111
[22] Wójtewicz S *et al* 2013 *J. Quant. Spectrosc. Radiat. Tr.* **130** 191
[23] Cygan A *et al* 2016 *J. Chem. Phys.* **144** 214202
[24] Lin H *et al* 2015 *J. Quant. Spectrosc. Radiat. Tr.* **161** 11
[25] Sironneau V, Hodges J T 2014 *J. Quant. Spectrosc. Radiat. Tr.* **152** 1
[26] Bielska K *et al* 23rd ICSLS Book of Abstracts, Wydawnictwo Naukowe UMK, Toruń, 2016, pp 116
[27] Domysławska J *et al* 23rd ICSLS Book of Abstracts, Wydawnictwo Naukowe UMK, Toruń, 2016, pp 122
[28] Stec K *et al* 23rd ICSLS Book of Abstracts, Wydawnictwo Naukowe UMK, Toruń, 2016, pp 136
[29] Zaborowski M *et al* 23rd ICSLS Book of Abstracts, Wydawnictwo Naukowe UMK, Toruń, 2016, pp 149

Comparison IR spectra of alanine $CH_3CH(NH_2)COOH$ and alanine $CD_3CH(NH_2)COOH$

Szczepan Brym
University of Warmia and Mazury, Faculty of Mathematics and Computer Science,
S oneczna 54, 10 -710 Olsztyn, Poland;
E-mail: szbrym@matman.uwm.edu.pl

Abstract. Comparison of spectra by ATR FTIR technique was made. For comparison the fingerprint region below 1500cm^{-1} was chosen. After replacement of hydrogen with deuterium in side chain the frequency shift of most of the bands to lower values in FTIR spectrum was observed.

1. Introduction

Alanine is one of optically active, unpolar aliphatic amino acids [1]. Lateral chain presented contains methyl group CH_3. Linear formula has form $CH_3CH(NH_2)$ COOH. Molecular weight 89,09 g/mol. Melting temperature 295^{0}C. After replacement of hydrogen with deuterium, the linear formula becomes $CD_3CH(NH_2)$ COOH with molecular weight 92,11 g/mol and melting temperature 314,5 ^{0}C. Structure of alanine is presented in figure 1 [2].

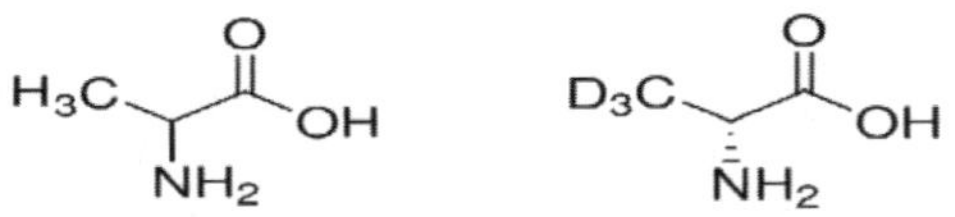

Alanine CH_3 Alanine CD_3

Figure 1: CH_3 and CD_3 forms of alanine molecules.

2. Experimental setup and results

FT-IR spectra of the solid compounds were obtained as attenuated total reflection (ATR), using Nicolet 6700 FT-IR spectrometer, equipped with a diamond crystals, XT-KBr beam splitter and a DlaTGS detector in the range from 500-4000cm^{-1}, with 64 scans and a spectra resolution of 4cm^{-1}, at a room temperature. For comparison the fingerprint region i.e. below 1500cm^{-1} was chosen. Figure 2 displays the 4000-550 cm^{-1} spectral region of the MIR spectra of the CH_3 and CD_3 alanine.

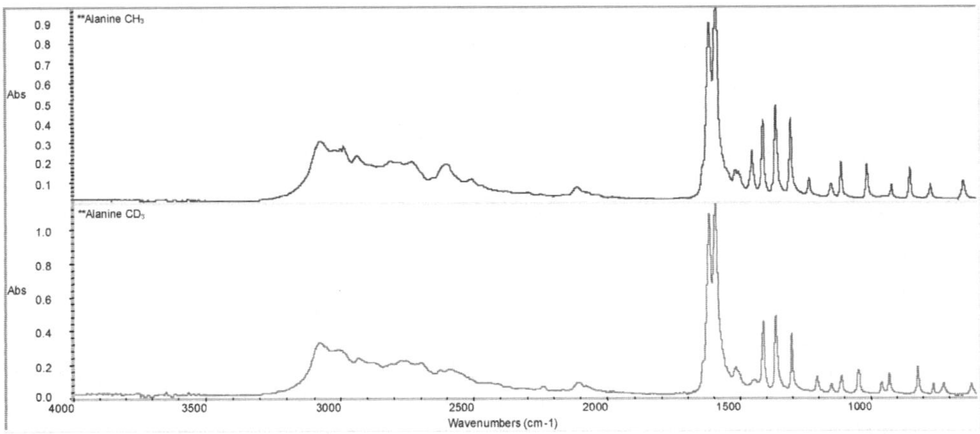

Figure 2: The MIR spectra of the CH_3 and CD_3 alanine.

Figure 3 displays the 1500-550 cm^{-1} spectral region of the MIR spectra of the CH_3 and CD_3 alanine.

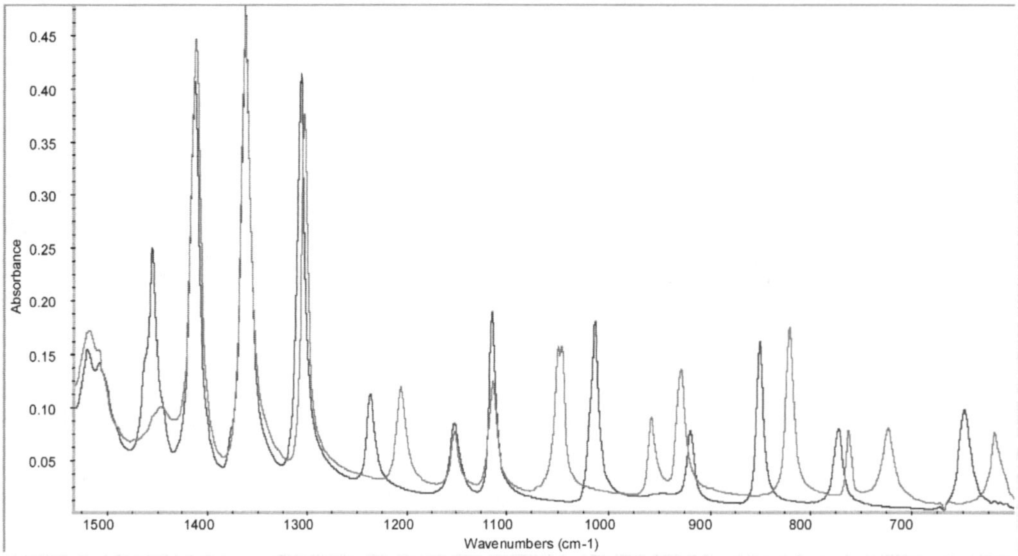

Figure 3: The spectra of the CH_3 and CD_3 alanine in the fingerprint region.

The corresponding frequencies and the band assignments are presented in table 1.

Table 1. Vibrational frequencies of alanine CH_3 and alanine CD_3 in the fingerprint region.

Alanine CH_3		Alanine CD_3	
Assignment	Exp. (MIR)	Assignment	Exp. (MIR)
こ$_{as}CH_3$	1455,87	こ$_{as}CD_3$	1446,75
ご$_{s}CH_3$	1413,12	ご$_{s}CD_3$	1412,11
そ$_{s}$COO	1363,15	そ$_{s}$COO	1363,14
こ$_{s}NH_2$	1307,77	こ$_{s}NH_2$	1304,02
ご$_{as}$CH	1236,97	ご$_{as}$CH	1206,66
ちNH_2	1152,39	ちNH_2	1152,22
そC-N	1114,71	そC-N	1113,34
ちCH_3	**1014,18**	ちCD_3	**1046,08**
ちNH_2	**919,32**	ちNH_2 (double)	**957,99**
そC-OO	**850,75**		**928,92**
そCN	772,25	そC-OO	**820,87**
ごCOO	648,97	そCN (double)	762,15
			723,11
		ごCOO	618,75

a) Frequencies in cm^{-1}; そstretching, こbending, どwagging, ちrocking, s-symmetric, as-asymmetric.
b) Bands in bold are presented in figure 4 below.

After replacement of hydrogen to deuterium in side chain in FTIR spectrum frequency shift of most of the bands to lower values was observed. Because the vibrations are complex, the frequency ratio of comparative bands did not exceed 1,1. The spectrum of alanine CH_3 in this field consists of separate single bands. After deuteration complex and double bands of diverse intensity were observed.

The profiles of these bands were examined. The selected bands and approximated Voigt profiles are presented in figure 4.

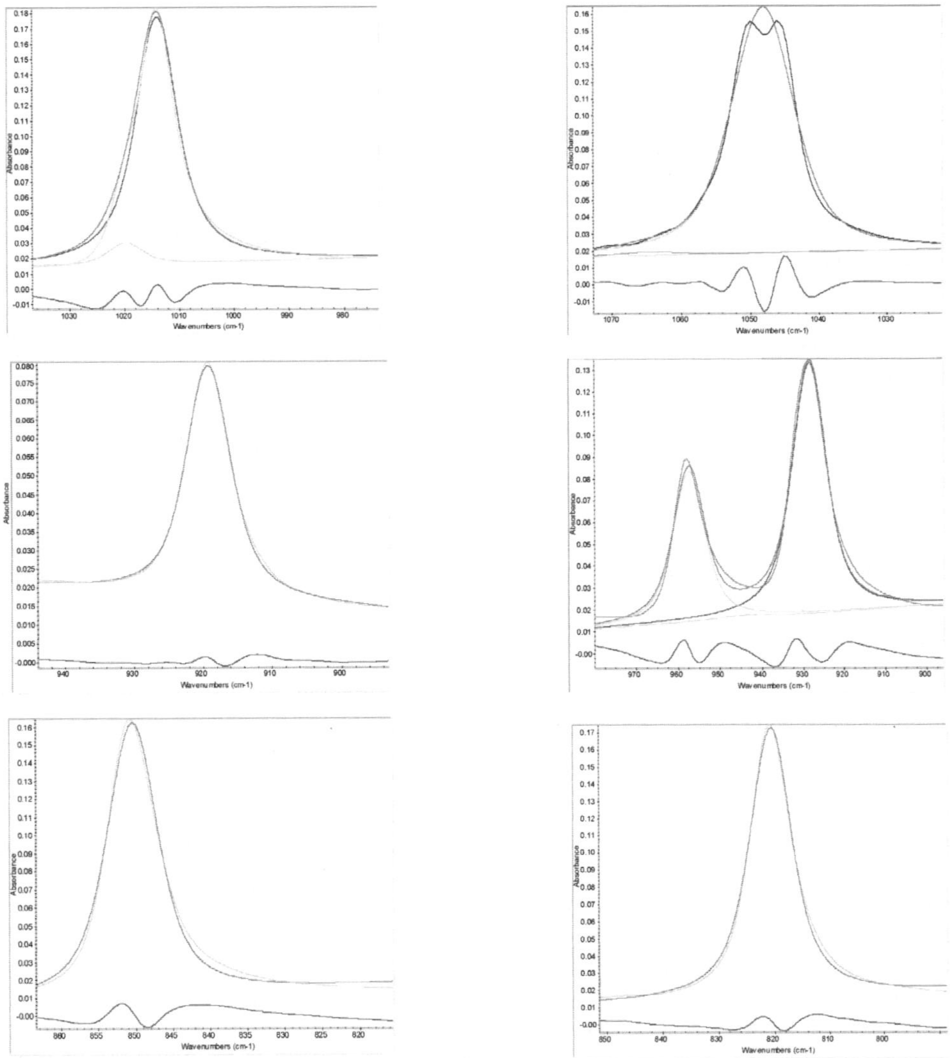

Figure 4: Selected bands of alanine CH_3 on the left with respective bands of alanine CD_3 on the right.

3. Conclusions

In FTIR spectrum the frequency shift was observed for most of the bands to lower values. We found the increase of vibrational frequencies for two bands. We also noticed that two vibrational frequencies turned into doublets in the spectrum after deuteration. Probably, this result is due to the Fermi resonance. In this work we also demonstrated that the shape of the spectral lines can be approximated by one or more of the Voigt profiles.

References

[1] Rosado M T S, Duarte M L R S and Rui F *Journal of Molecular Structure* 410-411(1997) p 343
[2] Lenarz W J and Lane M D (red.) *Encyclopedia of Biological Chemistry* 2004 Elsevier

Bound-free transitions in $H + H^-$ collisions

A Dadonova[1], **A Devdariani**[1,2], **E Dalimier**[3] **and P Angelo**[3]

[1] Department of Theoretical Physics and Astronomy, A.I. Herzen University, St. Petersburg 191186, Russia

[2] Department of Optics and Spectroscopy, St. Petersburg University, St. Petersburg 198904, Russia

[3] LULI - UPMC Univ Paris 06: Sorbonne Universites; CNRS, Ecole Polytechnique, CEA: Universit Paris-Saclay - F-75252 Paris cedex 05, France

E-mail: alladadonova@mail.ru

Abstract. The processes of photo-detachment or ionization of stable molecules are widely investigated in modern atomic physics. But the present work deals with the calculation of the photo-detachment cross-sections in the case of negative quasi-molecules formed during collisions. The reaction $H+H^- + \hbar\omega \rightarrow H+H+e$ has been taken as an example of bound-free transitions in single-active-electron quasi-molecules. For simplicity, we discuss the one-dimensional case in the frame of the zero-range potential model.

1. Introduction

The study of radiative transitions in collisional quasi-molecules, that is temporary molecules formed during atom or ion collisions, is one of the traditional areas in spectroscopy. Until recently the study has mainly been limited to bound-bound transitions between quasimolecular states. In particular, the spectral profiles produced by the reaction of charge exchange $H + H^- \rightarrow H^- + H + \hbar\omega$ was regarded in [1] as an example of asymptotically forbidden transition in quasi-molecules. The aim of the present work is to highlight the case of the bound-free transitions in quasi-molecules having in mind the recent progress in the study of the photo-effect in atoms and stable molecules, e.g. [2]. The specific reaction $H + H^- + \hbar\omega \rightarrow H + H + e$ has been taken into account as a typical example of quasimolecular photo-detachment or photo-ionization processes.

2. Bound-free transitions

The cross-section of photodetachment is proportional to the density of oscillator strengths which is the main characteristic of the dipole transitions

$$\sigma(\omega) = \frac{2\pi^2}{c}\frac{df_{bc}}{d\omega} = \frac{4\pi^2}{c}\omega_{bc}\left|X_{bc}\right|^2. \tag{1}$$

For simplicity, we discuss the one-dimensional case. Then the matrix element is

$$X_{bc} = \int \langle \Psi_b \left| x \right| \Psi_c \rangle \, dx. \tag{2}$$

In the LCAO approximation quasimolecular *gerade* and *ungerade* wave functions in the frame of the zero-range potentials model [3] are written as

IOP Conf. Series: Journal of Physics: Conf. Series **810** (2017) 012027 doi:10.1088/1742-6596/810/1/012027

$$\Psi_b^{g,u} = \sqrt{\frac{\alpha}{2}} \left[exp(-\alpha\,|x + R/2|) \pm exp(-\alpha\,|x - R/2|)\right]; \tag{3}$$

$$\Psi_c^{g,u} = \frac{1}{\sqrt{2\pi k}} \left[sin(k\,|x + R/2|) \pm sin(k\,|x - R/2|)\right]. \tag{4}$$

Then the matrix element dipole moments can be expressed by the formula:

$$X_{b=g,c=u} = \sqrt{\frac{\alpha k}{\pi}} \frac{4\alpha}{(\alpha^2+k^2)^2} cos^2\left(\frac{kR}{2}\right) + \sqrt{\frac{\alpha}{\pi k}} \frac{\alpha R}{2(\alpha^2+k^2)} sin\,(kR); \tag{5}$$

$$X_{b=u,c=g} = \sqrt{\frac{\alpha k}{\pi}} \frac{4\alpha}{(\alpha^2+k^2)^2} sin^2\left(\frac{kR}{2}\right) - \sqrt{\frac{\alpha}{\pi k}} \frac{\alpha R}{2(\alpha^2+k^2)} sin\,(kR). \tag{6}$$

Although the quasimolecular matrix elements increase with increasing R it does not lead to difficulties at large R. In this case quasimolecular potential energies of both states involved $U_u = U_g$, and it is obligatory to use the sum of amplitudes of two different ways of detachment [4]. Thus, the quasimolecular oscillator strength reduces to the oscillator strength for a single atomic ion at large distances.

The subsequent calculations of the detachment cross-sections of *gerade* σ_{gu} and *ungerade* σ_{ug} quasimolecular states have been made in the frame of the quasistatic approximation widely used in the theory of spectral line shapes, e.g. [5]. Then the main characteristic of the transition is the Condon point R_c that can be found from:

$$\omega = -U_{g,u}(R_c) + \frac{k^2}{2}. \tag{7}$$

The Figure 1 shows the results for the cross-sections in relative units as a functions of the frequency of light. The main feature is the shift of the detachment threshold in the direction of smaller frequencies compared with the threshold in the case of atomic ions, $2.7 * 10^{-2}$. The reason is that the *ungerade* quasimolecular term U_u is repulsive.

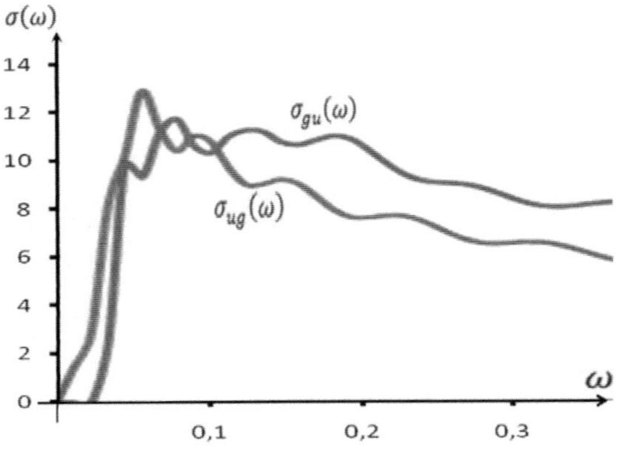

Figure 1. The cross-sections σ_{gu} and σ_{ug} of photo-detachment of *gerade* and *ungerade* quasimolecular states as functions of the frequency of light.

IOP Conf. Series: Journal of Physics: Conf. Series **810** (2017) 012027 doi:10.1088/1742-6596/810/1/012027

3. Conclusion

The comparisons and simple analytic formulas we have given should make the method presented a useful exploratory tool for the other detachment processes of this type. The main results obtained are as follows. First, collisions result in a lowering of the threshold of the electron detachment compared with the single ion. Second, the cross-sections are oscillating functions of the frequency of light as a consequence of interference under scattering of removed electron on two atoms. The results obtained may contribute to the cold and thermal plasma spectroscopies.

References

[1] Devdariani A Z, Dadonova A V 2015 *Russian Journal of Physical Chemistry B* **9(4)** 591-592.
[2] Amusia M Ya 1990 *Atomic Photoeffect* (New York London: Plenum Press).
[3] Demkov Yu N, Ostrovskii V N 1988 *Zero-range Potentials and their Applications in Atomic Physics* (New York: Plenum Pub Corp).
[4] Devdariani A Z 2015 *Optics and Spectr.* **119(3)** 333-337.
[5] Devdariani A Z 1999 *Optics and Spectr.* **86(6)** 853-858.

Radiative transitions in quasi-molecules Hg($6\,^3P_1 - 6\,^1S_0$) + Xe. The influence of buffer gas atom density on spectral line shape

A Z Devdariani[1,2], G M Grigorian[1], N A Kryukov[1], M G Lednev[3] and A L Zagrebin[3]

[1]St. Petersburg State University, Peterhof, Ul'janovskaya St. 3, 198504, St. Petersburg, Russia
[2]The Herzen State Pedagogical University of Russia, nab. rekiMoiki, 48, 191186, St. Petersburg, Russia
[3]Baltic State Technical University "VOENMEH" named after D.F. Ustinov, Krasnoarmeyskaya St. 1, 190005, St. Petersburg, Russia

E-mail: snbrn2@yandex.ru

Abstract. This study is aimed at the explanation of discrepancies in emission spectral profiles obtained by different authors for the excimer band HgXe(A^30^+) – HgXe(X^10^+). The calculations of the spectral profiles have been fulfilled for two limiting cases: for the high density and low density of buffer gas atoms. The results obtained let us to conclude that the discrepancies can be mainly caused by the recombination and relaxation processes leading to the population ro-vibrational states of the HgXe(A^30^+) excimer.

1. Introduction

The present work is devoted to the explanation of discrepancies in emission spectral profiles obtained in [1-3] for the excimer band HgXe(A^30^+) – HgXe(X^10^+) at the temperature T = 300 K. We suggest that the discrepancies are mainly caused by the recombination and relaxation processes leading to the population ro-vibrational states of the HgXe(A^30^+) excimer. This suggestion has been backing up by the calculations of the profiles in two limiting cases: for the high and low densities of buffer gas atoms.

The calculations are based on the experimental potential curves presented in [4, 5]. For the case of high density of buffer gas atoms the calculations have been perfomed in the framework of quasi-static approach [6]. For the case of low density the approach proposed in [7] has been used.

2. Interaction potential curves

The interaction potential curves obtained experimentally in [4, 5] for the excited and ground states and the corresponding difference potentials are presented on figure 1 and figure 2 respectivelly.

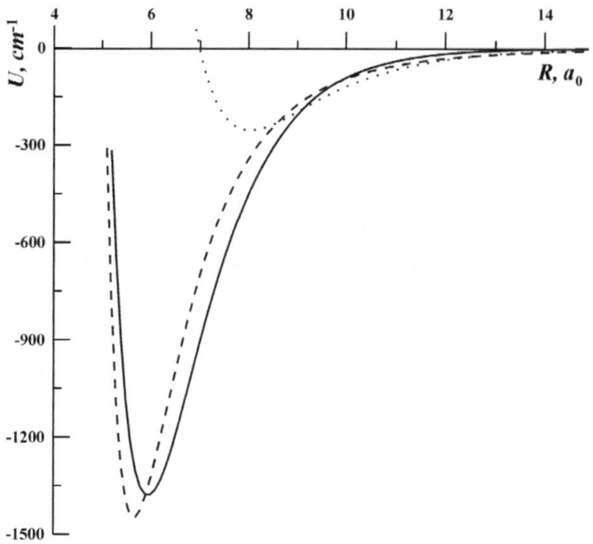

Figure 1. Interaction potential curves for the excited state $A^3 0^+$ obtained in [4] (dashed line) and [5] (solid line) and for the ground state $X^1 0^+$ obtained in [5] (dotted line).

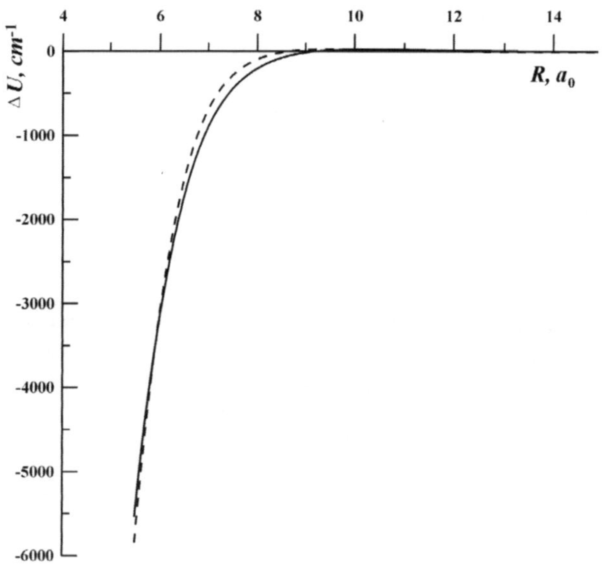

Figure 2. The difference potentials $\Delta U = U(A^3 0^+) - U(X^1 0^+)$ (dashed line – [4,5], solid line – [5,5]).

3. Emission spectra

The experimental and calculated emission spectra are presented on figure 3. The experimental spectra were obtained for different values N_{Xe} of the concentrations of xenon atoms. The calculated spectra are presented for the cases of high and low densities of xenon atoms.

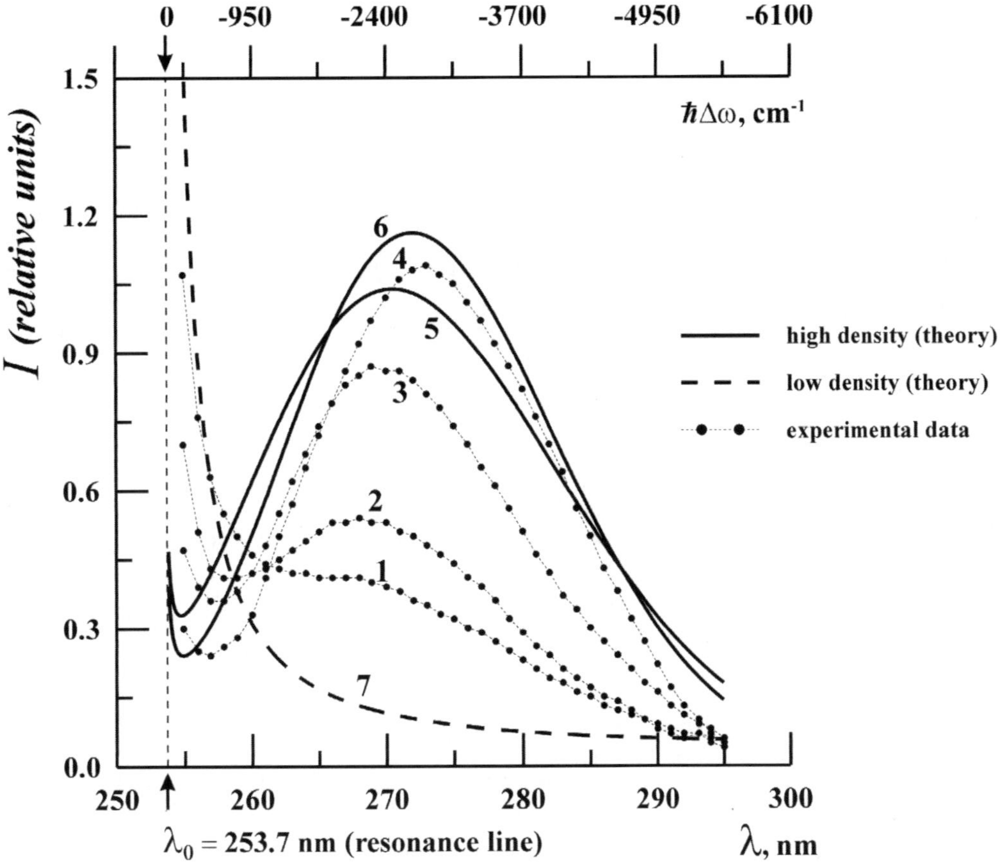

Figure 3. Experimental (lines 1 – 4) and calculated (lines 5 – 7) emission spectral profiles. Line 1 corresponds to $N_{Xe} = 3\cdot10^{17}$ cm^{-3} [1], line 2 – $N_{Xe} = 10.1\cdot10^{17}$ cm^{-3} [1], line 3 – $N_{Xe} \approx 10^{19}$ cm^{-3} [2], line 4 – $N_{Xe} \approx 1.1\cdot10^{20}$ cm^{-3} [3]. Lines 5 (on data [4] for A^30^+ and [5] for X^10^+) and 6 (on data [5] for A^30^+ and [5] for X^10^+) correspond to calculations for high density of Xe atoms. Line 7 (on data [4] for A^30^+ and [5] for X^10^+) corresponds to calculations for low density of Xe atoms.

The comparison leads to the conclusion that the experimental data obtained in [1] can be attributed to the case of the intermediate densities. The conditions of experiments [2, 3] are fairly close to the high density limit.

References

[1] Kryukov N A 1989 *Opt. Spektrosk.(USSR)* **66** 721
[2] Lam L K, Gallagher A and Drulligher R 1978 *J. Chem. Phys.* **68** 4411
[3] Nikiphorov E V *et al* 1976 *Opt. Spektrosk.(USSR)* **41** 339
[4] Grycuk T and Findeisen M 1983 *J. Phys.* B **16** 975
[5] Okunishi M, Nakazava H, Yamanouchi K and Tsuchiya S 1990 *J. Chem. Phys.* **93** 7526
[6] Gallagher A 1984 in: Excimer Lasers (*Top. Appl. Phys.* vol 30) ed Ch K Rhodes (Springer Berlin Heidelberg/Springer-Verlag) pp. 139-179
[7] Devdariani A Z and Zagrebin A L 1992 *Opt. Spectrosc.(USA)* **72** 309

Quasimolecular emission near the Xe($5p^5 6s$ $^{1,3}P_1 - 5p^6$ $^1S_0$) and Kr ($4p^5 5s$ $^{1,3}P_1 - 4p^6$ $^1S_0$) resonance lines induced by collisions with He atoms

O S Alekseeva[1], A Z Devdariani[2,3], G M Grigorian[2], M G Lednev[1] and A L Zagrebin[1]

[1]Baltic State Technical University "VOENMEH" named after D.F. Ustinov, Krasnoarmeyskaya St. 1, 190005, St. Petersburg, Russia
[2]St. Petersburg State University, Peterhof, Ul'janovskaya St. 3, 198504, St. Petersburg, Russia
[3]The Herzen State Pedagogical University of Russia, nab. reki Moiki, 48, 191186, St. Petersburg, Russia

E-mail: o.alek@rambler.ru

Abstract. This study is devoted to the theoretical investigation of the quasimolecular emission of Xe*-He and Kr*-He collision pairs near the Xe ($5p^5 6s$ $^{1,3}P_1 - 5p^6$ $^1S_0$) and Kr ($4p^5 5s$ $^{1,3}P_1 - 4p^6$ $^1S_0$) resonance atomic lines. The potential curves of the quasimolecules Xe($5p^5 6s$) + He and Kr($4p^5 5s$) + He have been obtained with the use of the effective Hamiltonian and pseudopotential methods. Based on these potential curves the processes of quasimolecular emission of Xe*+He and Kr*+He mixtures have been considered and the spectral distributions $I(\hbar\Delta\omega)$ of photons emitted have been obtained in the framework of quasistatic approximation.

1. Interaction potential curves

The potential curves of the interaction of excited atoms Kr($4p^5 5s$) and Xe($5p^5 6s$) with He atoms in the ground state were calculated in [1, 2] in the framework of the method of the effective Hamiltonian in the formulation [3] and the pseudopotential method [3]. Previously, these potential curves have been already applied for the calculations of absorption spectra of Xe + He and Kr + He mixtures and the results obtained are in good agreement with the experimental data [4]. For the ground states the potential curves determined from the experimental data in [5] have been used.

The interaction potential curves for excited and ground states and the difference potentials are shown on figure 1 and figure 2 for for Kr + He and Xe + He respectivelly.

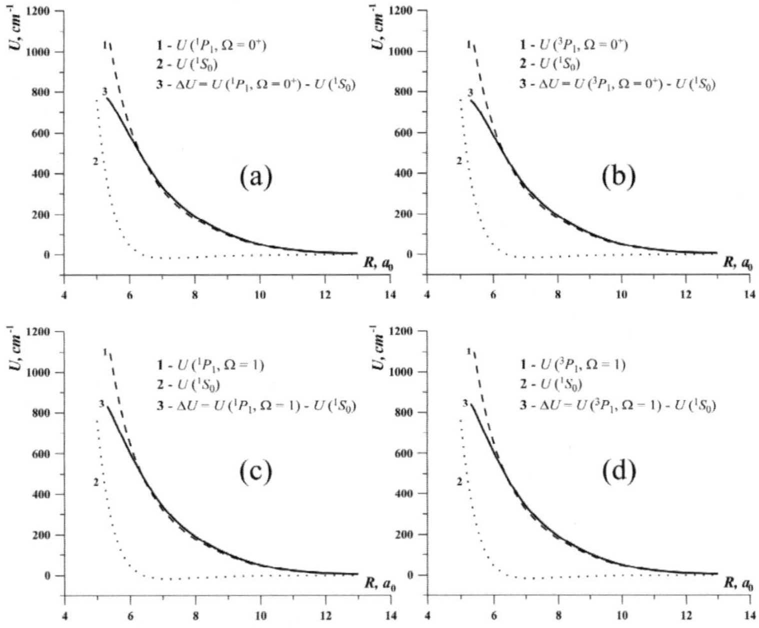

Figure 1. The interaction potential curves for the excited states $^1P_1\Omega=0^+$ (a), $^3P_1\Omega=0^+$ (b), $^1P_1\Omega=1$ (c) and $^3P_1\Omega=1$ (d) (lines 1, dashed), ground states (lines 2, dotted) and the corresponding difference potentials ΔU (lines 3, solid) for Kr-He quasimolecule.

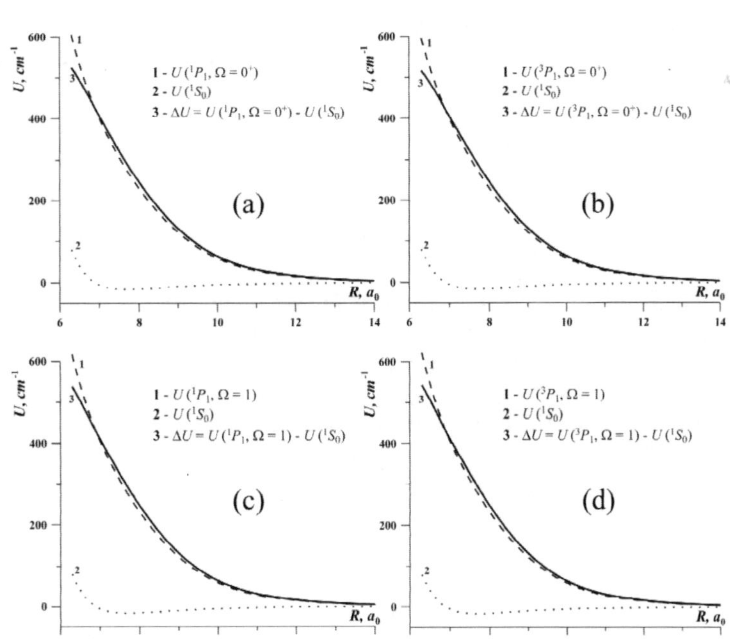

Figure 2. The interaction potential curves for the excited states $^1P_1\Omega=0^+$ (a), $^3P_1\Omega=0^+$ (b), $^1P_1\Omega=1$ (c) and $^3P_1\Omega=1$ (d) (lines 1, dashed), ground states (lines 2, dotted) and the corresponding difference potentials ΔU (lines 3, solid) for Xe-He quasimolecule.

2. Quasimolecular emission of Xe* + He and Kr* + He collision pairs

Based on the potential curves described in section 1 the processes of quasimolecular emission of Xe* + He and Kr* + He mixtures have been considered and the spectral distributions $I(\hbar\Delta\omega,\ \mathrm{cm}^5)$ of photons emitted have been obtained in the framework of quasistatic approximation [6]. The calculated emission spectra $\lg(I/I_0)$, where $I_0 = 10^{-35}\ \mathrm{cm}^5$, for the temperature T = 300 K are presented on figure 3 and figure 4 for Kr*+He and Xe*+He respectivelly.

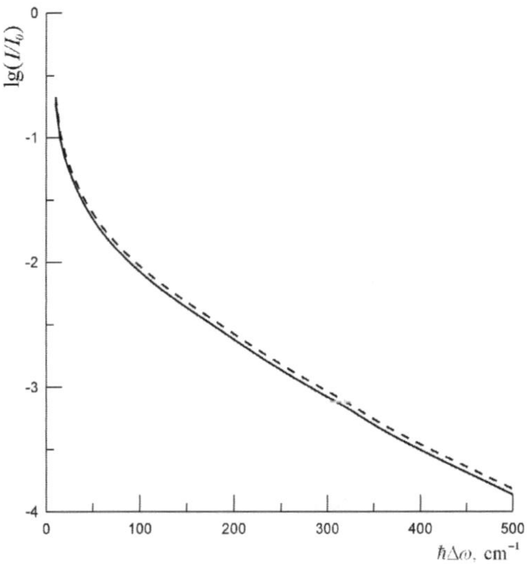

Figure 3. The emission spectra of Kr*+He near the resonance atomic lines $^1P_1 - {}^1S_0$ (solid line) and $^3P_1 - {}^1S_0$ (dashed line).

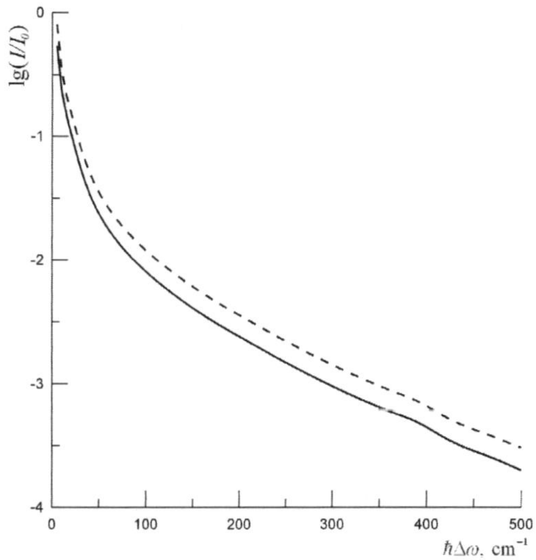

Figure 4. The emission spectra of Xe*+He near the resonance atomic lines $^1P_1 - {}^1S_0$ (solid line) and $^3P_1 - {}^1S_0$ (dashed line).

References

[1] Zagrebin A L and Pavlovskaya N A 1987 *Opt. Spectrosc.* **62 17**
[2] Zagrebin A L and Pavlovskaya N A 1989 *Opt. Spectrosc.* **66 465**
[3] Devdariani A Z, Zagrebin A L and Blagoev K B 1989 *Ann. Phys-Paris* **14** 464
[4] Alekseeva O S, Alekseev V A, Devdariani A Z, Lednev M G and Zagrebin A L 2010 *AIP Conf. Proc. (St. John's, Newfoundland, Canada)* vol 1290 (Melville, New York) p 231
[5] Keil M, Danielson L J and Dunlop P J 1991 *J. Chem. Phys.* **94** 296
[6] Gallagher A 1984 in: *Excimer Lasers (Top. Appl. Phys.* vol 30*)* ed Ch K Rhodes (Springer Berlin Heidelberg/Springer-Verlag) pp. 139-179

Speed-dependent Voigt profile parameters for oxygen B-band measured by cavity ring-down spectrometer referenced to the optical frequency comb

J Domysławska, S Wójtewicz, P Masłowski, A Cygan, K Bielska, R S Trawiński, R Ciuryło and D Lisak

Institute of Physics, Faculty of Physics, Astronomy and Informatics,
Nicolaus Copernicus University in Toruń, Grudziadzka 5, 87-100 Toruń, Poland

E-mail: jolka@fizyka.umk.pl

Abstract. We present the results of the line-shape analysis using the speed-dependent Voigt profile for the self-perturbed oxygen B-band resulting from the $b^1\Sigma_g^+ - X^3\Sigma_g^-(1 \leftarrow 0)$ transitions. The line-shape parameters data set contains the absolute positions, line intensities, pressure broadening, self-shift and speed-dependent line-shape coefficients.

1. Introduction

The Voigt profile (VP) is the most often used line-shape model for reproducing the atmospheric spectra. However, the use of the VP leads to systematic errors, even several percent, in the line intensities determination as well as collisional broadening and other line-shape coefficients. The data from our laboratory [1, 2, 3, 4] are so far the only complete measurements including all line-shape parameters determined simultaneously for oxygen B band. Line-shape analysis using models going beyond the VP revealed that taking into account the speed-dependent effects is sufficient to achieve sub-percent accuracy of line-shape parameters.

2. Experiment and data analysis

Self-broadened oxygen spectrum was measured at low pressure range by the PDH-locked FS-CRDS. The ring-down cavity was actively stabilized to a HeNe laser and the probe laser was locked to the cavity modes using the PDH method [5]. The CRD spectrometer was linked to an optical frequency comb (OFC) to enable absolute frequency measurements at every point of measured spectrum. Temperature and pressure of the gas sample were also measured at each point of the spectrum with relative uncertainties of 3.3×10^{-4} and 1×10^{-3}, respectively, and for analysis the mean values of these quantities were used. Natural isotopic abundance oxygen samples of high-purity (99.999%) were used in the measurements. The experimental data were analysed by means of the multispectrum fitting technique which reduces the numerical correlations between fitted line-shape parameters. For all line-shape models fitted to experimental profiles the Doppler width was set to the value corresponding to the gas temperature (approx. 0.95 GHz). The experimental setup details and specifics of the data analysis are described in subsequent papers [1, 2, 3, 4]. Besides the speed-dependent Voigt profile, data were analysed using several line profiles like the simple VP, profiles with Dicke

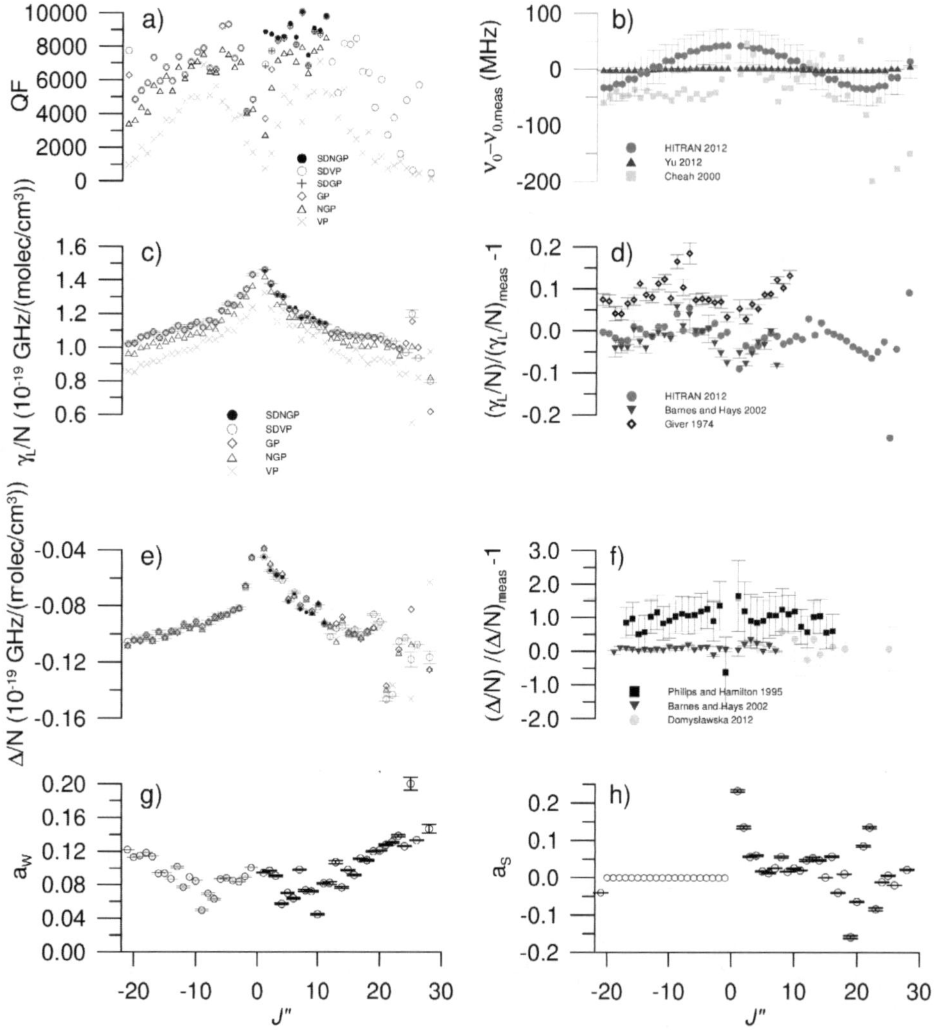

Figure 1. a) Quality-of-the-fit parameter for fitted profiles: VP, GP, NGP, SDVP, SDGP, SDNGP. b) – h) Line-shape parameters and comparison with other data. Details in text. J'' is the total angular momentum quantum number for the lower state. For clarity the P-branch is assigned with the negative J'' values.

narrowing: Galatry and Nelkin-Ghatak profiles, line profiles which simultaneously include Dicke narrowing and speed-dependent effect: speed-dependent Galatry and speed-dependent Nelkin-Ghatak profiles. For speed-dependence the quadratic and hypergeometric approximations were considered [1]. A review of the line-shape models can be found in Ref. [6].

3. Results and discussion

The numerical values of the fitted quadratic speed-dependent Voigt (qSDV) profile [7] for whole B band were collected and presented previously in Ref. [4]. Other profiles were thoroughly analysed in previous works for P branch [2] and R branch [1, 3]. Here we collected data and presented them on the graphs only. Figure 1a) shows the quality-of-the-fit (QF) parameter for fitted profiles. The most extensive analysis was performed for the first eleven lines of the R branch [1, 3]. It is clearly seen that the QF for VP is much lower than for other profiles, also NGP gives visibly worse QF values for every measured transition. The GP fits results in the QF values close to that of the qSDVP but lower in general, although it may not be clear in Fig. 1a). Only for the weaker lines which can be measured in higher pressure range the QF values are visibly different for various profiles. The results for more advanced profiles including Dicke narrowing and speed-dependent effects are not conclusive for stronger self-broadened transitions of the B band. Figure 1b) shows the comparison of line positions determined in our laboratory with other values [8, 9, 10]. In general they are consistent within 50 MHz while the combined standard uncertainty for our values is better than 0.5 MHz and for most of them is 170 kHz or better. Figures 1c) and 1d) show the results of the self-broadening coefficients and comparison with other data [8, 11, 12], respectively. It is clearly seen than the VP and NGP results are lower from other data by about 10% and 3%, respectively. The other data are different from ours by few percent or more. Figures 1e) and 1f) show the self-shift coefficients and comparison with previous data [13, 11, 14], respectively. Results for different profiles are consistent with each other except for the transitions from R branch with higher J" values due to the overlapping of the weak lines with the stronger ones. Figure 1g) and 1h) show values for the speed-dependent coefficients a_W and a_S. For the P branch fits were made with a_S set to 0 [2] except for the line with the highest J" value. No other data is available in the literature to compare with.

4. Conclusions

The quadratic speed-dependent Voigt profile was chosen to analyse experimental data. Line intensities, pressure width and shift coefficients and the speed-dependent parameters were determined with sub-percent accuracy. As shown previously [1, 2, 3, 4] the relatively simple qSDVP gives high quality of the fit in the range of investigated pressures and is consistent with IUPAC recommendation [15] of the line profile for atmospheric applications.

Acknowledgements

The research is part of the program of the National Laboratory FAMO in Toruń, Poland, and is supported by the Polish National Science Centre Projects nos. DEC-2012/05/D/ST2/01914, DEC-2013/11/D/ST2/02663, DEC-2014/15/D/ST2/05281 and DEC-2015-/17/B/ST2/02115. The research is also supported by the Foundation for Polish Science TEAM Project co-financed by the EU European Regional Development Fund.

References

[1] Domysławska J, et al. 2013 *J. Chem. Phys.* **139** 194312
[2] Wójtewicz S, et al. 2014 *J. Quant. Spectrosc. Radiat. Transfer* **144** 36–48
[3] Domysławska J, et al. 2015 *J. Quant. Spectrosc. Radiat. Transfer* **155** 22–31
[4] Domysławska J, et al. 2016 *J. Quant. Spectrosc. Radiat. Transfer* **169** 111–21
[5] Drever R W P, et al. 1983 *Appl. Phys.* B **31** 97–105
[6] Ciuryło R 1998 *Phys. Rev.* A **58** 1029–39
[7] Berman P R 1972 *J. Quant. Spectrosc. Radiat. Transfer* **12** 1331-42
[8] Rothman L S, et al. 2013 *J. Quant. Spectrosc. Radiat. Transfer* **130** 4–50
[9] Yu S, Miller C E, Drouin B J, Müller H S P 2012 *J. Chem. Phys.* **137** 024304
[10] Cheah S-L, Lee Y-P, Ogilvie J F 2000 *J. Quant. Spectrosc. Radiat. Transfer* **64** 467–82
[11] Barnes J E, Hays P B 2002 *J. Molec. Spectrosc.* **216** 98–104
[12] Giver L P, Boese R W, Miller J H 1974 *J. Quant. Spectrosc. Radiat. Transfer* **14** 793–802
[13] Phillips A J, Hamilton P A 1995 *J. Molec. Spectrosc.***174** 587–94
[14] Domysławska J, et al. 2012 *J. Chem. Phys.* **136** 024201
[15] Tennyson J, et al. 2014 *Pure Appl. Chem.* **86** 93143

Molecular dynamics simulations of collision-induced absorption: Implementation in LAMMPS

W Fakhardji and M Gustafsson

Applied Physics, Division of Materials Science, Department of Engineering Science and Mathematics, Luleå University of Technology, SE-97187 Luleå, Sweden

E-mail: wissam.fakhardji@ltu.se, magnus.gustafsson@ltu.se

Abstract. We pursue simulations of collision-induced absorption in a mixture of argon and xenon gas at room temperature by means of classical molecular dynamics. The established theoretical approach (Hartmann et al. 2011 J. Chem. Phys. **134** 094316) is implemented with the molecular dynamics package LAMMPS. The bound state features in the absorption spectrum are well reproduced with the molecular dynamics simulation in comparison with a laboratory measurement. The magnitude of the computed absorption, however, is underestimated in a large part of the spectrum. We suggest some aspects of the simulation that could be improved.

1. Introduction

Collision-induced absorption (CIA) can occur when interacting molecules acquire an electric dipole moment. The absorption happens even if the individual molecules are spectroscopically inactive. For example, argon (Ar) or xenon (Xe) atoms do not absorb in the microwave and far infrared spectral region. In an Ar–Xe gas mixture, however, the absorption due to interacting Ar–Xe pairs is observable. Note that the (homonuclear) Ar–Ar pairs and the Xe–Xe pairs do not contribute to the absorption since their symmetry is inconsistent with a dipole moment.

In planetary and cool stellar atmospheres CIA is often a major source of opacity. In the four gas giants of our solar system, the main atmospheric constituents are hydrogen and helium. CIA in those gases have been studied much both experimentally and theoretically [1]. Rigourous quantum mechanical calculations of CIA are possible for those low mass systems and they are an important complement to the excellent, but difficult and rare, lab measurements. In atmospheres of smaller planets more massive molecules like N_2, CH_4, and CO_2 are the most abundant. With molecular masses that high quantum mechanical calculations of absorption coefficients are challenging and, in fact, often not even necessary. Instead methods based on

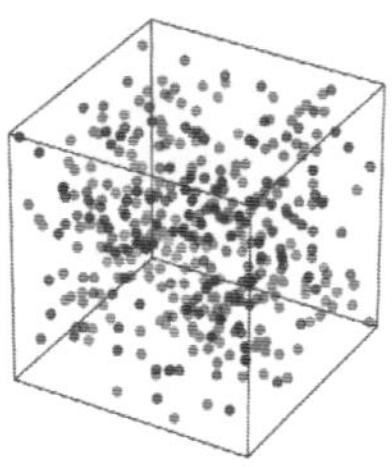

Figure 1. The molecular dynamics simulation is done for a cubic volume containing two atomic species. As the particles move around in the box during the simulation one has to somehow take care of those reaching the boundary. In MD this is commonly done by using periodic boundary conditions so that a particle "lost" through one box side instantly reenters the volume in the same position, but at the opposite wall, with unchanged momentum.

classical mechanics may be used, like the bimolecular approach of Ref. [2] or molecular dynamics (MD) [3]. The latter is a many-body approach and thereby has the advantage that it explores all phase space so that bound complexes, so-called dimers, can form and dissociate through three body encounters. All atomic and molecular pairs form dimers and they are generally abundant in a gas at atmospheric temperature and pressure conditions. In this work we test an implementation of the MD approach for a gas mixture of argon and xenon.

2. Theory and computational approach

The basic idea of MD is to simulate the motion of the particles in a section of the medium that is big enough so that macroscopic observables may be obtained, but small enough so that it is feasible to solve the equations of motion numerically. We are concerned with a gas mixture of Ar and Xe atoms in a cubic volume, V, as illustrated in Fig. 1. The particles interact with each other through pair potentials for the combinations Ar–Xe, Ar–Ar, and Xe–Xe, which are taken from Refs. [5, 6, 7], respectively. The calculations are done as illustrated in Fig. 2. The absorption coefficient [3]

$$\alpha(\omega, T) = \frac{\pi\omega}{3V\hbar c\epsilon_0} e^{-\hbar\omega/2k_BT}[1 - e^{-\hbar\omega/k_BT}]\, C(\omega) \tag{1}$$

is computed from the Fourier transform

$$C(\omega) = \frac{1}{\pi}\int_0^{\infty} cos(\omega t)\, \phi(t)\, dt \tag{2}$$

of the dipole autocorrelation function $\phi(t) = \langle \vec{\mu}(t) \cdot \vec{\mu}(0) \rangle$. $\vec{\mu}(t)$ is the time-dependent interaction-induced dipole moment for an Ar–Xe pair and $\langle ... \rangle$ indicates the canonical ensemble average over all pairs in the volume V (although the division by number of pairs, $\sim \rho_{\mathrm{Ar}}\rho_{\mathrm{Xe}}$, is taken care of later). The magnitude of $\vec{\mu}$ as a function of pair separation is displayed in Fig. 3. It is the model dipole that is used in this work and the ab inito data is shown to justify this choice. In Eq. (1) ω is the angular frequency, ϵ_0 is the vacuum permittivity, $\hbar$ is Planck's constant, k_B is Boltzmann's constant, and T is temperature.

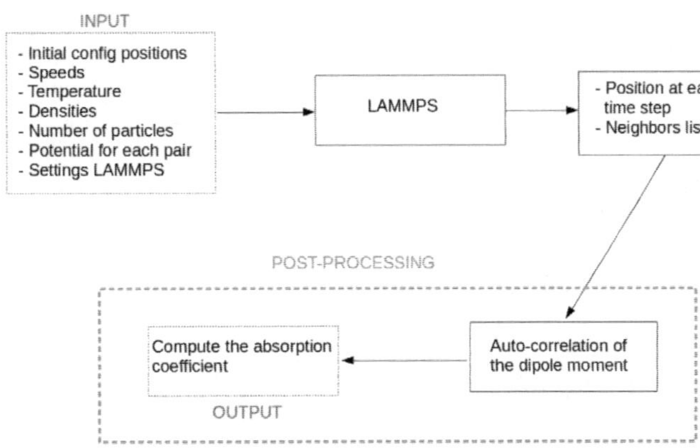

Figure 2. Schematic diagram showing how the MD simulations of collision-induced absorption coefficients are done in this work. The MD itself is done with the program LAMMPS [4] while the post-processing is done using an in-house python program.

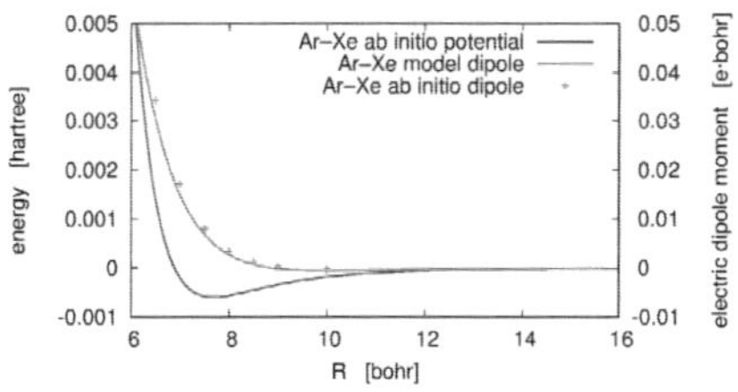

Figure 3. The Ar–Xe interaction potential [5] together with model [8] and *ab initio* [9] interaction-induced dipole moment, $\mu(R)$, versus the interatomic distance, R. The ab initio dipole is the MP2 calculation of Ref. [9] and here the values are displayed as the negative of those in the original work due to a difference in sign convention.

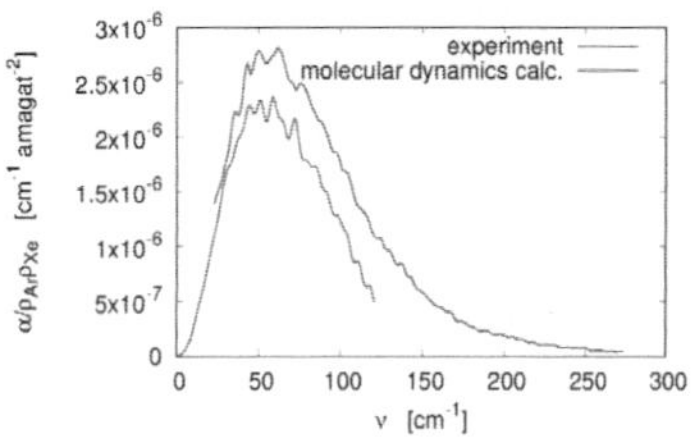

Figure 4. The collision-induced absorption coefficient at 292 K in a gas mixture of argon and xenon, normalized by the product of densities in units of amagat. The calculated data from the present molecular dynamics study is the lower curve. The experimental data from Ref. [8] (upper curve) is displayed for comparison.

We run the molecular dynamics simulations with the temperature held constant by applying a thermostat. Thermal equilibrium is achieved by allowing a thermalization period of 5 ps before trajectories are stored for the calculation of the spectrum. There are 10^5 atoms of each kind and the size of the box is chosen so that the total number density is 1 amagat, which equals about $2.7 \cdot 10^{19}$ atoms per cm^3. The densities of Ar and Xe are thus $\rho_{Ar} = \rho_{Xe} = 0.5$ amagat.

3. Results and discussion

Figure 4 shows a comparison between the present results and the laboratory measurement by Ref. [8]. The oscillatory features with widths on the order of 10 cm^{-1} are due to dimers that are dissociated by the radiation. These features are well produced in the calculation. There is a discrepancy, however, in the magnitude of the absorption, which is underestimated by around 20% by the calculation at the peak absorption. At higher frequencies the difference is larger. Likely there is room for improvement in the calculation and we will, for example, try with larger number of particles or time averaged autocorrelation function. We conclude that MD simulations of CIA with LAMMPS have great potential since the program is well optimized and documented, and has great flexibility.

Acknowledgments

This work was supported by the Grauduate School of Space Technology at Luleå University of Technology and the COST Action CM1405 Molecules In Motion (MOLIM).

References

[1] Frommhold L 1993 and 2006 *Collision-induced Absorption in Gases* (Cambridge, New York: Cambridge University Press)
[2] Levine H B and Birnbaum G 1967 *Phys. Rev.* **154** 86
[3] Hartmann J M, Boulet C and Jacquemart D 2011 *J. Chem. Phys.* **134** 094316
[4] Plimpton S 1995 *J. Comp. Phys* **117** 1 – 19 http://lammps.sandia.gov
[5] Aziz R A and van Dalen A 1983 *J. Chem. Phys.* **78** 2402 – 2412
[6] Aziz R A 1993 *J. Chem. Phys.* **99** 4518 – 4525
[7] Aziz R A and Slaman M 1986 *Molec. Phys.* **57** 825 – 840
[8] Grigoriev I M, Tonkov M V and Frommhold L 1998 *Phys. Rev. A* **58** 4978 – 4980
[9] Haskopoulos A and Maroulis G 2006 *J. Math. Chem.* **40** 233 – 241

Modified Complex Robert-Bonamy (MCRB) calculations of H_2O transitions broadened by H_2 for applications to planetary and exoplanet atmospheres

Robert R. Gamache[1] and Candice L. Renaud[1]

[1]Department of Environmental, Earth, and Atmospheric Sciences, University of Massachusetts Lowell, Lowell, MA, USA

Robert_Gamache@uml.edu

Abstract. Line shape parameters for hydrogen broadening of water vapor are needed to understand remote sensing measurements of planetary and exoplanet atmospheres. In order to address these needs, semiclassical calculations based on the Modified Complex Robert-Bonamy (MCRB) formalism were made. The intermolecular potential for the calculation is comprised of electrostatic, atom-atom (expanded to order 16 and rank 4), induction, and London dispersion terms. The trajectories were determined by numerical integration of the Hamilton's equations. The average over the Maxwell–Boltzmann distribution of velocities was performed by integration over 35 velocities corresponding to the temperature range 75K – 27000K. The formalism is complex valued yielding the half-width and line shift from a single calculation. The calculations are reported at 7 temperatures from 200 to 700 K. The half-width temperature dependence coefficient n was determined using the relation $\gamma(T)=\gamma(T_0)\left[T_0/T\right]^n$ with T_0=296K. The calculations are compared with the measurements of Brown and Plymate [JQSRT **56**, 263, 1996]. Future plans for further refinement of the intermolecular potential are discussed.

1. Introduction

Atmospheres surrounding planets in our solar system have been known and studied since the 19th century[1]. Some early studies of planetary atmospheres were the spectroscopic identification of methane in the atmospheres of the giant planets[2], carbon dioxide on the terrestrial planets[3], and the spectroscopic detection of an atmosphere on Titan[4]. Starting from these early observations, the basic physics and chemistry of planetary atmospheres have been established.

In the mid 1990s exoplanets orbiting sun-like stars were first observed using radial velocity detection. Because of this detection technique many of the exoplanets discovered in the first few years orbited exceedingly close to the host star. Astronomers tell us that statistically every star in the Milky Way Galaxy should have at least one planet[5] and that small rocky planets are extremely common[6,7]. The Milky Way Galaxy has 100 billion stars, and our Universe has upwards of 100 billion galaxies. Thus, based on sheer probability the chance for life elsewhere seems inevitable. Currently there are over 3000 known exoplanets; see http://exoplanet.eu/catalog.php for details.

A variety of methods[8] were developed in the past two decades to discover and characterize exoplanets. Part of the development are techniques to study exoplanet atmospheres. Spectra have been measured for a handful of exoplanets and broadband spectrophotometry has been used to study dozens more. For many researchers the ultimate goal is the search for habitable exoplanets. The exoplanet atmosphere is the only way to infer whether or not a planet is habitable or likely inhabited.

Over fifty years ago it was realized that signs of life could be recognized on a distant planet by remotely sensing gases in the planet's atmosphere[9,10]. Gases produced by life, called biosignature gases, can accumulate in a planet atmosphere to detectable levels. These gases provide a means by which we can deduce the possible existence of life on an exoplanet. Thus, the planetary atmosphere is our window into temperatures, habitability indicators, and biosignature gases.

The list of possible biosignature gases is growing and there is a need to determine which molecules provide the best opportunities for detection. This task is accomplished by modeling using the radiative transfer equations. As light travels from the parent star through the atmosphere of the exoplanet, it acquires molecular features from the exoplanet due to the composition, temperature, and pressure structure of the atmosphere.

Currently measurements are being made by space borne satellites, e.g. the Hubble Space Telescope (HST) and the Spitzer Space Telescope (SST), and by ground based telescopes, such as the Keck II 10-m telescope, the Cerro Tololo Inter-American Observatory, or the 10.4 m Gran Telescopio Canarias. These platforms have instruments that operate at low to moderate resolution. For example on the HST the Near Infrared Camera and Multi-Object Spectrometer (NICMOS) has resolving power ($R=\Delta\lambda/\lambda$) ~ 15[11], the Space Telescope Imaging Spectrograph (STIS) has R=500[12], and the Wide Field Camera 3 (WFC3) has R=130[12]. The Infrared Spectrograph on the Spitzer Space Telescope comprises four separate spectrograph modules covering the wavelength range from 5.3 to 38 μm with spectral resolutions, R = 90 and 600[13].

In 2018 the James Webb Space Telescope (JWST) will be launched. JWST (jwst.stsci.edu) instruments include the Near-Infrared Spectrograph (NIRSpec) and the Mid Infrared Instrument (MIRI), which will measure the infrared line emission by gaseous molecules such as H_2O, CO, CO_2, OH and a growing list of organics. These instruments have a much improved resolving power (R~2700) that will separate individual lines in complex molecular bands and allow for accurate measurements of weak lines critical for constraining chemical models, including those of rare isotopologues.

While the current measurements are of moderate resolution, it is important to note that the interpretation of these spectra are and have been done using line by line radiative transfer models, hence all spectroscopic parameters are needed. Madhusudhan and Seager[14] developed a temperature and abundance retrieval method for exoplanet atmospheres based on parametric pressure–temperature (P -T) profile coupled with line-by-line radiative transfer, hydrostatic equilibrium, and energy balance, along with prescriptions for non-equilibrium molecular composition and energy redistribution. This method has become the standard and has been used by many to model exoplanet atmosphere spectra[15-20]. However, in the line by line model much of the line shape data are poorly estimated.

In order to understand the measured spectra, the spectral parameters for the gases in the atmosphere must be known. Of the spectroscopic parameters the line shape is the least well known. Here we address the line shape of water vapor broadened by hydrogen as determined via semi-classical calculations.

2. Modified Complex Robert-Bonamy formalism

Semi-classical line shape calculations using the MCRB[21] formalism were made for 274 ν_2 transitions to compare with the measurements of Brown and Plymate[22]. The intermolecular potential is comprised of electrostatic components, an atom-atom potential, and the vibrational dependence of the isotropic potential, which uses the induction and London dispersion potentials. Water is a light asymmetric rotor with typical energy gaps for collisionally induced transitions in the range of 0 to 500 cm^{-1}. Hydrogen is a diatomic molecule with a B value of 61.264895 cm^{-1}. Figure 1 shows the average energy gaps as a function of J. The figure suggests that the H_2O and H_2 energy gaps will seldom match leading to mostly off-resonance collisions. This fact hints that the atom-atom potential will play a large role in the determination of the line-shape parameters.

The initial atom-atom potential (pot 00) used combination rules to determine the atom-atom coefficients. Four additional runs were made increasing each atom-atom coefficient by 10%, pot 01-04, to determine the effect of the atom-atom coefficients on the half-widths. Figure 2 shows the difference between the pot 01-04 calculations and pot 00 versus an energy ordered Index.

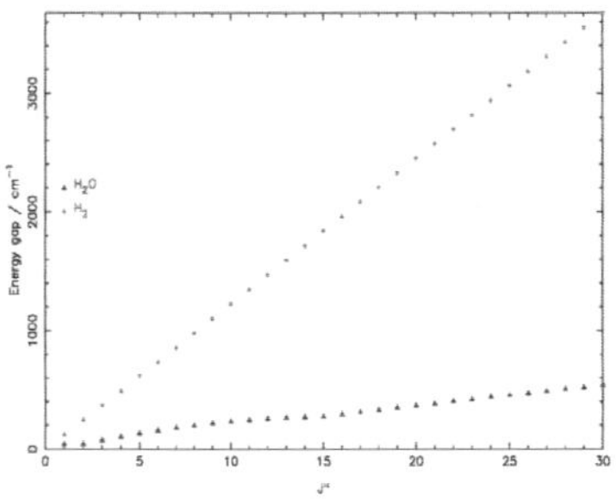

Figure 1 Average energy gaps for collisionally induced transitions for H_2O (blue) and H_2 (red)

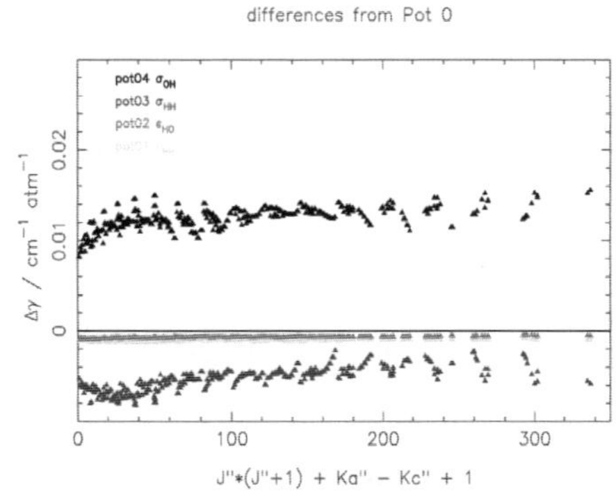

Figure 2 difference between the pot 01-04 calculations and pot 00 versus an energy ordered index.

The fact that there is structure in Figure 2 suggest adjusting the atom-atom coefficients will affect the computed half-widths differently for each transition. Ten iterations were made. In the final potential, the epsilon values were not changed; σ_{HH} is 8.36% lower and σ_{OH} is 10.% higher than the pot 00 values. Figures 3 and 4 show the comparison of the half-width measurements of Brown and Plymate[12] with the pot 00 and pot 10 calculations, respectively. The pot 00 comparison has an average percent difference of -59.75 and a standard deviation of 23.3%. The pot 10 comparison has an average percent difference of -0.11 and a standard deviation of 8.2%.

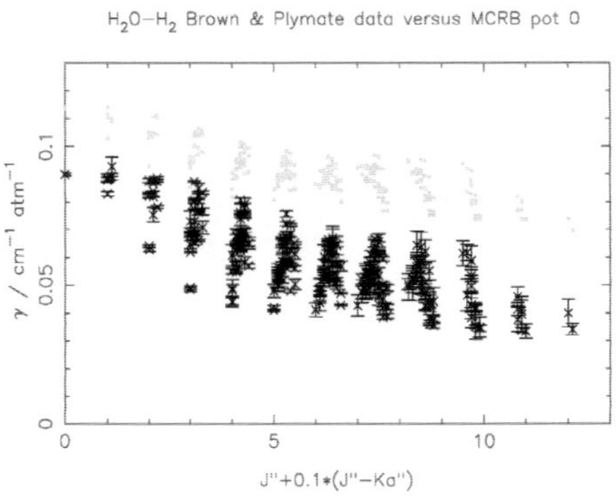

Figure 3 H_2O-H_2 B&P data versus MCRB pot 00

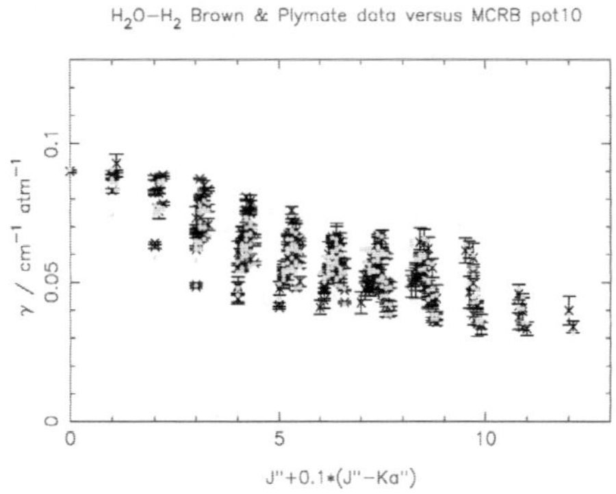

Figure 4 H_2O-H_2 B&P data versus MCRB pot 10

The intermolecular potential will be further refined using results of an intercomparison of all measurements of H_2O-H_2 broadening, the *family of transitions* idea of Brown and Plymate[11], and the partner transition and smooth variation rules of Ma *et al.*[23].

The authors are pleased to acknowledge support of this research by the National Science Foundation through Grant Nos. AGS-1156862 and AGS-1622676 . Any opinions, findings, and conclusions or recommendations expressed in this material are those of the author(s) and do not necessarily reflect the views of the National Science Foundation.

References

[1] Chalis J 1863 MNRAS **23**, 231
[2] Adel A, Slipher VM 1934 Phys. Rev. **46** 902
[3] Adel A 1937 ApJ. **85** 345
[4] Kuiper G 1944 ApJ. **100** 378
[5] Cassan A, et al. 2012 Nature **481**(7380) 167
[6] Fressin F, et al. 2013 Astrophys J **766** 81
[7] Marcy G W, et al. 2014 Proc Natl Acad Sci USA **111** 12655
[8] Seager S 2011 *Exoplanets*, ed Seager S (Univ of Arizona Press, Tucson, AZ)
[9] Lederberg J 1965 Nature **207**(992) 9
[10] Lovelock JE, Nature **207**(997), 568, 1965
[11] Sivaramakrishnan R M 2000 Proc. SPIE, UV, Optical, and IR Space Telescopes and Instruments **4013** 386
[12] Huitson C M 2013 MNRAS; **434** 3252
[13] Houck J R , 2004 The Astrophysical Journal Supplement Series; **154** 18
[14] Madhusudhan N, Seager S. 2009 The Astrophysical Journal **707** 24
[15] Antígona L K, Subhanjoy S 2011 The Astrophysical Journal **733** 35
[16] Ranjan S, et al. 2014 Ap. J. **785** 148
[17] Stevenson K B, et al. 2014 Astronomical Journal **147** 161
[18] Swain M R, Vasisht G, Tinett G 2008 Nature **452** 329
[19] Swain M R, et al. 2009 The Astrophysical Journal **704** 1616
[20] Swain M R, et al. 2009 Ap. J. **690** L114
[21] Ma Q, Tipping R H, Boulet C 2007 JQSRT **103** 588
[22] Brown L R, Plymate C 1996 JQSRT **56** 263
[23] Ma Q, Tipping R H, Lavrentieva N N 2011 Mol. Phys. **109** 1925

A study of atmospheric aerosol optical properties over Alexandria city- Egypt

E E Kohil, I H Saleh and Z F Ghatass

Department of Environmental Studies, Institute of Graduate Studies and Research, Alexandria University, 163 Horrya Avenue, P.O. 832 Shatby, Alexandria, Egypt

E-mail: z_ghatass@yahoo.com

Abstract. Aerosols are minute particles suspended in the atmosphere. When these particles are sufficiently large, we notice their presence as they scatter and absorb sunlight. They scatter and absorb optical radiation depending upon their size distribution, refractive index and total atmospheric loading. Aerosol optical depth (AOD) was measured at Alexandria city (31° 16′ N, 30° 01′ E and 21 m above sea level) using hand-held microprocessor-based sun photometer "MICROTOPS II". AOD is studied at five different wavelengths from 380 to 1020 nm during the period from Aug-2015 to Feb-2016. Precipitable water column (PWC) is estimated from the measurements of solar intensity at 936 and 1020 nm. Diurnal, monthly and seasonal variation of AOD and water vapor content was studied during the study period. The seasonal variation of AOD has high value (0.416) in summer and low value (0.176) in winter at wavelength of 380 nm. The changes in the PWC have been found to be correlated with changes in AOD. This is supported by the observed increase of AOD with relative humidity (RH) values.

1. Introduction

Aerosols play an important role in earth's radiation budget, air quality and environmental health [1-3]. Aerosols are tiny particles in solid or liquid phase suspended in the air. They occur over a wide range extended from 10^{-2} µm to about 10^{2} µm; the size range 0.1-1 µm are most effective in attenuating sunlight [4]. There are two sources of aerosols; natural source and anthropogenic source. Natural source such as volcanic activity, windblown dust, sea spray, convective and general circulations produce regional-scale effects of modulating background aerosol, while the anthropogenic sources of aerosols are combustion, industrial activities, transport and mining to regional-scale differences in aerosol properties. One of the most important optical properties of aerosol used in radiative transfer calculation is aerosol optical depth (AOD). It is defined as the attenuation of direct solar radiative passing through the atmosphere by scattering and absorption due to aerosols [5]. The aim of this study is to measure AOD and PWC by using MICROTOPS II sunphotometer (Solar light co.) at Alexandria City. Probably the local aerosols are a mixture of aerosols originating from windblown dust, vehicular sea bubbles burst, as the region is a coastal and crowded with population as well as it is near from the desert.

2. Materials and method

2.1. Site description

This study was carried out at Alexandria city. Alexandria is a coastal city (31° 16' N, 30° 01' E and 21m above sea level) and it is an urban area. It has 40% of total industry in Egypt. It extends along the coast of the Mediterranean Sea in the north-central Egypt. It has a pleasant Mediterranean climate. August is the hottest month has an average temperature reaching 31°C, while January is the coldest month has an average temperature reaching to 18°C. The summer season is hot and the average of the humidity varies from 58.2 to 48.8%. The site of the study is an urban area, so it is very crowded and the pollution load is high compared with rural areas.

2.2. Methodology

MICROTOPS II is a five channel hand-held sunphotometer which is used for measuring AOD easily, accurately and dependably. The MICROTOPS II sunphotometer provides the aerosol optical thickness at the standard wavelengths 380, 440, 500, 936, and 1020 nm. It also measures direct solar irradiance at the five wavelengths and PWC at 936 and 1020 nm [6].GPS Model Garmin (72H) was used. Accurate location parameters can be automatically transferred from the GPS receiver to the MICROTOPS II. MICROTOPS II works only when the sky is clear otherwise an error will occur, so clear days were chosen (no clouds or mist in the sky). The measurements were carried out one day per week during the study period (Aug-2015 to Feb-2016). The diurnal and monthly variations were calculated during the study period. Meteorological data (humidity and wind speed and direction) were downloaded from the internet.

3. Results and discussion

Measuring Aerosols distributed within a column of air from earth's surface to the top of the atmosphere was made by MICROTOPS II which calculates the aerosol optical depth (AOD) values at each wavelength based on the channels, its extraterrestrial constant, atmospheric pressure (for Rayleigh scattering), time and location. Sun distance correction is automatically applied. All calculations of AOD are based on the Boughuer- Lambarert-Beer law at the following formula,

$$\tau_\lambda = \left(\frac{\ln(V_{o\lambda}) - \ln(V_\lambda * \mathrm{SDCORR})}{M}\right) - (\tau_{R\lambda}) * \frac{P}{P_o}$$

Where λ refers to the selected wavelength, ln ($V_{o\lambda}$) refers to the AOD calibration constant of the selected wavelength, V_λ is the signal intensity at the selected wavelength in (mV), SDCORR is the mean Earth-sun distance correction, M is the optical air mass, $\tau_{R\lambda}$ is Rayleigh optical depth at the selected wavelength and P and P_o are the atmospheric pressure at the study region and sea level pressure in (mbar) respectively [6].

PWC is measured through two wavelengths, the first one is 936 nm which water absorption band and the second is 1020 nm which is no water absorption band. The following equation that measures PWC based on Bouguer- Lambarert- Beer law and it is measured in cm scale [7].

$$W = \left(\frac{\ln V_{o(936)} - \ln V_{(936)} - \tau_{(936)} * M}{K * M^b}\right)^{1/b}$$

Where K and b are constants numerically derived for the filter.

3.1. Diurnal variation

'Figure 1' shows the diurnal variation of aerosol optical depth at five different wavelengths (380, 440, 500, 936 and 1020 nm), while 'Figure 2' shows the diurnal variation of PWC at wavelengths 936 and 1020 nm during the study period.

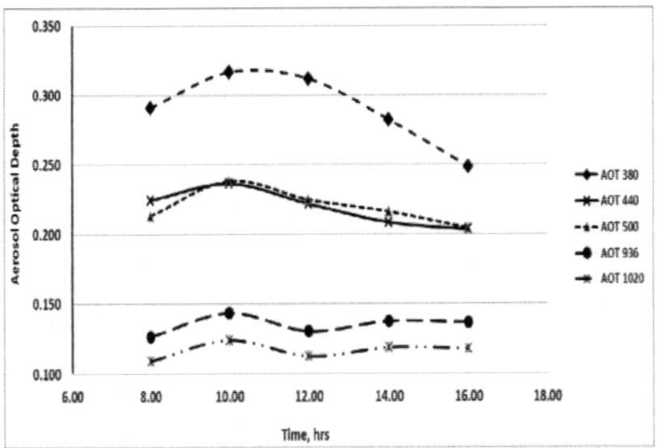

Figure 1. Diurnal variation of AOD at the five wavelengths during the study period

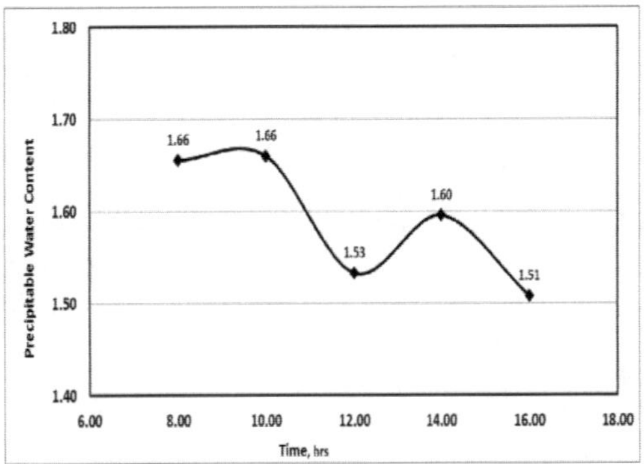

Figure 2. Diurnal variation of PWC during the study period

The maximum value of AOD was at 10 am while the minimum value was 4 pm. This trend is consistently seen at all wavelengths. AOD values also show comparatively high values at 10 am, then a slight decrease, but the values increased again, this may be due to increasing of the temperature and the human activity as well. 'Figure 2' shows the that the PWC along the day has variable values as it is very high at the morning (0.166 at 8 and 10 am) and very low at the afternoon (0.151 at 4 pm), this may be due to raise in the temperature, so water vapor evaporated.

3.2. Monthly Variation

Monthly values of AOD were calculated during the study period at mentioned wavelengths as shown in 'Figure 3'. It can be seen that the highest values was at Aug-15 (0.416 at 380 nm) while the lowest values was at Jan-16 (0.176 at 380 nm). The trend was decreasing regularly till Jan-16. High values of AOD during the summer may be due to increasing of the heat of the surface of the earth, because of the increasing of the temperature, which makes evaporation to molecules and this molecules raise in the atmosphere and this, was observed during the high temperature periods. The lowest values of AOD were obtained in winter at Jan-16 and this may be due to cloud scavenging and rain wash out processes and decreasing the temperature and this ensures the previous interpretation.

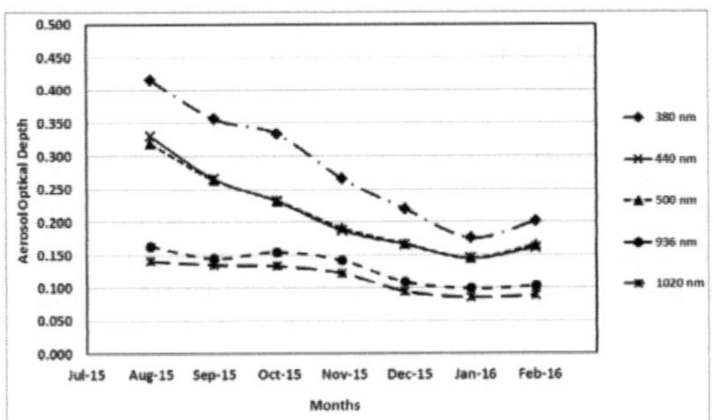

Figure 3. Monthly variations of AOD at the five wavelengths along the study period

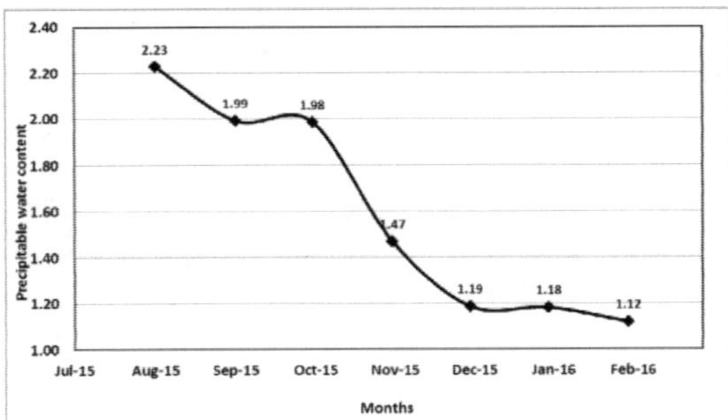

Figure 4. Monthly variation of mean values of PWC along the period

'Figure 4' shows the monthly variation of precipitable water content for the study period. The highest value was at Aug-15 (2.23 cm), while the lowest value was Feb-15 (1.12 cm). Trend of precipitable water content is as well as the AOD trend because both are related to digital voltage (V) and its calibration constant $\ln(V_0)$. Generally the precipitable water content values are high in summer and low in winter this is may be due to hygroscopic growth of AOD and the humidity is very high in summer which lets some liquid particles suspended in the atmosphere.

3.3. AOD Spectral dependence and Ångstrom parameters

AOD spectral dependence was studied during the study period at the five wavelengths as shown in 'Figure 5'. This study observed that AOD values are inversely proportional with the wavelength. There is an enhancement in AOD values between the two wavelengths; 440 and 500 nm; this is may be due to water vapor interference, clouds or sunlight [8].

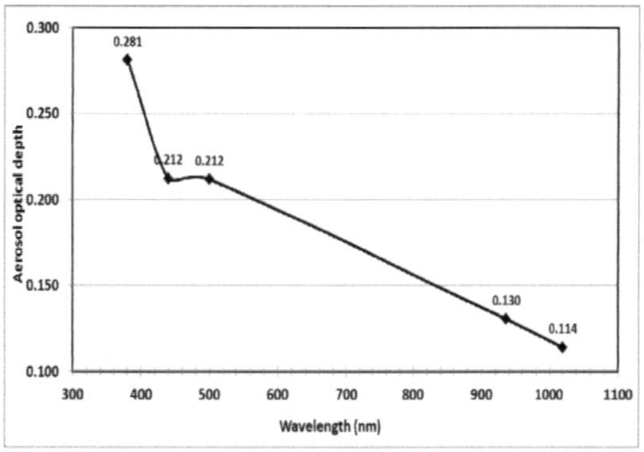

Figure 5. AOD Spectral dependence for the study period

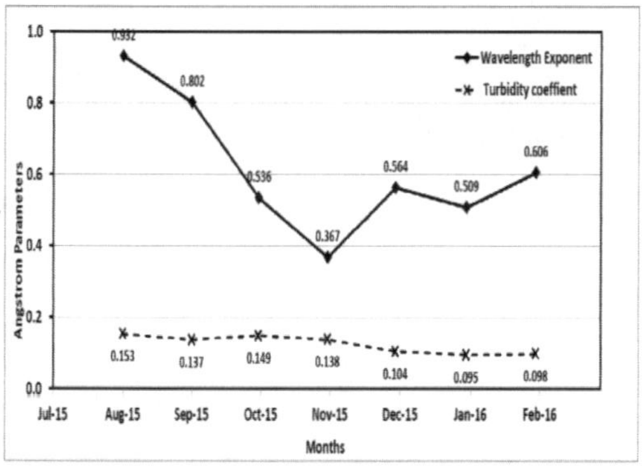

Figure 6. Ångstrom parameters; wavelength exponent and turbidity coefficient during the study period

Ångstrom parameters; i.e. wavelength exponent α and turbidity coefficient β, were calculated as shown in 'Figure 6'. In this study α are less than 2.0 along the study period, so the aerosol is coarse along this period but differs from month to another. Turbidity coefficient (β) values are varied between 0.1 and 0.2 during Aug-2015 to Nov-2015, so the atmosphere is not clear during this period, but β values are less than 0.1 during the period from Dec-15 to Feb-16, so the atmosphere is clear as expected because of rains, this interpretation is due to standard Ångstrom parameters [9].

3.4. Meteorological conditions

It was observed from this study that all meteorological conditions affect the AOD values and PWC values. Temperature, humidity and wind speed and direction is directly proportional, but the atmospheric pressure is inversely proportional with the value of AOD and PWC. Meteorological data; humidity, wind speed and direction were collected from the internet [10], but the temperature and the atmospheric pressure were collected by the instrument itself.

4. Conclusion

Aerosols are tiny particles suspended in the air in solid or liquid phase. Aerosol values are high during the summer months, but AOD values are low during the winter and the post monsoon rains due to rain wash out. PWC values tend to follow the AOD values, as it is high during the summer due to raising of the humidity and low during the winter. Meteorological conditions; temperature, pressure, humidity and/or wind speed and direction, have an effect on the values of AOD and PWC. There is an enhancement in the spectral dependence between the two wavelengths; 400 and 500nm because of interference with water vapor, clouds or sunlight according to NASA, earth observatory.

References

[1] Prospero J M Charlson R J Mohnen B Jaenicke R Delany A C Mayers J and Rahn K 1983 *J. Geophys. & Sp. Phys.* **21** pp 1607-1629

[2] Satheesh S K and Ramanathan V 2000 *J. Letters to Nature* **405** pp 60-63

[3] Pinker R T Pandithurai G Holben B N Dubovik, O and Aro T O 2001 *J. Geophys. Res.* **106** pp 22,923-22,930

[4] Ranjan R R Ganguly N D Joshi H P and Iyer K N 2007 *Ind. J. Rad. & Sp. Phys.* **36** pp 27-32

[5] Ranjan R R Joshi H P and Iyer K N 2007 *J. Aer. & Air Qu. Res.* **7 (1)** pp 33-45

[6] Solar Light Co. 2003 *User's Guide of MICROTOPS II* Philadelphia USA **5.5** p 5-15

[7] Morys M Forest M Mims III Hagerup S Anderson S E Baker A Kia J and Walkup T 2001 *J. Geophys. Res.* **106** pp 14,573–14,582

[8] NASA (2007). *http://earthobservatory.nasa.gov/IOTD/view.php?id=7469*

[9] Akpootu D O and Momoh M 2013 *Nig. J. basic & app. Sci.* **21(1)** pp 70-78

[10] Website: *"www.wundrground.com"*

Collisional shift of hyperfine line for thallium in an atmosphere of the buffer inert gases

A V Glushkov, V F Mansarliysky, O Yu Khetselius, A V Ignatenko, A V Smirnov and G P Prepelitsa

Odessa State Environmental University, L'vovskaya str.15, Odessa-16, 65016, Ukraine

E-mail: dirac13@mail.ru

Abstract. A relativistic approach, based on the relativistic many-body perturbation theory (PT) with using the optimized Dirack-Kohn-Sham wave functions sets, is applied to computing the hyperfine structure collision shift for the thallium atom in atmosphere of the inert gases (A=He,Ar, Xe). Data for the collisional shifts of the Tl-A system are presented and compared with other available ones.

1. Introduction

This paper goes on our work on theoretical studying hyperfine lines shift and broadening for heavy atoms in an atmosphere of the buffer gases (see, for example [1-5]). Let us remind that studying the collisional shifts and broadening of the hyperfine structure lines for heavy elements (alkali, alkali-earth, lanthanides, actinides and others) in an atmosphere of inert gases is one of the important and actual topics of the modern quantum chemistry, atomic optics and spectroscopy [6,7]. Besides, calculation of the hyperfine structure line shift and broadening allows one to check the quality of wave functions and study a contribution of relativistic effects in two-center (multi-center) atomic systems. From the applied science point of view, the mentioned physical effects form a basis for creating an atomic quantum measure of frequency. The detailed non-relativistic theory of the hyperfine line collisional shift and broadening for simple elements (alkali elements etc) was developed by many authors (see Refs. [1,6-8]). However, consideration of heavy elements faces serious difficulties related with account for relativistic and correlation effects. In our paper a new relativistic optimized approach, based on the gauge-invariant perturbation theory (PT) with using the optimized Dirac-Kohn-Sham wave functions basis's and the exchange PT (modified version EL-HAV [1,7]), is applied to computing the hyperfine line collision shift for thallium atom in atmosphere of the inert gases. The basic expressions for the collision shift are taken from the kinetic theory of spectral lines [1,6].

2. Hyperfine lines shift for system "alkali atom- inert gas"

As the method of computing hyperfine lines shift is earlier presented in details [1-4], here we are limited only by the key topics. To calculate the collision shift of hyperfine structure spectral lines one could use the following expression known from kinetic theory of spectral line form (see [6]):

$$f_p = \frac{\Delta}{p} = \frac{4\pi N_b}{p} \int_0^\infty e^{-U(R)/kT} \left[1 + g(R)\right] \delta\omega(R) R^2 dR \tag{1}$$

Here *U(R)* is an effective central interatomic interaction potential of a system A-B (for example, A=Tl, B=He;*)*; *T* is a temperature, w_0 is a frequency of the hyperfine structure transition in the isolated active atom; $d\omega\ (R)=Dw\ (R)/w_0$ is the relative local shift of the hyperfine structure lines, which is arisen due to the disposition of the active atoms on distance *R*, $\{1+g(R)\}$ is the temperature factor. To calculate interatomic interaction potential, the exchange PT method (modified version EL-HAV; see [1-3,7]) is used. Within exactness to 2nd order terms on potential of Coulomb interaction of the valent electrons and atomic cores one can write:

$$\delta\omega(R)=\frac{S_0}{1-S_0}+\Omega_1+\Omega_2-\frac{C_6}{R^6}\left(\frac{2}{\overline{E}_a}+\frac{1}{\overline{E}_a+\overline{E}_B}\right), \tag{2}$$

where C_6 is the van der Waals constant for the A-B interaction; *I*, $E_{Ia,b}$ are the ionization potential and excitation energy for atoms A,B correspondingly; S_0 is the overlapping integral. The values Ω_1, Ω_2 in Eq.(2) are the non-exchange and exchange non-perturbation sums of the first order correspondingly:

$$\Omega_1=\frac{2}{N(1-S_0)\rho_0}\sum_k \frac{\langle\Phi_0'(1)|H_{CT}'|\Phi_k'(1)\rangle V_{k0}}{E_0-E_k},$$

$$\Omega_2=\frac{2}{N(1-S_0)\rho_0}\sum_k \frac{\langle\Phi_0'(1)|H_{CT}'|\Phi_k'(1)\rangle U_{k0}}{E_0-E_k}, \tag{3}$$

$$\rho_0=\langle\Phi_0^1(1)|H_{CT}'|\Phi_0'(1)\rangle/\langle\Phi_0^1(1)|\Phi_0'(1)\rangle$$

where H'_{CT} is the operator of hyperfine interaction, *N* is the total number of electrons taken into account in calculation; E_k, $\Phi_k'(1)=F_{k_a}'(1)\varphi_{k_b}(2...N)$ — energy and non-symmetrized wave function of state $k=\{k_a,k_b\}$ for isolated atoms A and B. The non-exchange matrix element of the Coulomb interatomic interaction is as follows: $V_{ko}=\langle\Phi'_k(1)|V(1)|\Phi'_0(1)\rangle$. Correspondingly the exchange matrix element is: $U_{k0}=\sum_{i=2}^{N}\langle\Phi_k'(1)|V(i)|\Phi_0'(i)\rangle$. Operator *V (i)* is (for the system Tl-He) as follows:

$$V(i)=U_{SCF}(r_{a3})+U_{SCF}(r_{a4})-2U_{SCF}(R)+\frac{1}{r_{bi}}, \tag{4}$$

where $U_{SCF}(r)$ is the self-conjunctive field, created by the thallium tomic core.
Let us return to consideration of the van der Waals constant C_6 for the interatomic A-B interaction. As a rule, one uses the approximate values for the van der Waals constant C_6 etc. Often the sufficiently great mistake in determination of the van der Waals constants provides non-high accuracy of the inter-atomic potentials calculation and further inaccuracies. The van der Waals constant may be written as [1]:

$$C_6(L,M)=C_{6,0}(L)-\frac{3M^2-L(L+1)}{(2L-1)(2L+3)}\cdot C_{6,2}(l), \tag{5}$$

where $C_{6,0}$ *(L)* is the isotropic component of the interaction and $C_{6,2}$ *(L)* is the component corresponding to the $P_2(\cos\theta)$ term in the expansion of the interaction in Legendre polynomials. The coefficients $C_{6,0}$, $C_{6,2}$ may be expressed in terms of the scalar and tensor polarizabilities evaluated at imaginary frequencies [8]. Usually (see [6]) the non-relativistic Hartree-Fock sets of the wave functions are used. More correct approach is based on using the relativistic Dirac-Fock (DF) wave functions or wave functions as the solutions of the Dirac equations with model potential (the Ivanov-Ivanova type) [2-10]). Here paper we used the set of the relativistic functions, generated by the Dirac-Kohn-Sham equation with the optimized exchange-correlation functional (see details in [11-19]).

IOP Conf. Series: Journal of Physics: Conf. Series **810** (2017) 012034 doi:10.1088/1742-6596/810/1/012034

3. Results

In table 1 we list our data for the shift f_p (Hz/Torr) for the Tl-He, Ar, Xe pairs. The observed value of the line shift (T=700K) and other theoretical data, obtained with using the standard DF, optimized DF methods [2,3,6] are listed too. A difference between our theoretical data and other calculation data is explained by using the different sets of wave functions and different PT schemes. For other temperatures there are no quite precise data. It is obvious that using the gauge-invariant optimized sets of the wave functions and correct version of the exchange PT will be necessary in a case of the heavy elements in an atmosphere of more heavy inert gases.

Table 1. The shift f_ρ (Hz/Torr) of the Tl hyperfine line for the pairs: Tl-He, Kr, Xe (T=700 K).

System	Tl-He	Tl-Ar	Tl-Xe
Experiment	130 ± 30	-490±20	-1000±80
Theory [6]	155.0	-850.0	-1420.0
Theory [2]	137.2	-504	-1052
Theory [3]	135.4	-501	-1044
This work	136.6	-503	-1048

References

[1] Malinovskaya S V, Glushkov A V, Khetselius O Yu, Mischenko E V, Loboda A V and Svinarenko A A 2009 *Int. J. Quant. Chem.* **109** 3031

[2] Khetselius O Yu, Glushkov A V, Gurnitskaya E P, Loboda A V, Mischenko E V, Florko T A and Sukharev D E 2008 *Spectral Line Shapes* (AIP) **15** 231

[3] Khetselius O Yu, Florko T A., Svinarenko A A, Tkach T B 2013 *Phys.Scr.* **T153** 014037

[4] Florko T , Ambrosov S , Svinarenko A , Tkach T 2012 *J. of Phys.: Conf. Ser.* **397** 012037

[5] Glushkov A V, Khetselius O Yu, Lopatkin Y M, Florko T A, Kovalenko O A, Mansarliysky V F 2014 *J. of Phys.: Conf. Ser.* **548** 012026

[6] Batygin V V and Sokolov I M 1983 *Opt. Spectr.* **55** 30

[7] Kaplan I G 1985 *Theory of intermolecular interactions* (Moscow: Nauka)

[8] Chi X, Dalgarno A, Groenenborn G C 2007 *Phys.Rev.A.* **75** 032723

[9] Glushkov A V, Khetselius O Yu, Svinarenko A A 2013 *Phys. Scr.* **T153** 014029

[10] Glushkov A V and Ivanov L N 1992 *Phys. Lett.* A **170** 33; 1993 *J. of Phys. B* **26** L379

[11] Glushkov A V, Khetselius O Yu and Lovett L, 2009 *Adv. in Theory of Atomic and Molecular Systems Dynamics, Spectroscopy, Clusters, and Nanostructures, Ser: Progress in Theor. Chem.& Phys.*, ed P Piecuch, J Maruani, G Delgado-Barrio, S Wilson (Springer) **20** 125

[12] Malinovskaya S V, Glushkov A V, Khetselius O Yu, Svinarenko A A, Mischenko E V, Florko T A 2009 *Int.Journ.Quant.Chem.* **109** 3325

[13] Malinovskaya S V, Glushkov A V, Khetselius O Yu, Loboda A V, Lopatkin Yu, Nikola L, Svinarenko A A, Perelygina T 2011 *Int.Journ.Quant.Chem.* **111** 288

[14] Svinarenko A A 2014 *J.of Phys.: Conf. Ser.* **548** 012039

[15] Svinarenko A A, Khetselius O Yu, Buyadzhi V V, Florko T A, Zaichko P A, Ponomarenko E 2014 *J. of Phys.: Conf. Ser.* **548** 012048

[16] Serga I N, Dubrovskaya Yu V, Kvasikova A S, Shakhman A N, Sukharev D E 2012 *J. of Phys.: Conf. Ser.* **397** 012013

[17] Malinovskaya S V, Dubrovskaya Yu V, Vitavetskaya L A 2005 *AIP Conf. Proc.* **796** 201

[18] Glushkov A V, Malinovskaya S V 1988 *Rus. J. Phys. Chem.* **62**(1) 100

[19] Glushkov A V, Khetselius O Yu, Malinovskaya S V 2008 *Europ. Phys. Journ. ST* **160** 195; 2008 *Mol. Phys.* **106** 1257

IOP Conf. Series: Journal of Physics: Conf. Series **810** (2017) 012035 doi:10.1088/1742-6596/810/1/012035

VIPA spectrometer calibration and comb-cavity locking schemes comparison for sensitive and accurate frequency comb spectroscopy

G Kowzan[1], K F Lee[2], M Borkowski[1], P Ablewski[1], S Wójtewicz[1], K Stec[1], D Lisak[1], M E Fermann[2], R S Trawiński[1], P Masłowski[1]

[1]Institute of Physics, Faculty of Physics, Astronomy and Informatics, Nicolaus Copernicus University in Toruń, ul. Grudziadzka 5, 87-100 Toruń, Poland
[2]IMRA America, Inc., 1044 Woodridge Ave., Ann Arbor, MI 48105, USA

E-mail: gkowzan@fizyka.umk.pl

Abstract. In direct frequency comb spectroscopy with a VIPA spectrometer, the resolution of the spectrometer is usually insufficient to resolve the comb modes. Thus, one can either filter the modes, reducing the density of spectral elements or cope with the inability to uniquely identify spectral elements with individual comb modes by calibrating the spectrometer itself. Here, we present a way to make use of the inherent frequency accuracy of a stabilized frequency comb to calibrate the spectrometer. We also present a comparison between two commonly used schemes to stabilize the coupling between a frequency comb and cavity resonances: the Pound-Drever-Hall locking scheme and the swept coupling scheme.

In direct frequency comb spectroscopy (DFCS) the comb beam interacts directly with a gas sample and its broadband spectrum is dispersed to retrieve the absorption profile [1–3]. Being a laser spectroscopy technique, it offers high sensitivity [4] and high resolution [5], but also combines acquisition time reaching down to tens of microseconds with broad spectral bandwidth, limited by the spectrum of the comb itself. Owing to this combination of desirable features, it has found applications in chemical kinetics experiments [6], cold spectroscopy of complex molecules [7], ultrafast and sensitive pump-probe spectroscopy [8] and biomedical process monitoring [9], among others. Moreover, frequency conversion through nonlinear processes, enabled by the high peak powers of the mode-locked lasers, is frequently used to transfer these advantages to spectral regions from ultraviolet to mid-infrared [10–12].

Among the varied approaches to dispersing the broadband femtosecond optical frequency comb (OFC) spectrum [13–17], virtually-imaged phased array (VIPA) spectroscopy is unique in providing tens of nanometers coverage, microsecond-level acquisition time and spectral element density sufficient to measure the shapes of individual resonances [18–23]. The drawback of this approach is the inability to uniquely identify spectral elements with individual comb modes, which necessitates some form of spectrometer calibration. Here we present a calibration scheme which makes use of the inherent stability of the OFC and a comparison between OFC-cavity coupling schemes commonly used in cavity-enhanced DFCS experiments.

The comparison of the coupling schemes and VIPA spectrometer calibration was performed with the setup shown in Fig. 1. The OFC is generated by an Er:fiber-based mode-locked laser with the repetition rate (f_{rep}) of 250 MHz. The coupling of the comb into the cavity is accomplished

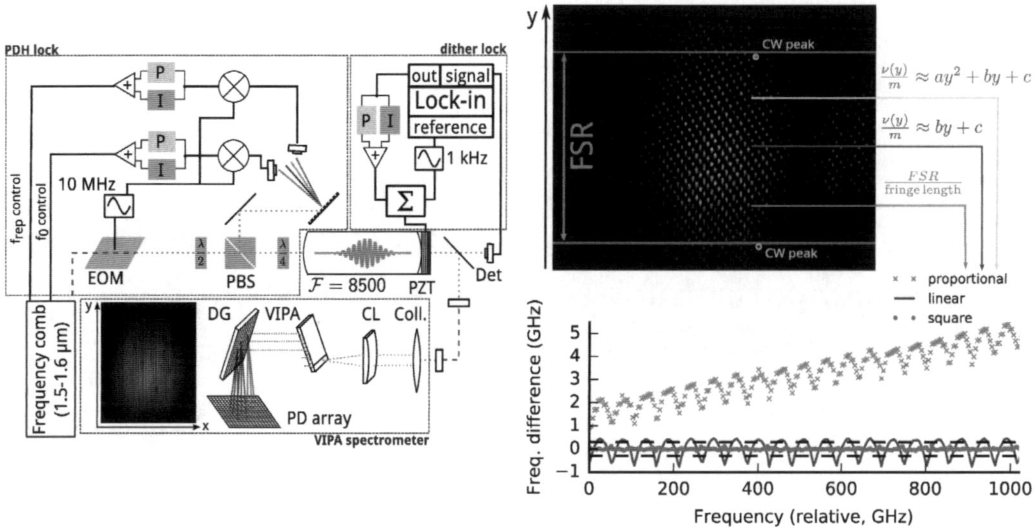

Figure 1. Left: Schematic of the spectrometer system. Solid black lines are electrical connections, dashed blue lines are fiber links and dotted red lines are free-space beams. EOM, electro-optic modulator; PZT, piezoelectric transducer; Det, InGaAs photodetector; f_0, offset frequency; f_{rep}, repetition rate; $\frac{\lambda}{2}$, half-wave plate; $\frac{\lambda}{4}$, quarter-wave plate; PBS, polarizing beam splitter; $\mathcal{F}$, Finesse; Coll., collimator; CL, cylindrical lens; DG, diffraction grating; PD array, HgCdTe photodetector array. Right: Upper part of the figure illustrates the construction of the frequency axis (details in the text). Lower part shows differences between frequencies corresponding to peak positions of comb teeth, as determined by the calibrated frequency axis, and the actual frequencies of comb teeth. The region between the dashed lines covers twice the frequency span corresponding to a single pixel, approximately equal to 600 MHz.

either with the Pound-Drever-Hall (PDH) locking scheme [24, 25] or with the dither coupling scheme [26]. In the PDH coupling scheme the OFC produced by the laser is tightly locked to cavity resonances. This is achieved by producing two error signals from different regions of the spectrum and using them to lock the f_{rep} and the offset frequency (f_0) to the cavity [27]. One of the error signals is fed to an electrooptic modulator inside the laser resonator and a piezoelectric transducer, upon which the resonator mirror is mounted, to lock the f_{rep} of the OFC; the second error signal is fed to the pumping diode current driver to lock the f_0 of the OFC. In the swept coupling scheme the OFC is stabilized to an Rb frequency reference and the absorption cavity resonances follow the comb lines. The absorption cavity length is modulated in the vicinity of comb-cavity resonance with a 1 kHz sine wave and the cavity transmission signal is recorded by a photodetector. Lock-in detection of the transmission signal referenced to the modulation signal is proportional to cavity length mismatch and constitutes a good error signal. This signal is added to the modulation signal and stabilizes the coupling of the comb to the cavity. The transmitted beam is coupled into a fiber and the collimated output is introduced into the VIPA spectrometer [28].

The VIPA etalon [29] used here has free spectral range (FSR) of 53.4 GHz and resolution of 600 MHz. The broadband spectrum of the OFC is folded by the etalon into a single stripe comprising multiple overlapping interference orders; the diffraction grating (DG) is used to separate these orders and produce a two-dimensional diffraction image. The dispersion of the grating is important inasmuch as it resolves the orders, so for the spectrometer calibration procedure only the VIPA dispersion law [29] is important:

$$\frac{\nu(y)}{m} = g(y) = 2ct\left(n_r \cos\theta_v - \frac{\tan\theta_v \cos\theta_{vn} y}{f} - \frac{\cos\theta_v y^2}{2n_r f^2}\right)^{-1}, \tag{1}$$

where y is the spatial direction in which the dispersion occurs in the detection plane (as shown on the right of Fig. 1), c is the speed of light in vacuum, ν is the optical frequency, m is the interference order, t is the etalon thickness, n_r is the index of refraction, f is the focal length of imaging lens, and θ_v and θ_{vn} are the incident and the propagation angle inside the etalon, respectively. It is clear from the equation that the dispersion not only changes with the interference order, but also changes within the order. This also implies that the FSR is fringe-dependent and that resolution is a function of wavelength.

Since in the dither scheme we can couple the OFC spectrum with the cavity resonances, while stabilizing it to an Rb frequency reference, we use the frequency comb to calibrate the spectrometer. The comb modes are filtered by the cavity in the Vernier scheme [19] and resolved by the spectrometer, as shown in the top right part of Fig. 1. The mode number of a single comb tooth is identified by measuring the beat of the OFC with a CW laser, which is at the same time imaged on the camera and whose frequency is measured independently with a wavelength meter. Three different dispersion law models are then fitted to the known frequencies of imaged comb modes and the frequency axis obtained from the calibration is compared with the frequencies of the comb lines. The proportional model, which neglects the order-dependence and the square term fares the worst, showing a drift of the frequency axis and parabolic distortions in every order; the linear model removes the drift, but keeps the parabolic distortions; the square model agrees with the actual comb frequencies within the uncertainties caused by the comb mode position fitting procedure.

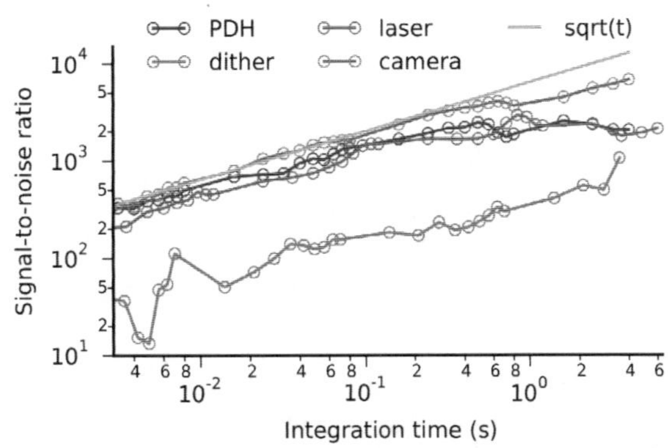

Figure 2. Signal-to-noise ratio of a region in the most intense part of the spectrum for the PDH (blue) and dither (red) locking schemes, amplified frequency comb (green) and camera (magenta).

The calibrated spectrometer was used to compare noise properties of acquired spectra in both coupling schemes, for amplified OFC bypassing the cavity and for camera noise. Signal-to-noise ratio (SNR) used for the comparison was calculated by: dividing two spectra taken with the same gas mixture, incident spectrum, and averaging time; selecting a region with the highest light intensity; and taking the reciprocal of the standard deviation of this reduced array. The noise-equivalent absorption (NEA) was calculated analogously, except in the last step the standard deviation was divided by the effective optical path length (L_{eff}). The L_{eff} for small-absorption cavity-enhanced measurements is well-approximated by $L_{\text{eff}} = \beta\mathcal{F}L/\pi$, where $\beta = 2$ for the PDH lock and $\beta = 1$ for the dither lock [26, 30, 31]. Figure 2 shows that the PDH lock noise closely follows the amplified OFC's behavior. The dither locking scheme starts with very high noise, but shows quick improvement. This is caused by high shot-to-shot variability in measured power and spectrum. The large variance means that the incident power has to be kept lower than in the PDH case, which reduces the dynamic range and the maximum obtainable SNR. Another point of comparison between the schemes is the NEA (per spectral element (p.s.e) [32]). For the PDH lock, the minimum NEA of 9.9×10^{-10} cm^{-1} and 7.4×10^{-11} cm^{-1} Hz$^{-1/2}$ p.s.e. (6085 elements) is obtained for an integration time of 2×1.6 s (2×17 s measurement time). For the dither lock, the minimum NEA of 5.3×10^{-9} cm^{-1} and 1.5×10^{-9} cm^{-1} Hz$^{-1/2}$ p.s.e (3320 elements) is obtained for an integration time of 2×2.8 s (2×133 s measurement time). The significant difference between measurement and integration time is caused by the high level of thermal noise of HgCdTe photodetectors, preventing longer integration times.

Acknowledgements
Research is supported by the National Science Centre, Poland, Project No. 2015/18/E/ST2/00585. P.M. and G.K. are supported by the Foundation for Polish Science HOMING PLUS project.

References
[1] Adler F, Thorpe M J, Cossel K C and Ye J 2010 *Annu. Rev. Anal. Chem.* **3** 175–205
[2] Foltynowicz A, Masłowski P, Ban T, Adler F, Cossel K C, Briles T C and Ye J 2011 *Faraday Discuss.* **150** 23–31
[3] Masłowski P, Cossel K C, Foltynowicz A and Ye J 2014 Cavity-Enhanced Direct Frequency Comb Spectroscopy *Cavity-Enhanced Spectroscopy and Sensing* (*Springer Series in Optical Sciences* vol 179) ed Gagliardi G and Loock H P (Berlin, Heidelberg: Springer Berlin Heidelberg) p 271 ISBN 978-3-642-40002-5 978-3-642-40003-2
[4] Foltynowicz A, Ban T, Masłowski P, Adler F and Ye J 2011 *Phys. Rev. Lett.* **107** 233002
[5] Maslowski P *et al.* 2016 *Phys. Rev. A* **93** 021802
[6] Bjork B J *et al.* 2016 *arXiv:1608.07321 [physics]* (*Preprint* 1608.07321)
[7] Spaun B, Changala P B, Patterson D, Bjork B J, Heckl O H, Doyle J M and Ye J 2016 *Nature* **533** 517–520
[8] Reber M A R, Chen Y and Allison T K 2016 *Optica* **3** 311
[9] Gołkowski M, Gołkowski C, Leszczynski J, Plimpton S R, Masłowski P, Foltynowicz A, Ye J and McCollister B 2012 *Plasma Sci. IEEE Trans. On* **40** 1984–1991
[10] Adler F, Masłowski P, Foltynowicz A, Cossel K C, Briles T C, Hartl I and Ye J 2010 *Opt. Express* **18** 21861
[11] Leindecker N, Marandi A, Byer R L, Vodopyanov K L, Jiang J, Hartl I, Fermann M and Schunemann P G 2012 *Opt. Express* **20** 7046
[12] Khodabakhsh A, Ramaiah-Badarla V, Rutkowski L, Johansson A C, Lee K F, Jiang J, Mohr C, Fermann M E and Foltynowicz A 2016 *Opt. Lett.* **41** 2541
[13] Mandon J, Guelachvili G and Picqué N 2009 *Nat. Photonics* **3** 99–102
[14] Thorpe M J, Moll K D, Jones R J, Safdi B and Ye J 2006 *Science* **311** 1595–1599
[15] Khodabakhsh A, Johansson A C and Foltynowicz A 2015 *Appl. Phys. B* **119** 87–96
[16] Coddington I, Newbury N and Swann W 2016 *Optica* **3** 414
[17] Gambetta A, Cassinerio M, Gatti D, Laporta P and Galzerano G 2016 *Sci. Rep.* **6** 35541
[18] Thorpe M J, Balslev-Clausen D, Kirchner M S and Ye J 2008 *Opt. Express* **16** 2387
[19] Diddams S A, Hollberg L and Mbele V 2007 *Nature* **445** 627–630
[20] Fleisher A J, Bjork B J, Bui T Q, Cossel K C, Okumura M and Ye J 2014 *J. Phys. Chem. Lett.* **5** 2241–2246
[21] Gresh D N, Cossel K C, Zhou Y, Ye J and Cornell E A 2016 *J. Mol. Spectrosc.* **319** 1–9
[22] Klose A, Ycas G, Cruz F C, Maser D L and Diddams S A 2016 *Appl. Phys. B* **122** 1–8
[23] Scholten S K, Anstie J D, Hébert N B, White R T, Genest J and Luiten A N 2016 *Opt. Lett.* **41** 1277
[24] Black E D 2001 *Am. J. Phys.* **69** 79–87

[25] Drever R W P, Hall J L, Kowalski F V, Hough J, Ford G M, Munley A J and Ward H 1983 *Appl. Phys. B* **31** 97 105

[26] Thorpe M J and Ye J 2008 *Appl. Phys. B* **91** 397–414

[27] Foltynowicz A, Masłowski P, Fleisher A J, Bjork B J and Ye J 2012 *Appl. Phys. B* **110** 163–175

[28] Xiao S and Weiner A M 2004 *Opt. Express* **12** 2895

[29] Xiao S, Weiner A M and Lin C 2004 *Quantum Electron. IEEE J. Of* **40** 420–426

[30] Kowzan G *et al.* 2015 *Photonics Lett. Pol.* **7** pp–78

[31] Romanini D, Ventrillard I, Mejean G, Morville J and Kerstel E 2014 Introduction to Cavity Enhanced Absorption Spectroscopy *Cavity-Enhanced Spectroscopy and Sensing* (*Springer Series in Optical Sciences* vol 179) ed Gagliardi G and Loock H P (Berlin, Heidelberg: Springer Berlin Heidelberg) p 1 ISBN 978-3-642-40002-5 978-3-642-40003-2

[32] Gherman T and Romanini D 2002 *Opt. Express* **10** 1033

IOP Conf. Series: Journal of Physics: Conf. Series **810** (2017) 012036 doi:10.1088/1742-6596/810/1/012036

Cavity Ring-Down Spectroscopy measurements of Acetone concentration

G Revalde[1,3], J Alnis [1,2], E Nitišs [4], K Blušs[2], K Grundšteins[1]

[1]Ventspils University College, Inzenieru str. 101, Ventspils, Latvia;
[2] Institute of Atomic Physics and Spectroscopy, University of Latvia, Skunu str. 4, Riga, Latvia;
[3]Institute of Technical Physics, Faculty of Materials Science and Applied Chemistry, Riga, Technical university, P.Valdena 3,Riga, Latvia;
[4] Institue of Solid State Physics, Kengaraga str., Riga, Latvia.

E-mail: gitar@latnet.lv

Abstract. The Cavity Ring-Down Spectroscopy system for medical applications is created and first results for detection of acetone are shown. The acetone concentration was measured in the air. In the system a pulsed 1 kHz 266 nm DPSS laser with pulse length of 10 ns is coupled in a gas cell with high reflectivity mirror.

1. Introduction

Acetone in human breath gas has been established as a biomarker for type-1 diabetes (T1D) [1] and different other diseases. Normal human breath contains many volatile organic compounds (VOC) in low concentration. The Cavity Ring-Down spectroscopy (CRDS) is sensitive technique to detect VOC in the air or human breath in real time however there are no commercial systems to detect broad range of VOCs. We report in this publication the first measurements of acetone concentration by the CRDS system build in Latvia.

2. Results and discussion

The acetone concentration in the air was measured using the Cavity Ring-Down Spectroscopy (CRDS) technique. System is assembled on the optical breadboard. The picture of the CRDS system is shown in Fig.1. In the vacuum part we use Teflon pipes, metal valves, Teflon thread *Locktight 55*, three stage oil-free membrane pump *Ilmvac*, 5.5 purity grade nitrogen. Resonator can be evacuated to 0.2 mbar. In the system a pulsed 1 kHz 266 nm DPSS laser with pulse length of 10 ns is coupled in a gas cell with high reflectivity mirrors attached to both ends. The cell operates as a resonator in which the laser pulse decays with a specific decay rates.

The decay rate is determined by the distance between the mirrors, mirror reflectivity coefficient and absorptivity of the gas at 266 nm.

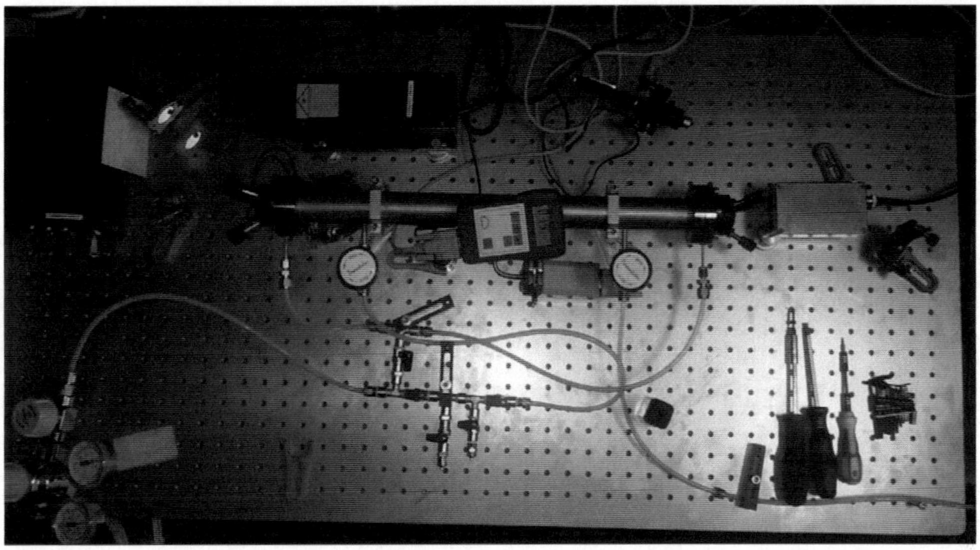

Figure 1. The set-up of the cavity ring-down system (CRDS) for acetone concentration measurements in air.

An example of pulse decay for empty cell (vacuum of 0.3 mbar) is shown in Fig. 2. The light intensity is captured using a high-speed avalanche photodiode. Afterwards an exponential fit in the form f(t)=A*exp(-t/τ)+B is done, where the A is the amplitude of the signal, B is the intercept and τ is the decay rate. In the experimental data curve small oscillations are visible. These are caused by different modes which travel at different speeds in the resonator.

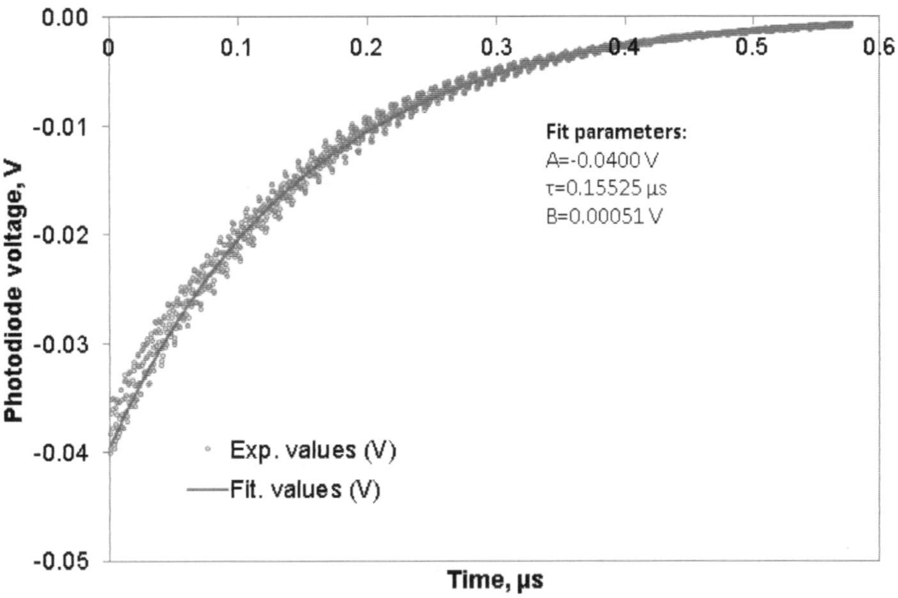

Figure 2. The decay of laser pulse in the resonator cell measured at low pressure of 0.3 mbar.

The decay rate was captured as the cell was consecutively filled to 1000 mbar with air with low amount of acetone (30 ppm and 5 ppm respectively), high purity nitrogen gas as well as vacuumed (air pressure of around 0.3 mbar) cell. The obtained experimental data are evident in Fig. 3.

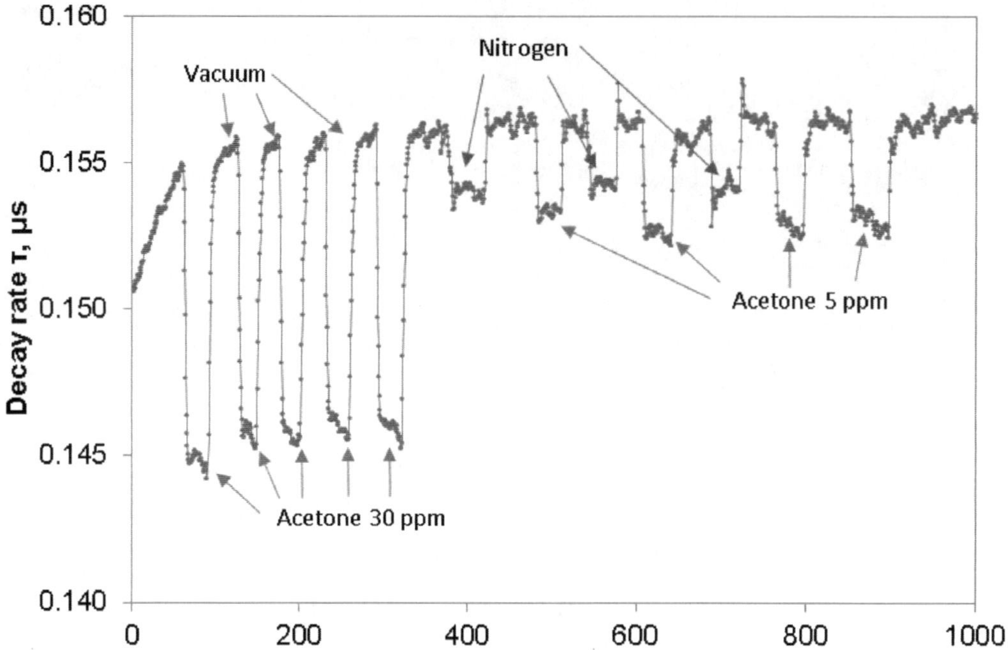

Figure 3. The ringdown decay rate signal for Acetone of 30 ppm and 5 ppm concentration, captured in time.

3. Conclusion

The Cavity Ring-Down Spectroscopy (CRDS) system for medical applications was created and first results for detection of acetone were received. The possibility to measure acetone concentration in the air at the level of 5 ppm was shown.

Acknowledgements

This research has been partly supported by the National Research Programme "The next generation of information and communication technologies" (NexIT).

References

[1] C.Wang and A. B Surampudi, 2008, *Measurement Science and Technology* **19,** 105604

IOP Conf. Series: Journal of Physics: Conf. Series **810** (2017) 012037 doi:10.1088/1742-6596/810/1/012037

A modified formula for the Voigt spectral line profile

G D Roston, M F Ahmed, M S Helmi and O S Mahran
Department of physics – Faculty of science - Alexandria University-Egypt

Abstract. A modified formula for the Voigt spectral line profile has been obtained. The Voigt formula has been transferred from the integral formula to a second order differential equation. It was found that the results of the obtained formula are in a perfect agreement with the corresponding published results which are obtained by different numerical calculation methods for a wide range of the line damping parameter (a) for $(0 \le a \le 200)$. Also, four special formulas are obtained at different ranges of the line profile, Gaussian profile at (a → 0), Lorentzian profile at (a → ∞), line center at (x → 0) and line wings at (x→ ∞).

1. Introduction

The spectral line pressure broadening can be described using the Voigt profile which is a convolution of both Doppler and Lorentzian profiles which is given by:

$$V(a, x) = \frac{a}{\pi}\int_{-\infty}^{\infty} \frac{e^{-(y)^2}}{a^2 + (x-y)^2}\, dy \qquad (1)$$

Where (a) is the spectral line damping parameter which is given by

$$a = \sqrt{ln2}\cdot\frac{\Delta\nu_L}{\Delta\nu_D} \qquad \text{and} \qquad x = 2\sqrt{ln2}\,\frac{(\nu_0-\nu-\Delta)}{\Delta\nu_D}$$

Here, $\Delta\nu_L$ *and* $\Delta\nu_D$ are the Lorentzian and Doppler widths and Δ is the pressure shift of the spectral line with respect to the unperturbed frequency ν_0. Many accurate tables of the Voigt profile [1-4] have been developed for some values of the parameter (***a***). G. D. Roston and F. S. Obeid [4] obtained an analytical formula for this profile as a solution of the second-order differential equation using homogeneous and non-homogeneous solutions for some values of (**a**) $(0 \le a \le 1)$, but the obtained formula obtained by [4] contain two terms which should be solved numerically. The aim of this work is to convert those two numerical terms to analytical terms to obtain a total analytical formula for the Voigt spectral line profile.

2. Theoretical

According to Roston and Obaid [4] and Andersen [5], the Voigt profile has been converted from the integral formula to the second- order differential equation as:

$$\frac{\partial^2 V(a,x)}{\partial x^2} + 4x\frac{\partial V(a,x)}{\partial x} + (4a^2 + 4x^2 + 2)V = \frac{4a}{\sqrt{\pi}} \qquad (2)$$

The solution of Eq. (2) is obtained by [6] as

$$V(x) = e^{a^2-x^2}cos(2ax)erfc(a) - \frac{2}{\sqrt{\pi}}e^{-x^2}$$

$$[\cos(ax)\int_0^x sin\,(2ax)\;e^{x^2} - sin(2ax)\int_0^x \cos(2ax)\,e^{x^2}dx] \qquad (3)$$

Where **erfc** is the complementary error function. Eq. (3) was not totally analytical, the second and the third terms should be solved numerically. For that, the two integrals must be converted to analytical formulas as follows:

(1). $\int_0^x sin(2ax)\ e^{x^2}dx = \frac{1}{2i}\int_0^x e^{x^2}\left(e^{2iax} - e^{-2iax}\right)dx$

$$= [\frac{1}{2i}\cdot\frac{1}{2}\sqrt{\pi}e^{a^2}erfi(x+ia) - \frac{1}{2i}\cdot\frac{1}{2}\sqrt{\pi}e^{a^2}erfi(x-ia)]_0^x$$

$$= \frac{1}{4}\sqrt{\pi}\,e^{a^2}\,[erf(a-ix) + erf(a+ix) - 2erf(a)] \qquad (4)$$

(2). $\int_0^x cos(2ax)\ e^{x^2}dx = \frac{1}{2}\int_0^x e^{x^2}\left(e^{2iax} + e^{-2iax}\right)dx$

$$= [\frac{1}{2}\cdot\frac{1}{2}\sqrt{\pi}e^{a^2}erfi(x+ia) + \frac{1}{2}\cdot\frac{1}{2}\sqrt{\pi}e^{a^2}erfi(x-ia)]_0^x$$

$$= \frac{i}{4}\sqrt{\pi}\,e^{a^2}\,[erf(a-ix) - erf(a+ix)] \qquad (5)$$

By inserting Eq. (4) and Eq. (5) in Eq. (3) we get:

$$V\,(a,x) = erfc(a)\,e^{a^2-x^2}cos(2ax) - \frac{2}{\sqrt{\pi}}e^{-x^2}\{\frac{1}{4}\sqrt{\pi}\,e^{a^2}cos(2ax)\,[erf\,(a\text{-}ix)$$

$$+ erf\,(a+ix) - 2erf\,(a)] - \frac{i}{4}\sqrt{\pi}\,e^{a^2}\,sin\,(2ax)\ [erf\,(a\text{-}ix) - erf\,(a+ix)]\}$$

$$= e^{a^2-x^2}\{erfc\,(a)\,cos(2ax) - \frac{1}{2}[\,cos(2ax)\,[erf\,(a\text{-}ix) + erf\,(a+ix)$$

$$- 2erf\,(a)] - isin(2ax)\,[erf\,(a\text{-}ix) - erf\,(a+ix)]]\} \qquad (6)$$

The two functions $erf\,(a\text{-}ix)$ and $erf\,(a+ix)$ given by Abramowitz and Stegun [6] as:

1. $erf\,(a\text{-}ix) = erf\,(a) + \frac{e^{-a^2}}{2\pi a}[(1 - \cos(2ax)) - i\sin(2ax)]$

$$+ \frac{2}{\pi}e^{-a^2}\sum_{n=1}^{\infty}\frac{e^{-n^2/4}}{n^2+4a^2}[\,f_n(a,x) - iR_n(a,x)] + \varepsilon(a,x)$$

2. $erf\,(a+ix) = erf\,(a) + \frac{e^{-a^2}}{2\pi a}[(1 - \cos(2ax)) + i\sin(2ax)]$

$$+ \frac{2}{\pi}e^{-a^2}\sum_{n=1}^{\infty}\frac{e^{-n^2/4}}{n^2+4a^2}[\,f_n(a,x) + iR_n(a,x)] + \varepsilon(a,x)$$

The functions $f_n(a,x)$, $R_n(a,x)$ and $\varepsilon(a,x)$ are given by Abramovitz and Stegun [6] as:

$f_n(a,x) = 2a - 2a\ cosh\,(nx)\ cos\,(2ax) + nsinh\,(nx)\ sin\,(2ax)$

$R_n(a,x) = 2a\ cosh\,(nx)\ sin\,(2ax) + n\ sinh\,(nx)\ cos\,(2ax)$, $\quad \varepsilon(a,x) \approx 10^{-16}\ |erf(a+ix)|$.

Introducing the expansions of erf $(a\text{-}ix)$ and erf $(a+ix)$ in Eq. (6) we get:

$$V(a,x) = e^{-x^2}\{erfc\,(a)\cos(2ax)e^{a^2} + \frac{1}{2\pi a}(\cos^2(2ax) + \sin^2(2ax)) - \frac{1}{2\pi a}(cos(2ax)$$

$$+ \frac{2}{\pi}\sum_{n=1}^{\infty}\frac{e^{-n^2/4}}{n^2+4a^2}\,[sin(2ax)\,R_n(a,x) - cos(2ax)f_n(a,x)]\} \qquad (7)$$

The obtained formula is in agreement with the formula obtained by Limandri et al [7]. Substituting the values of $f_n(a,x)$ and $R_n(a,x)$ in Eq. (7), we get

$$V(a,x) = e^{-x^2}\{erfc(a)\, e^{a^2}\cos(2ax) + \frac{1}{2\pi a}(1-\cos(2ax))$$

$$+\frac{4a}{\pi}\sum_{n=1}^{\infty}\frac{e^{-n^2/4}}{n^2+4a^2}[\cosh(nx)-\cos(2ax)] \quad (8)$$

This is the final formula for the Voigt profile, which is valid in the wide range of spectra [$0 \leq a \leq 5$ and $-10 \leq x \leq 10$], but for $a > 5$, erfc (a) in the first term of Eq. (8) is very small and approaches to zero. For that, the expansion of [$erfc\ (a)\, e^{a^2}$] given by [1] can be used as:

$$erfc\,(a)\, e^{a^2} = \frac{1}{\sqrt{\pi}}\frac{1}{a}\sum_{n=0}^{\infty}\frac{(-1)^n(2n)!}{n!(2a)^{2n}} \quad (9)$$

By substituting Eq. (9) in Eq. (8), we have

$$V(a,x) = e^{-x^2}\{\frac{1}{\sqrt{\pi}}\frac{1}{a}\sum_{n=0}^{\infty}\frac{(-1)^n(2n)!}{n!(2a)^{2n}}\cos(2ax) + \frac{1}{2\pi a}(1-\cos(2ax))$$

$$+\frac{4a}{\pi}\sum_{n=1}^{\infty}\frac{e^{-n^2/4}}{n^2+4a^2}[\cosh(nx)-\cos(2ax)]\} \quad (10)$$

This formula is valid for larger values of (a) [$5 < a \leq 15$ and $-10 \leq x \leq 10$].

Special cases at different ranges of the line profile

1-At ($a \simeq 0$) , $V(0, x) = e^{-x^2}$ (Doppler profile) .

2-At ($a \simeq \infty$), $V(a \to \infty, x) = \frac{1}{\sqrt{\pi}}\frac{a}{a^2+x^2}$ (Lorentzian profile)

This formula is valid for [$15 < a \leq 200$ and $-200 \leq x \leq 200$].

3-At the line center where ($\nu = \nu_0$) and $x = 0$, $V(a, 0) =$ erfc (a) e^{a^2}

4-At the line wing where ($x = \infty$), $V(a, x \to \infty) = \frac{1}{\sqrt{\pi}}\frac{a}{x^2}$

3. Results and discussions

3.1. The effect of (n) parameter

As the Voigt profile (Eqs.8,10) is dependent on the number of summation (n) which effects on the shape and the accuracy of the spectral line, It was found that the best fitting between the calculated profiles and the corresponding numerical profiles takes place at (n =25), and for n > 25 does not make any variations. For small values of (a) [0.01 - 4] we get a good fitting at of (n) [7 - 20] as shown in Fig.(1) with the fitting polynomial equation given by:

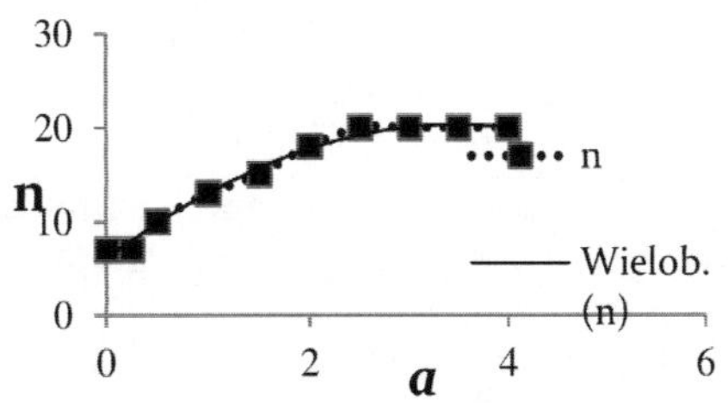

Figure (1): The relation between the values of (n) and the damping parameter (a) [0.01 – 4] and the fitting polynomial equation

$$n = 6.02 + 8.214a - 1.179a^2.$$

3.2 Comparison between the obtained and the corresponding published results

The comparison between the results of the obtained formulas and the corresponding published data of [1-3] are illustrated in the table.

Table: The comparison between the obtained results and Mofreh R.Zaghloul [1], Mamedov [2] and Z.Shippony and W.G.Read[3].

x	This work	Zaghloul [1]
At damping parameter (*a* = 0.01)		
1	0.368702417397766	0.368702416670236
2	0.0206200654455691	0.0206200653731694
3	0.000908830706741581	0.000908830706010574
4	0.000392604421617868	0.00039260442161698
5	0.00030237707114623	0.000240803391951175
At damping parameter (a = 0.5)		
1	0.354900332867578	0.354900332432228
2	0.103358823741367	0.103358823749883
3	0.037126366054692	0.037126366055133
4	0.019224945518739	0.019224945518740
5	0.011900325522594	0.011900325522594
At damping parameter (a = 5)		
1	0.106797738398065	0.106797738687967
2	0.0964981126066414	0.0964981125848205
3	0.0829877379769018	0.082987737977024
4	0.0692362095804915	0.0692362095804913
5	0.056965439888177	0.0569654398881766
At damping parameter (a = 15)		
1	0.0373653474578327	0.0373653478628073
2	0.0368810149294909	0.0368810149383951
3	0.0361008874912074	0.0361008874910956
4	0.035062152853902	0.0350621528539018
5	0.0338107397543061	0.033810739754306

x	This work	Mamedov [2]
At damping parameter (a = 0.01)		
1.00E-09	9.88815461046343E-01	9.88815461046342E-01
5.00E-05	9.88815458602020E-01	9.88815458602018E-01
1.00E-04	9.88815451269048E-01	9.88815451269048E-01
5.00E-02	9.86374167467079E-01	9.86374167467078E-01
2	2.06200654455691E-02	2.06200654455695E-02
9	7.09844469661236E-05	7.09844886591770E-05
At damping parameter (a = 0.5)		
1.00E-09	6.15690344192926E-01	6.15690344192925E-01
5.00E-05	6.15690362791042E-01	6.15690343294561E-01
1.00E-04	6.15690418585387E-01	6.15690340599466E-01
5.00E-02	6.14792754945695E-01	6.14792754945695E-01
2	1.03358823741367E-01	1.03358823741366E-01
9	3.53780588850978E-03	3.53780778549224E-03
At damping parameter (a = 5)		
1.00E-09	1.10704637733069E-01	1.10704637793912E-01
5.00E-05	1.10704637722967E-01	1.10704637715343E-01
1.00E-04	1.10704637692662E-01	1.10704637694591E-01
5.00E-02	1.10694536873867E-01	1.10694536864910E-01
2	9.64981126066414E-02	9.64948274000898E-02
9	2.68729312688368E-02	2.68729317409139E-02

x	(This work)	Z.Shippony and W.G.Read [3]
At damping parameter (a = 1)		
1	0.304744205256913	0.3047442052569126
2	0.140239581366278	0.1402395813662779
3	0.065317777289047	0.0653177772890470
4	0.0362814564899886	0.0362814564899886
5	0.023003132594060	0.0230031325940600
At damping parameter (a = 2)		
1	0.218492615274891	0.2184926152748907
2	0.147952759512016	0.1479527595120158
3	0.0927107664264434	0.0927107664264433
4	0.0596869296104459	0.0596869296104459
5	0.0406436763334944	0.0406436763334944
At damping parameter (a = 3)		
1	0.129888159930841	0.1298881599308405
2	0.112139477902116	0.1121394779021160
3	0.0909339041947653	0.0909339041947653
4	0.0715704334263653	0.0715704334263653
5	0.0559973771425239	0.0559973771425239
At damping parameter (a = 4)		
1	0.164261136392986	0.1642611363929863
2	0.130757469669849	0.1307574696698486
3	0.0964025055830446	0.0964025055830445
4	0.0697909616496483	0.0697909616496483
5	0.0512259965673866	0.0512259965673866
At damping parameter (a = 5)		
1	0.106797738398065	0.1067977383980653
2	0.0964981126066414	0.0964981126066414
3	0.0829877379769018	0.0829877379769018
4	0.0692362095804915	0.0692362095804914
5	0.056965439888177	0.0569654398881770

It is seen from the last tables that the results of the obtained formulas are in good agreement with the corresponding theoretical and numerical values.

4. Conclusions

1-A modified formulas for the Voigt profile Eq. (8, 10) makes the calculation of the Voigt profile easier in a wide range of the spectral line profile.

2-Using the obtained formulas, four special formulas were also obtained for [(a $\rightarrow$ 0) Gaussian profile, (a $\rightarrow\infty$) Lorentzian profile , for (x $\rightarrow$0) at line center and (x $\rightarrow$ ∞) at line wing].

References

[1] M.R. Zaghloul, On the calculation of the Voigt profile: a single proper integral with a damped sine integrand, Mon. Not. R. Astron. Soc. (2007):375:1043-1048.

[2] B.A. Mamedov, Analytical evaluation of the Voigt function using binomial coefficients and Incomplete Gamma functions, Mon. Not. R. Astron. Soc. 2008: 387(4):1622-1626.

[3] Z. Shippony, W.G. Read, A highly accurate Voigt function algorithm, J. Quant. Spectrosc. Radiat. Transfer. 1993:50(6):635-646.

[4] G.D. Roston, F.S. Obaid, Exact analytical formula for Voigt spectral line profile, J. Quant. Spectrosc. Radiat. Transfer. 2005:94:255-263.

[5] T. Andersen, The differential equations of the Voigt function, J. Quant. Spectrosc. Radiate. Transfer. 1978:19:169-171.

[6] M. Abramovitz, I.A. Stegun, Handbook of Mathematical Functions, Ninth printing, Dowe Publications, Inc., New York, (1970).

[7] S.P. Limandri, R.D. Bonetto, H.O. Di Rocco, J.C. Trincavelli, Fast and accurate expression for the Voigt profile function, Application to the determination of uranium M line widths, Spect. Acta.B. 2008:63:962-967.

An algorithm for calculating the contour Voigt and its improvement and refinement for some ranges of parameters

A Y Sukhanov[1]

[1] Laboratory of Lidar Methods, V.E. Zuev Institute of Atmospheric Optics SB RAS, 1, Academician Zuev square, 634021 Tomsk, Russia

e-mail: say@iao.ru

Abstract. We present an approximation Voigt contour for some parameters intervals such as the interval with y less than 0.02 and absolute value x less than 1.6 gives a simple formula for calculating and relative error less than 0.1%, and for some of the intervals suggetsted to use Hermite quadrature.

In [1] was presented the method for calculating the Voigt contour, in this, it refined and improved. Voigt profile contour used for calculations of the absorption lines and takes into account the effect of Doppler broadening and collision effects [2], this intractable integral with infinite boundaries of integration:

$$\chi(\nu) = \frac{x_0 y}{\pi} \int_{-\infty}^{\infty} \frac{\exp(-t^2)}{y^2 + (x-t)^2} dt, \quad (1)$$

where x, $y \geq 0$ - parameters of Voigt conotour depending on wavelength and broadening effects, $x_0 = \frac{S}{\gamma_D}\left(\frac{\ln 2}{\pi}\right)^{1/2}$, $y = \frac{\gamma_L}{\gamma_D}(\ln 2)^{1/2}$, $x = \frac{\nu - \nu_0}{\gamma_D}(\ln 2)^{1/2}$, γ_D, γ_L - Doppler and Laurence half width, $\gamma_D = \frac{\gamma_0}{c}\left(\frac{2kT}{m}\ln 2\right)^{1/2}$, T - temperature, m - molecular mass, k - Boltzmann constant, ν_0 - absorption line frequency.

Different algorithmic approaches to calculate the integral in (1) are used, some of them described in [3-7]. Here it offers improved relative error for $|x| < 1.6$ and $y < 0.02$ range with an error of less than 0.1% compared with the error <0.5% at first algorithm realization and for $y < 0.001$ less than 0.001% by approximation given:

$$f(x,y) = \frac{\exp\left(\frac{-x^2}{y+1}\right)}{(0.351y + 0.3183)y - (1.4x^4 - 0.38x^2)y^2 0.093} \quad (2)$$

It approximate following function:

$$\varphi(x,y)=\int_{-\infty}^{\infty}\frac{\exp(-t^2)}{y^2+(x-t)^2}dt \tag{3}$$

Relative errors estimated for different parameter y are presented at figure 1, to determine the coefficients of the approximation formula and its type used a genetic algorithm [8].

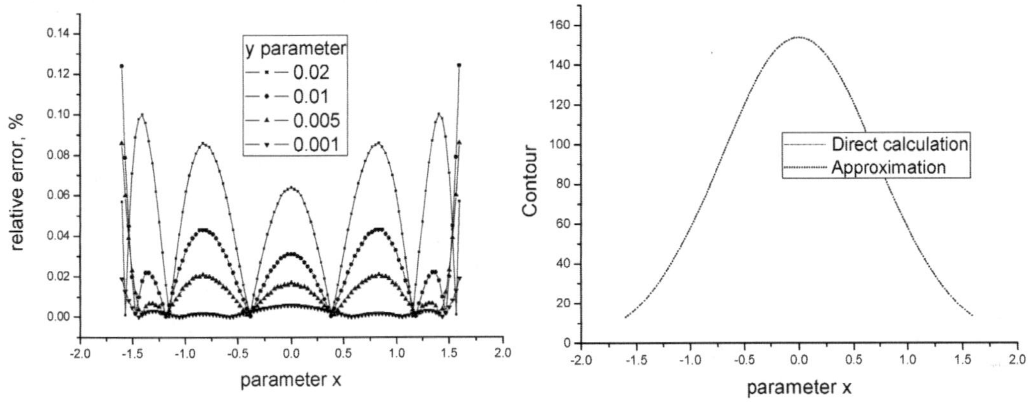

Figure 1. – Relative errors of contour approximation for $|x|<1.6$ and $y<0.02$

Approximation based on orthogonal polynomials Hermite:

$$f(x,y)=\sum_{i=1}^{10}\frac{A_i}{y^2+(x-Ax_i)^2},\quad (y\geq 1.0)\vee(|x|>3.55\wedge y>0.01)\vee|x|>5.000001 \tag{4}$$

where $A_i.Ax_i$ - precalculated weight Hermite quadrature, values listed in Table 1.

Table 1. precalculated weight Hermite quadrature values

Index number	*Ax*	*A*
1	3.4361583	7.641e-6
2	2.5327307	1.344e-3
3	1.7566830	0.034
4	1.0366104	0.240
5	0.3429013	0.611
6	-0.3429013	0.611
7	-1.0366104	0.240
8	-1.7566830	0.034
9	-2.5327307	1.344e-3
10	-3.4361583	7.641e-6

For $|x|>5.000001$ error is less than 0.001%, for $y\geq 0.7$ and $y\leq 1.0$ and $|x|<3.55$ error less than 0.5%, for $y\geq 1.0$ and $|x|<3.55$ less than 0.1%, and for $|x|>3.55$ less than 0.01% (figure 2).

For the intervals $y<0.01$ and $|x|>3.55$ and $|x|<5.000001$ is recommended to use direct calculation using recursive procedure dividing each grid spacing if necessary and check the relative error of calculation in the division of the interval for calculating the integral for 2, 4 and 8 parts, or use another methods [4-7], unfortunately this algorithm with a small stack size for small $y<0.000001$ can not enough memory, this range has features in which, for example, a standard Mathcad integration algorithm is not working properly.

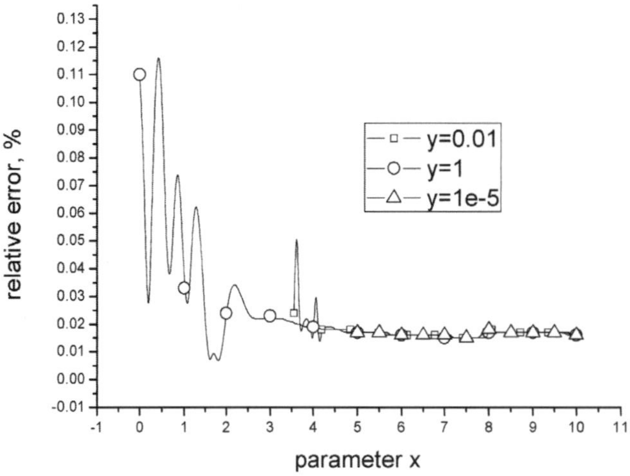

Figure 2. Relative errors of contour approximation for $|x|>5.000001$, for $y=0.01$ and $x\geq 3.55$, for $y=1.0$ and $|x|>0$

It is suggested to use a linear approximation of integral in expression (1) on a uniform x and y grid for another cases, the accuracy will depend on the amount of memory allocated for the approximation coefficients. Computation speed does not depend on grid spacing, as the respective numbers of coefficients are linear based on it, but it requires estimation and preservation of the approximation coefficients and a large amount of memory for them (for example if $|x|<1.6$ and $y>0.02$ and $y<0.7$ it is need more 100 points for x and more than 2000 points for y with relative error <0.01%).

References

[1] Sukhanov A Ya 2015 *The XVIIIth Symp. and School on High Resolution Molecular Spectroscopy* (Tomsk: Publishing House of IAO SB RAS) P 119
[2] Zuev V E, Makushkin Yu S, Ponomarev Yu N 1987 *Spectroscopy of the Atmosphere* (Leningrad: Gidrometeoizdat) p 248
[3] Mitsel A A, Ptashnik I V, Milyakov A V 2000 *Atmospheric and oceanic optics* **13** 1051-55
[4] Humlíček J 1982 *Journal of Quantitative Spectroscopy and Radiative Transfer* **27** 437–44
[5] Kuntz M, Höpfner M 1999 *Journal of Quantitative Spectroscopy and Radiative Transfer* **63** 97–114
[6] Letchworth K L Benner D C 2007 *Journal of Quantitative Spectroscopy & Radiative Transfer* **107** 173–192
[7] Amamou H, Ferhat B, Bois A 2013 *American Journal of Analytical Chemistry* **4** 725-31
[8] Holland J H 1992 *Adaptation in Natural and Artificial Systems* (Cambridge MA: MIT Press. Second edition)

Super Lorentzian effects on the wings of self-broadened HCl and of HCl diluted in Ar

H Tran[1], J-M Hartmann[1], G Li[2], V Ebert[2,3,4]

[1]Laboratoire Interuniversitaire des Systèmes Atmosphériques (LISA, UMR CNRS 7583), Université Paris Est Créteil, Université Paris Diderot, Institut Pierre-Simon Laplace, 61 avenue du Général de Gaulle, 94010 Créteil, France
[2]Physikalisch-Technische Bundesanstalt, Bundesallee 100, 38116 Braunschweig, Germany
[3]Physikalisch Chemisches Institut, U Heidelberg, INF 253, 69116 Heidelberg
[4]Center of Smart Interfaces, TU Darmstadt, Petersenstraße 32, Darmstadt 64287, Germany

E-mail: ha.tran@lisa.u-pec.fr

Abstract. Super-Lorentzian effects in the troughs between HCl lines were observed long time ago [Varanasi et al., J Quant Rad Transfer, Vol. 12, pag. 857, 1972]. The observed spectral shape was then modelled by using an empirical law and there was no explanation about the mechanisms underlying these super-Lorentzian effects. In this work, new spectra of pure HCl and HCl diluted in Ar have been measured using a high resolution Fourier Transform spectrometer, for pressure from 6 to 10 bars. Spectra of pure HCl and HCl in Ar have been also computed using classical molecular dynamics simulations (CMDS). First comparisons between CMDS-calculated spectra and measured ones, for regions at the troughs between HCl lines, show that the observed super-Lorentzian behaviour is correctly reproduced by the calculations. These results thus open the paths for the determination of the origin of these super-Lorentzian effects.

1. CMDS calculations and experimental spectra

Detailed description of our classical molecular dynamic simulations for HCl can be found in [1], thus only the main features are recalled here. In our CMDS calculations, a large number of molecules ($>10^6$) are treated. They are placed in cubic boxes with periodic boundary conditions [2]. Their initializations are perfomed as follows: the center of mass position and molecular axis orientation are randomly initialized. For the translational and angular velocity, their orientations are randomly initialized while Maxwell-Boltzmann distributions are used for their modulus. The time evolution of the system is then computed using the classical equation, i.e. for the molecule i :

$$m\dot{\vec{v}}_i = \vec{f}_i \quad \text{and} \quad \mathrm{I}\dot{\vec{\omega}}_i = \vec{\tau}_i \tag{1}$$

Where the force f and the torque τ are computed from the potential gradient summed over all nearby collision partners $\sum_{j\neq i} \vec{\nabla} V_{i,j}$. The intermolecular potential is taken from Refs. [3,4]. These calculations thus provide the center of mass positions and velocities as well as the orientations and angular velocity of all molecules at all times. The auto-correlation function of the dipole moment carried by the molecular axis and thus the corresponding spectrum can thus be computed using the Fourier-Laplace transform. Note that the same requantization procedure as in Ref. [1] has been used in order to associate a rotational quantum number J_m to a value of ω_m, i.e. $\frac{1}{2}I\omega_m{}^2 \approx \frac{\hbar^2}{2I}J_m(J_m+1)$.

Spectra of pure HCl and HCl diluted in Ar in the 2-0 band have been measured at room temperature at pressures ranging from 6 to 10 bars, using a Fourier transform spectrometer. The detailed description of the experimental setup can be found in [5]. Spectra of HCl diluted in Ar in the 1-0 band, measured at room temperature and pressures up to 350 bars of Ref. [6] are also used for the comparison with our calculations.

2. First results and discussions

In order to analyze both measurements and CMDS-calculated spectra, different following steps have been carried: (i) at low pressure range, line parameters such as line position, line integrated intensity and line broadening have been retrieved from fits of measured (or calculated) spectra using the Voigt profile, (ii) using these data, we then calculate the corresponding spectra at higher pressure (more than 6 bars) using the Lorentzian profile. From these spectra, we determine the absorption at the troughs between HCl P and R branches lines, which will be denoted $a_{Lorentz}$. Absorption at the same regions but determined from measured (CMDS-calculated) spectra are denoted a_{obs} (a_{CMDS}). The super-Lorentzian effect is then determined by calculating the ratio $a_{obs}/a_{Lorentz}$ ($a_{CMDS}/a_{Lorentz}$).

Figure 1 show the $a_{obs}/a_{Lorentz}$ ratio obtained for three pressures: 6, 8 and 10 bars for pure HCl in the 2-0 band. We can observe that this ratio is constant with pressure as expected since absorption in the wings is proportional to the squared pressure [7]. The same analysis procedure has been performed for all measured and CMDS-calculated spectra.

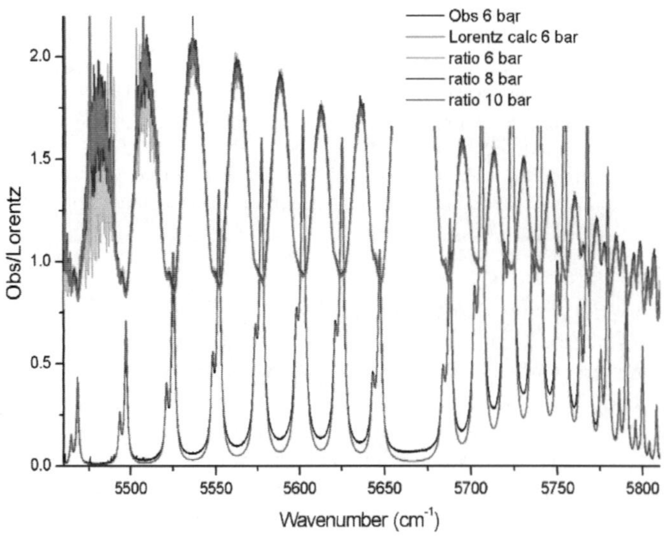

Figure 1: Comparison between spectra calculated using the Lorentz profile and measurements at 6 bars and the $a_{obs}/a_{Lorentz}$ ratio at various pressures for pure HCl.

Figure 2 shows the comparison between the measured and CMDS-calculated results for HCl diluted in Ar in the 1-0 band where a very good agreement can be observed. Work is now in progress for the 2-0 band region and for self-broadening HCl. We also plan to perform similar study for HCl in

CO_2 for which the observed super-Lorentzian effect is much larger. Finally, the different physical mechanisms contributing to this super Lorentzian behavior will be determined and analyzed using CMDS.

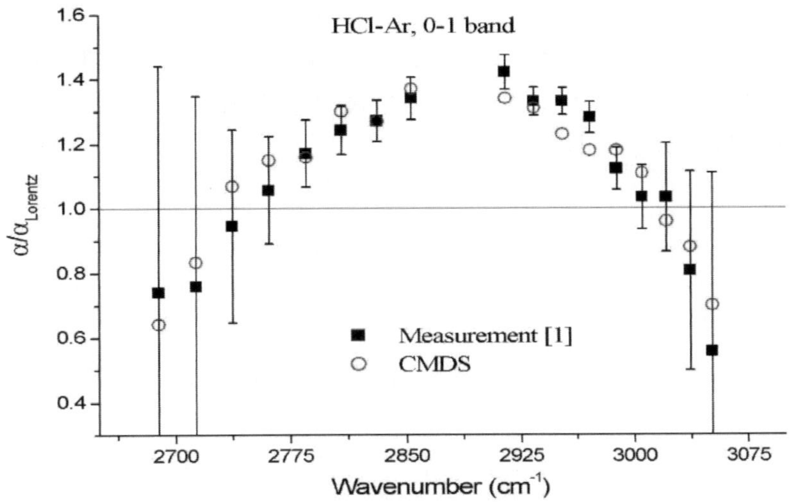

Figure 2: Comparison between measured and CMDS-calculated results

References

[1] H. Tran, J. L. Domenech 2014 *J. Chem. Phys.* **140** 084308
[2] M. P. Allen and D. J. Tildesley 1987 *Computer Simulation of Liquids.* (Clarendon, Oxford)
[3] A. Medina, J. M. M. Roco, A Calvo Hernandez, A.Velasco 2004 *J. Chem. Phys.* **121** 6353
[4] P. Hoang, P. Joubert, D. Robert 2001 *Phys. Rev. A* **65** 012507
[5] Li G, Serdyukov A, Gisi M, Werhahn O and Ebert V 2015 *J. Quant. Spectrosc. Radiat. Transfer* **165** 76-87
[6] Boulet C, P. M. Flaud, J. M. Hartmann 2004 *J. Chem. Phys.* **120** 11053
[7] J.-M. Hartmann, C. Boulet, and D. Robert 2008 *Collisional effects on molecular spectra. Laboratory experiments and models, consequences for applications* (Elsevier, Amsterdam)

Influence of spontaneous emission transfer on the resonances of interaction of laser radiation with alkali atoms confined in an extremely thin cell

G Todorov[1], V Polischuk[2], A Krasteva[1], A Sargsyan[3], S Cartaleva[1] and T Vartanyan[2]

[1]Institute of Electronics, Bulgarian Academy of Sciences, 72, 1784 Sofia, Bulgaria
[2]ITMO University, St. Petersburg, 197101, Russian Federation
[3]Institute for Physical Research, National Academy of Sciences of Armenia, Ashtarak-2, Armenia

E-mail: Tigran.Vartanyan@mail.ru

Abstract. The nonlinear absorption and fluorescence spectra of ^{133}Cs vapours confined to the extremely thin cell are computed in the framework of perturbation theory regarding to the intensity of the laser light illumination. The amplitude as well as the sign of the longitudinal alignment of the ground state is shown to be defined by the spontaneous decay of the upper level. The sign and the magnitude of the nonlinear absorption resonance on the closed optical transition are shown to be defined by this process as well.

1. Intoduction

The sub-Doppler spectroscopy of alkali atom vapour was strongly advanced by the invention of the Extremely Thin Cells (ETC) [1]. Using the conventional spectroscopic methods of registration, a number of peculiarities in the absorption, reflection and fluorescence spectra were observed in these cells, which are not present in commonly used (cm-size) cells. The main reason for the formation of the sub-Doppler structures in the spectra is the strong spatial anisotropy in the ETC environment. Collisions with the cell walls induce loss of atomic phase memory and excitation quenching. Hence, the atoms departing from the cell walls undergo a transient regime of excitation before they acquire the steady state polarization that corresponds to their velocity as well as to the laser field intensity and detuning. Nonlinear resonances in absorption and fluorescence are also influenced by velocity selective optical pumping [2].The present work gives a description of the interaction of resonant laser fields with alkali atoms in an ETC with the full account of the hyperfine sublevel degeneracy using iteration procedure over the laser field intensity.

The problem of determination of the nonlinear atomic polarization was solved for arbitrary values of the total angular momenta of the resonance levels for excitation with linearly polarized laser light. Using the previously developed methodology [3-6], the equations of motion for the statistical operator in the irreducible tensor operator (ITO) representation were solved, taking into account the transient processes started by the atomic collisions with the cell walls as well as the spontaneous transfer of coherence and population from the upper level to the ground state [7].

2. Theory

The developed theoretical approach is based on the full account of transient polarizations induced by the atom-wall collisions in extremely thin cells as well as nonlinear optical effects and magnetic sublevel degeneracy. Let us start with the introduction of the statistical operator ρ_q^{κ} $(\rho = f, \varphi, \xi)$ that defines the populations (k=0) and alignments (k=2) of the upper (f_q^{κ}) and the lower (φ_q^{κ}) levels as well as the optical coherence

$$\xi_q^{\kappa}(x,v,t) = \xi_q^{\kappa}(x, v)\exp\{-i\,\omega_{las}\,t\}, \qquad (1)$$

where x is the atomic coordinate along the propagation direction of the laser beam, which is perpendicular to the cell windows, t is time, v is the projection of the atomic velocity on the x axis, and ω_{las} .is the laser frequency. In the rotating wave approximation one obtains time-independent equations of the form [8]

$$v\frac{\partial}{\partial x} f_q^{\kappa} + \gamma_f^{\kappa} f_q^{\kappa} = ih^{-1}(2F_{\varphi}+1)^{-1/2}\sum_{\kappa' q' Q} E_{-Q}(x)\, C_{qq'Q}^{\kappa\kappa'}\left[d\xi_{q'}^{\kappa'}(x,v) + d^{*}(\xi_{-q'}^{\kappa'}(x,v))^{*}(-1)^{\kappa+\kappa'+q'} \right]$$
$$+\left(2F_f+1\right)\gamma_f^{\kappa} N_f W(v)\delta_{\kappa 0} \qquad (2a)$$

$$v\frac{\partial}{\partial x}\varphi_q^{\kappa} + \gamma_{\varphi}^{\kappa}\varphi_q^{\kappa} = (-1)^{\kappa} ih^{-1}(2F_{\varphi}+1)^{-1/2}\sum_{\kappa' q' Q} E_{-Q}(x)(-1)^{\kappa'} B_{qq'Q}^{\kappa\kappa'}\left[d\,\xi_{q'}^{\kappa'}(x,v) + d^{*}(\xi_{-q'}^{\kappa'}(x,v))^{*}(-1)^{\kappa+\kappa'+q'} \right]$$
$$+\left(2F_{\varphi}+1\right)\gamma_{\varphi}^{\kappa} N_{\varphi} W(v)\delta_{\kappa 0} + \Gamma_{F_f F_{\varphi}}^{\kappa} f_q^{\kappa} \qquad (2b)$$

$$v\frac{\partial}{\partial x}\xi_q^{\kappa}(x,v) + \left[\gamma_{\xi}^{\kappa} + i(\omega_0 - \omega_{las})\right]\xi_q^{\kappa}(x,v) = ih^{-1}(2F_f+1)^{-1/2} d^{*} \times$$
$$\sum_{\kappa' q' Q} E_{-Q}(x)\left[S_{qq'Q}^{\kappa\kappa'} f_{q'}^{\kappa'} + (-1)^{\kappa+\kappa'} R_{qq'Q}^{\kappa\kappa'}\varphi_{q'}^{\kappa'} \right] \qquad (2c)$$

where

$$C_{qq'Q}^{\kappa\kappa'} = (-1)^{2F_{\varphi}+q'}(2F_f+1)^{1/2}(2\kappa'+1)\begin{Bmatrix} \kappa' & 1 & \kappa \\ F_f & F_f & F_{\varphi} \end{Bmatrix}\begin{pmatrix} \kappa' & 1 & \kappa \\ -q' & Q & q \end{pmatrix} \qquad (3a)$$

$$R_{qq'Q}^{\kappa\kappa'} = (-1)^{2F_{\varphi}+q'}(2F_f+1)^{1/2}(2\kappa'+1)\begin{Bmatrix} \kappa' & 1 & \kappa \\ F_f & F_{\varphi} & F_{\varphi} \end{Bmatrix}\begin{pmatrix} \kappa' & 1 & \kappa \\ -q' & Q & q \end{pmatrix} \qquad (3b)$$

while $B_{qq'Q}^{\kappa\kappa'}$ and $S_{qq'Q}^{\kappa\kappa'}$ may be obtained out of $C_{qq'Q}^{\kappa\kappa'}$ and $R_{qq'Q}^{\kappa\kappa'}$ via substitution $F_f \rightarrow F_{\varphi}$. $d \equiv (F_f \| d \| F_{\varphi}) \equiv \| d_{F_f F_{\varphi}} \|$ is the reduced dipole moment of the transition.

The populations ρ_0^0 and the alignments ρ_0^2 $(\rho = f, \varphi)$ of both the ground and the excited states were obtained in the second order of the perturbation theory and integrated over the cell thickness

$$^{(2)}\rho_0^{\kappa_2}(v) = \frac{1}{l}\int_0^l \left[^{(2)}\rho_0^{\kappa_2}(x, v>0) + {}^{(2)}\rho_0^{\kappa_2}(x, v<0)\right]dx, \quad \rho = f, \varphi\ . \qquad (4)$$

The linearly polarized laser light induces the population difference of magnetic sublevels of the ground state that results in the longitudinal alignment φ_0^2. Spontaneous emission from the upper level [7-9] accounted for by the term $\Gamma_{F_f F_{\varphi}}^{\kappa} f_q^{\kappa}$ in Eq. (2b) partially returns the atoms back and leads to the changes of alignment as well. According to the results of detailed computations, accounting for the spontaneous emission leads to the sing reversal of the alignment on the closed transition

$F_\varphi = 4 \rightarrow F_f = 5$. Open transitions are affected as well but less strongly. The amplitudes of the resonances depend on the values of the relaxation constants (Fig. 1).

Numerical integration of ${}^{(2)}f_0^{\kappa_2}(v)$ over the Maxwell velocity distribution $W(v)$ with the thermal velocity v_T gave the intensity of the unpolarized fluorescence in the given direction $\vec{n}$

$$ {}^{(2)}I_{f\varphi}(\vec{n}) = K_0 \left\| d^2_{F_\varphi F_f} \right\| \left[\frac{{}^{(2)}f_0^0}{3(2F_f+1)^{1/2}} + 5(-1)^{F_f+F_\varphi} \begin{Bmatrix} 1 & 1 & 2 \\ F_f & F_f & F_\varphi \end{Bmatrix} \left[\sum (-1)^q \, {}^{(2)}f_q^2 \, T^2_{-q}(\vec{n}) \right] \right] \quad (5) $$

Here $T^2_{-q}(\vec{n})$ is the observation tensor [10].

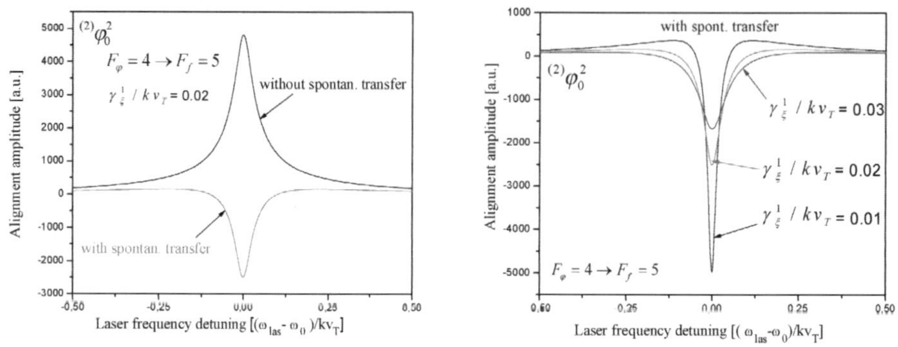

(a) (b)

Figure 1. (a) Inversion of the longitudinal alignment resonance with the account for the spontaneous decay on the closed transition $F_\varphi = 4 \rightarrow F_f = 5$, (b) influence of the spontaneous relaxation rate on the amplitude of the inverted resonance.

In the first and the third orders of perturbation theory the computed values of the optical coherences ${}^{(1)}\xi_0^1(x,v) = {}^{(1)}\xi_0^1(x,v>0) + {}^{(1)}\xi_0^1(x,v<0)$ and ${}^{(3)}\xi_0^1(x,v) = {}^{(3)}\xi_0^1(x,v>0) + {}^{(3)}\xi_0^1(x,v<0)$ were used to obtain the transmitted field amplitudes [4,5]

$$ E^1_{tr} = (-1)^{F_f - F_\varphi}(2F_\varphi + 1)^{-1/2} 2\pi i k d \int_0^l \exp[-ikx')] \int_{-\infty}^{+\infty} W(v) [{}^{(1)}\xi_0^1(x,v) + {}^{(3)}\xi_0^1(x,v)] \, dv \, dx' \quad (6) $$

Finally, the intensity of the transmitted light was obtained as the squared modulus of the sum of the incident light and the fields ${}^{(1)}E^1_{tr}$ and ${}^{(3)}E^k_{tr}$ induced by the vapor polarization

$$ {}^{(\Sigma)}I_{tr} \propto |E_0|^2 \left[1 + 2\,\mathrm{Re}\left(\frac{{}^{(1)}E^1_{tr}}{E_0} + \frac{{}^{(3)}E^\kappa_{tr}}{E_0} \right) \right] \quad (7) $$

Let us stress that contrary to the previous publications, relaxation rates γ_ρ^k $(\rho = f, \varphi, \xi)$ were assumed to be different. They were varied to reach the best fit to the experimental data.

3. Comparison with the experiment

The results of the computations were used for interpretation of the experimental findings. Fig. 2 plots the fluorescence spectra of ^{133}Cs vapor confined to the extremely thin cells of the thickness $l=(3/2)\lambda$, where $\lambda = 852,35$ nm corresponds to the cesium D_2 line $6s\ ^2S_{1/2} \rightarrow 6s\ ^2P_{3/2}$.

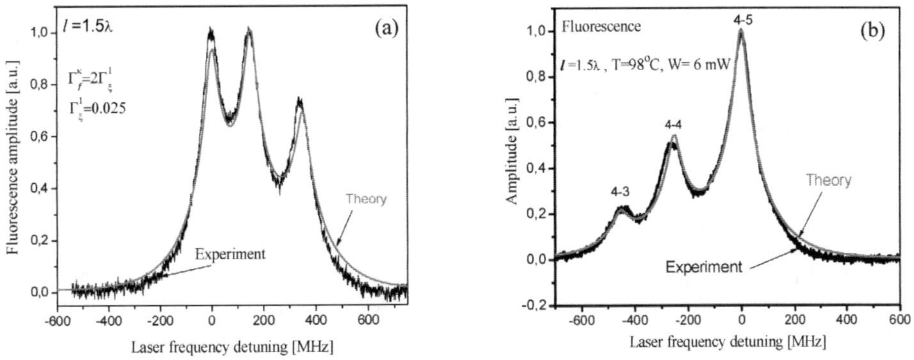

Figure 2. Theoretical and experimental fluorescence spectra of [133]Cs vapor confined to the extremely thin cells of the thicknesses *l*=(3/2)λ. (a) hyperfine transitions $F_\varphi = 3 \rightarrow F_f = 2, 3, 4$, (b) hyperfine transitions $F_\varphi = 4 \rightarrow F_f = 3, 4, 5$.

Fig. 3 plots the absorption spectra of cesium vapor in the same extremely thin cells. Saturated absorption resonances are well reproduced by the calculated nonlinear term ${}^{(3)}E^k_{tr}$. The enhanced absorption that is clearly seen in the case of *l*=(3/2)λ is due to the Dicke-like transient polarizations [4]. This linear effect dominates and obscures the influence of the nonlinear resonances and of the spontaneous transfer.

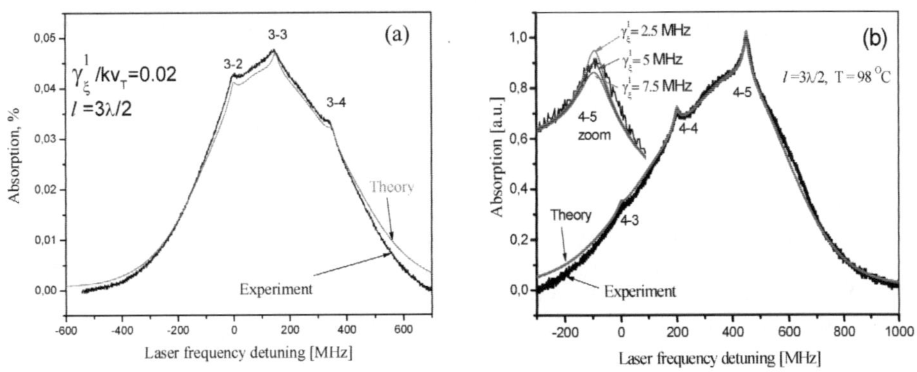

Figure 3. Theoretical and experimental absorption spectra of [133]Cs vapour confined to the extremely thin cells of thicknesses *l*=(3/2)λ.(a) hyperfine transitions $F_\varphi = 3 \rightarrow F_f = 2, 3, 4$, (b) hyperfine transitions $F_\varphi = 4 \rightarrow F_f = 3, 4, 5$

The role of the spontaneous transfer is shown in pure form in Fig. 4 for the cell that is 6λ thick. The detailed theoretical considerations have shown that absorption is enhanced only on the closed transition $F_\varphi = 4 \rightarrow F_f = 5$. This effect is the direct consequence of the sign reversal of the longitudinal alignment. The amplitude of this resonance depends on the relaxation constants and may be used for their estimations by comparing with the experimental data.

4. Conclusion

Steadily increasing demand for miniaturized optical devices leads to the growing importance of the theories that are able to account for atom-wall collisions and to predict the spectral lines shapes in the extremely thin cells. Conventional approaches based on simplified models failed to give quantitative description of the optical processes in the extremely thin cells. Simultaneous account for the transient polarizations and optical pumping effects are known to be the indispensible ingredients of the theoretical models.

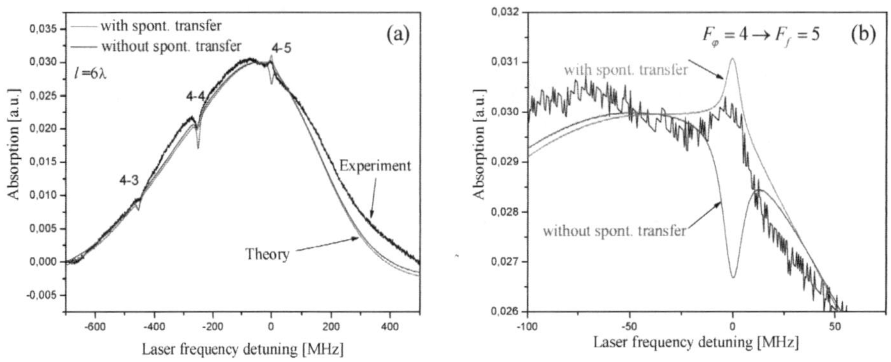

Figure 4. Theoretical and experimental absorption spectra of ^{133}Cs vapours confined to the extremely thin cell of the thicknesses l=6λ; (a) overview of hyperfine transitions $F_\varphi = 4 \rightarrow F_f = 3, 4, 5$; (b) detailed absorption spectrum associated with the closed transition $F_\varphi = 4 \rightarrow F_f = 5$.

In this contribution we showed that accounting for the alignment in the ground state induced by the spontaneous emission transfer is also of crucial importance. In particular, the sing of the nonlinear absorption resonance on the closed transition can not be obtained in accord with the experimental data unless the theory accounts for the spontaneous emission transfer and the right sing of alignment in the ground state.

Aknowledgements
This work was supported by the Russian–Armenian bilateral project RFBR 15-52-05030, 15 RF-024; and by the Russian Ministry of Education and Science 2014/190, 074-U01. AK acknowledges the project DFNP-188/14.05.2016 under scientific program "Assistance for young scientists", BAS. G.T. thanks also Dr D. Slavov for the assistance in the numerical calculations.

References

[1] Sarkisyan D, Bloch D, Papoyan A and Ducloy M 2001 *Opt. Comm.* **200** 201-208.
[2] Dey S, Ray B, Ghosh P N, Cartaleva S and Slavov D 2015 *Opt. Comm.* **356** 378-388.
[3] Vartanyan T A 1985 *Sov. Phys. JETP* **61** 674-677.
[4] Vartanyan T A and Lin D L 1995 *Phys. Rev.* A **51** 1959-1964.
[5] Vartanyan T A and Lin D L 1998 *Eur. Phys. J.* D **1** 217-221.
[6] Andreeva C, Cartaleva S, Petrov L, Saltiel S M, Sarkisyan D, Varzhapetyan T, Bloch D and Ducloy M 2007 *Phys. Rev.* A **76** 013837.
[7] Ducloy M and Dumont M 1970 *Le journal de Physique* **31**, 419-427.
[8] D'yakonov M I and Perel V I 1966 *Opt. Spek.* **20**, 472-480.
[9] Krasteva A, Slavov D, Todorov G and Cartaleva S 2013 *Proc. of SPIE* **8770**, 87700N.
[10] Alexandrov E B., Chaika M P and Khvostenko G I 1993 Interference of Atomic States (Springer-Verlag Berlin Heidelberg)

IOP Conf. Series: Journal of Physics: Conf. Series **810** (2017) 012041 doi:10.1088/1742-6596/810/1/012041

The ambiguity in determination of interatomic potential of diatomic molecule

T Urbańczyk[1], **J Koperski**[1], **A Pashov**[2]

[1] Smoluchowski Institute of Physics, Jagiellonian University, prof. S. Łojasiewicza 11, 30-348 Kraków, Poland

[2] Department of Physics, Sofia University, 5 James Bourchier Boulevard, 1164 Sofia, Bulgaria

E-mail: tomek.urbanczyk@uj.edu.pl

Abstract. A method of estimation of errors of interatomic potential parameters is presented. It can be employed in order to obtain errors of parameters of analytical potentials. The method works even for correlated parameters of the interatomic potential *e. g.*, parameters dependent on the vibrational frequency ω_e' and anharmonicity $\omega_e'x_e'$. Exemplary results of application of the method to interatomic potential represented by a Morse function are presented.

1. Introduction

The main goal of this study is to show a method for assessment of errors of interatomic potentials of diatomic molecules. For a simplicity, here we consider molecular states which potentials can be described using a Morse function. The inspiration for this work lays in observation that for a given molecular spectrum (especially without resolved rotational structure) there exist a large number of significantly different combination of molecular constants: vibrational frequency ω_e' and anharmonicity $\omega_e'x_e'$ (or equivalently large numbers of Morse potentials with different parameters D_e' and β') for which we can obtain an appropriate simulation of the experimental spectrum [1]. Figure 1 presents values of so-called agreement parameter P, which describes an agreement between simulated and experimental spectra in function of the ω_e' and $\omega_e'x_e'$ for the $B^31(5^3P_1)\leftarrow X^10^+(5^1S_0)$ transition in CdAr complex [2]. Detailed description of the procedure of constructing of the plot can be found in Ref. [3]. It was observed that each combination of molecular constants for which parameter P is larger than 15, leads to a simulation which satisfactorily agrees with the result of experiment. To demonstrate the fact, in Fig. 2 we present the experimental spectrum and three exemplary simulations that were conducted for pairs of values of molecular constants ω_e' and $\omega_e'x_e'$ depicted in Fig. 1 with points A, B and C. One can see that each of these simulations reconstructs satisfactorily the experimental spectrum, especially if we neglect the rotational structure. It is also worth to mention that during the analysis of molecular spectra one should also consider a correlation between value of the equilibrium distance R_e' and the transition dipole moment (TDM) in case when the TDM is included in the simulation as both the factors have influence on the intensities of vibrational transitions.

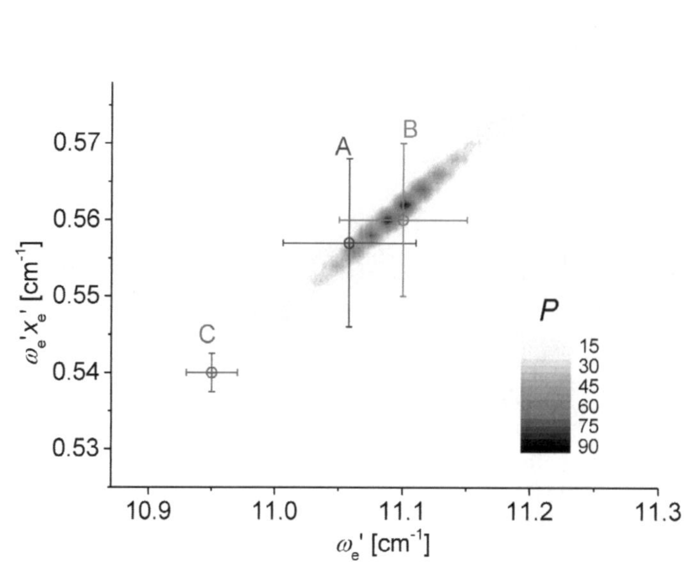

Figure 1. Agreement plot for the $B^31(5^3P_1)\leftarrow X^10^+(5^1S_0)$ transition in CdAr. A (red): the result of Birge-Sponer (B-S) analysis, B (blue): the result of correlation plot analysis, C (green): an example of vibrational constants ω_e' and $\omega_e'x_e'$ beyond the error bars resulted from the B-S analysis which provides a high simulation-to-experiment agreement; compare with Fig. 2.

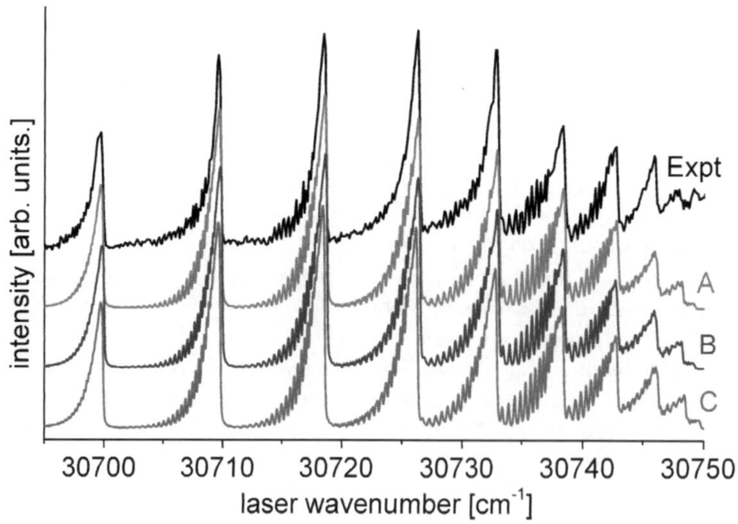

Figure 2. Simulation of the $B^31(5^3P_1)\leftarrow X^10^+(5^1S_0)$ transition in CdAr. Experimental (Expt: black) and simulated (A: red, B: blue and C: green) spectra of the transition under consideration. The simulated spectra A, B and C were obtained for values of vibrational constants ω_e' and $\omega_e'x_e'$ depicted in Fig.1 with points A, B and C, respectively.

2. Estimation of errors of interatomic potential parameters

To estimate an error of a given potential parameter of interatomic potential, we propose the following approach. In the first step, one of the initial parameters of the potential (the one for which we search for the error) is changed by a small fraction (*i.e.*, a proposed error) and then fixed. In the second step, all other parameters of the potential are refitted (using a gradient descent method) to obtain the best possible agreement between the new simulation (performed using LEVEL program [4]) and the experimental spectrum. In the last step, an assessment of the agreement between new, temporary simulation and the experimental spectrum is performed. The result of the assessment is described by an χ^2 coefficient:

$$\chi^2 = \frac{1}{N}\sum_{i=1}^{N}\left(I_i^{sim} - I_i^{expt}\right), \tag{1}$$

where I_i^{sim} and I_i^{expt} are intensities of the simulated and the experimental spectra, respectively, measured for the *i*-th point of the spectrum, and N is the total number of points in the spectrum. If the χ^2 calculated according to Eq. (1) is not significantly different from the coefficient χ_0^2 calculated for initial parameters of the potential (*e.g.*, $\chi^2 \leq 1.2\chi_0^2$), we assume that the error should be larger, so we repeat the whole procedure until the χ^2 exceeds certain limit. As an example, we tried to assess the uncertainty of a Morse potential parameters for the excited state of the $A^30^+(5^3P_1) \leftarrow X^10^+(5^1S_0)$ transition in CdAr. To do this, we simulated the reference spectrum for parameters of a Morse potential from Ref. [5] (R_e'=3.510[Å], D_e'=314.88[cm^{-1}], β'=1.4632[Å^{-1}]). In the error estimation procedure, we used the reference spectrum instead of the real experimental spectrum to perform the proposed calculations. The results of the error assessment are presented in Table 1.

Table 1. Estimated values of errors of parameters of a Morse potential representing the $A^30^+(5^3P_1)$ state in CdAr. Performing the estimation we assumed that the temporary simulation is indistinguishable from the reference spectrum if $\chi^2 \leq 1.2\chi_0^2$.

Parameter	Value	Estimated error
R_e' [Å]	3.510	0.005
D_e' [cm^{-1}]	314.88	0.35
β' [Å^{-1}]	1.4632	0.0025

3. Conclusions

In this work we present a method for estimation of errors of analytical interatomic potential parameters which can be used even when the potential parameters are correlated. Possibly, after small modifications, the method can be used for pointwise interatomic potentials.

Acknowledgements
This work was supported by the National Science Centre Poland under grant number UMO-2015/17/B/ST4/04016.

References

[1] Urbańczyk T, Koperski J 2014 *Mol. Phys.* **112** 2486–94
[2] Urbańczyk T, Koperski J 2014 *Chem. Phys. Lett.* **591** 64–68
[3] Urbańczyk T, Koperski J 2015 *Chem. Phys. Lett.* **640** 82–6
[4] LeRoy R J 2002 *LEVEL 8.0 A Computer Program for Solving the Radial Schrödinger Equation for Bound and Quasibound Levels* (University of Waterloo Chemical Physics Research Report) CP-655R
[5] Urbańczyk T, Strojecki M, Koperski J 2011 *Chem. Phys. Lett.* **503** 18–24

IOP Conf. Series: Journal of Physics: Conf. Series **810** (2017) 012042 doi:10.1088/1742-6596/810/1/012042

Ultra accurate measurements and *ab initio* calculations of collisional effects in pure D_2.

Mikołaj Zaborowski*[1], Piotr Wcisło[1], Franck Thibault[2], Szymon Wójtewicz[1], Agata Cygan[1], Grzegorz Kowzan[1], Piotr Masłowski[1], Daniel Lisak[1] and Roman Ciuryło[1]

[1] Institute of Physics, Faculty of Physics, Astronomy and Informatics, Nicolaus Copernicus University, Grudziadzka 5, 87-100 Torun, Poland

[2] Institut de Physique de Rennes, UMR CNRS 6251, Université de Rennes 1, Campus de Beaulieu, Bât.11B, F-35042 Rennes, France

E-mail: *zaborowski@fizyka.umk.pl

Abstract. We present our experimental spectra of the very weak S(2) transition from the 2–0 band of molecular deuterium, measured with a frequency-stabilized cavity ring-down spectroscopy (FS–CRDS) assisted by the optical frequency comb (OFC). Experimental collisional broadening and shifting are compared with results of *ab initio* quantum scattering calculations.

1. Introduction

With recent developments in the theoretical calculations of molecular state energies of H_2 and its isotopologues [1,2] the transition frequencies with uncertainties exceeding the level of 10^{-3} cm^{-1} for the first overtone band 2-0 became available [3]. Such predictions provide an opportunity to test relativistic and quantum electrodynamics corrections. At this level of accuracy, the line-shape effects, including asymmetry, affect the uncertainty of the determination of H_2, HD and D_2 line positions in the Doppler limit [4].

We have measured the self-broadened spectral line shapes of the very weak S(2) transition of deuterium in the 2–0 band up to the pressure of about 1500 Torr. The collisional effects on the spectral line shape are described using an advanced *ab initio* model. The line has been measured using the well established frequency-stabilized cavity ring-down spectroscopy (FS–CRDS) with the optical frequency axis provided by an optical frequency comb (OFC) [5,6]. Results from the experiment are compared with our *ab initio* quantum scattering calculations of the collisional broadening and shifting coefficients.

2. Experimental setup

The frequency-stabilized cavity ring-down spectrometer used in the experiment has been previously described in [7,8,9]. Here we only provide the details of the recently developed thermal stabilization of the cavity, see Fig. 1.

In our thermal stabilization system four QC–127–1.4–8.5MD *Quick-Cool* Peltier elements, mounted in the top of an insulating box, were used to provide most of the heating or cooling of the optical cavity. ICK PGA 21 x 21 *Fischer Elektronik* heatsinks as well as EB40201S2-000U-999 *SUNON* fans are mounted on both sides of the Peltier devices. Thermoelectric modules are current-driven by a PI controller with an accuracy of 0.01 mA . Depending on the temperature of the cavity, a significant gradient is observed along the cavity. Therefore, three sets of *Minco* thermofoil, silicone-based heaters are attached to different parts of the cavity. Another PI controller (working independently of the main one) is used to make fine local changes of temperature and eliminates the temperature gradient in the cavity.

Three TCS610 *Wavelength Electronics* temperature sensors powered via a 4-wire circuit are

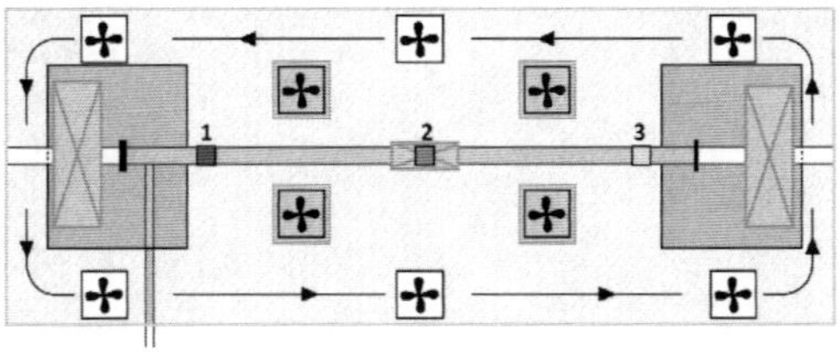

Figure 1. Thermal stabilization of the cavity. Four Peltier cooling/heating modules (orange squares on the top center part above the cavity) effect the major changes of the temperature. For the fine corrections three heaters are directly attached to the cavity (green rectangles). All heaters are thermally isolated from the air next to the cavity and from the thermistors. This way it is possible to reduce temperature differences at the points 1 (violet), 2 (red) and 3 (green) where thermistors are placed.

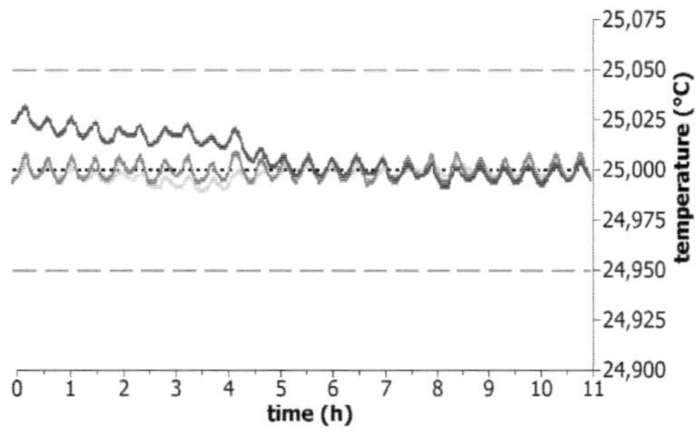

Figure 2. Temperature graph for the stabilization at the temperature of 25°C at the points 1 (violet), 2 (red) and 3 (green) near the cavity, see Fig. 1. With small gradient at the beginning, all temperatures stabilize to the required values in several hours. Orange lines show the deviation from the setting point by 50 mK, which corresponds to the accuracy of the measurement.

arranged at the two sides and at the center of the cavity. They are calibrated with a *Fluke* 5640 Series thermistor at an accuracy of 10 mK at room temperature and 50 mK in the full range from 15°C to 60°C. The heating or cooling is sped up by six EB40201S2-000U-999 *SUNON* fans mounted around the cavity, which also reduce temperature gradient of the gas in the optical cavity. In addition, an electronic system was built to automatically shut down the power supply for the Peltier modules when any heatsink temperature is higher than a set point.

The absolute frequency of the cw-ECDL laser is determined by measuring the heterodyne beat of the CW laser and an optical frequency comb. Both the OFC and the RF signal driving the AOM are referenced to the UTC(AOS) primary frequency standard, provided by the OPTIME network [10,11] from AstroGeodynamic Observatory in Borowiec (Poland) [12,13].

3. Preliminary results and discussion

The spectra were recorded in several pressures between 380 Torr and 1483 Torr and at a temperature of 15.2°C. The average spectrum over 5 scans for the highest pressure is shown in Fig. 3. The measured spectral line shapes were fitted with the Nelkin-Ghatak profile (NGP) [14] which, beyond the Voigt profile approximation, takes into account also the velocity-changing collisions leading to Dicke narrowing. The significant improvement in the quality of the fit (QF) [15], as compared to the VP fit, confirms a large contribution of Dicke narrowing to the line shape. However, the asymmetry of the line is clearly visible. It is a manifestation of the speed-dependence of the collisional shift. This can be confirmed by the flat residuals when the speed-dependent Nelkin-Ghatak profile (SDNGP) [16] is fitted (see Fig. 3). We used the NGP for the estimation of experimental values for pressure broadening and shifting coefficients as well as unperturbed frequency of investigated transition.

The *ab initio* approach was used to compare experimental pressure broadening and shifting

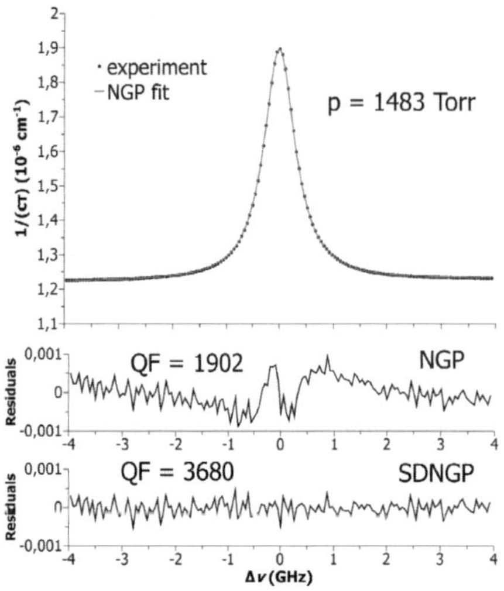

Figure 3. Experimental spectrum of the S(2) 2–0 line of D_2 averaged over 5 scans, recorded with the FS–CRDS technique. Below residuals of the NGP fit (with QF=1902) and the SDNGP fit (with QF=3680).

coefficients with theoretical ones. We calculated the scattering S-matrix by solving the close-coupling equations for the quantum scattering problem. From the S-matrix we determined the real and imaginary parts of the generalized spectroscopic cross sections, which provide the collisional broadening γ_L/p and shifting δ/p, respectively. The quantum scattering calculations were performed on potential energy surfaces (PES) given by Hinde [17]. The frequency of velocity-changing collisions $\nu_{\rm VC}$ is calculated from the diffusion coefficient taken from Ref. [18].

The theoretical values were compared with the experimental ones obtained from the analysis with the NGP, see Table 1. The values calculated this way show only 6.3% disagreement between the collisional widths but 33% disagreement in collisional shift. This discrepancy in shift, caused by fitting an oversimplified profile neglecting speed-dependent effects, could be explained by non-linear dependence of the collisional shift on pressure [4]. A similar explanation can be given for the large relative difference between the experimentally determined frequency of the velocity-changing collisions $\nu_{\rm VC}$ and the calculated one [18].

Finally, we estimated the unperturbed frequency of the S(2) transition from D_2 2–0 band to be equal to 187104.2909(5) GHz. Adding a correction value of 13.2 MHz resulting from the non-linear dependence of the collisional shift on pressure, estimated for a different transition in H_2 [4], to our transition frequency leads to a reasonable agreement with the results of both Mondelain *et al.* [3] — 187104.29951(50) GHz, and Komasa *et al.* [19] — 187104.301(3) GHz.

It is clear that the line-shape model applied in this preliminary characterisation is insufficient. Therefore, a more physical model must be used to correctly describe the recorded spectra. The velocity-changing collisions could be described, for instance, by the hard-sphere approximation of the *ab initio* potential [20,21]. This will be important for further work on verification of quantum electrodynamics in molecular systems [19] as well as for searching for new physics beyond the Standard Model [22].

Table 1. Comparison between our theoretical and experimental (form the NGP fit) values of the collisional width γ_L/p, collisional shift δ/p and frequency of the velocity-changing collisions ν_{VC}.

	γ_L/p (10^{-3} cm^{-1}/amg) (FWHM)	δ/p (10^{-3} cm^{-1}/amg)	ν_{VC}(GHz) for 1483 Torr
calculated	7.21 (this work)	-1.58 (this work)	1.01 (Song [17])
measured (this work)	7.69(5)†	-2.232(7)‡	1.52(2)†

† Obtained from the spectra recorded at 1483 Torr and at 15.2°C.
‡ Obtained from the linear fit performed with the spectra recorded in four pressures (380 Torr, 703 Torr, 1119 Torr and 1483 Torr) at 15.2°C.

Acknowledgments

The project is supported by the National Science Centre, Poland, Projects Nos. 2015/19/D/ST2/02195, DEC-2013/11/D/ST2/02663, 2015/17/B/ST2/02115 and DEC-2012/05/D/ST2/01914. The research is part of the program of the National Laboratory FAMO in Torun, Poland. It is also supported by the COST Action CM1405 MOLIM. The financial support provided by the French-Polish "Programm Hubert Curien" POLONIUM program is acknowledged. P. M. and G. K. are supported by the Foundation for Polish Science HOMING PLUS project.

References

[1] Pachucki K and Komasa J 2010 *Phys. Chem. Chem. Phys.* **12** 9188
[2] Pachucki K and Komasa J 2015 *J. Chem. Phys.* **143** 034111
[3] Mondelain D, Kassi S, Sala T, Romanini D, Gatti D and Campargue A 2016 *J. Mol. Spectrosc.* **326** 5–8
[4] Wcisło P, Gordon I E, Cheng C-F, Hu S-M and Ciuryło R 2016 *Phys. Rev.* A **93** 022501
[5] Domysławska J, Wójtewicz S, Lisak D, Cygan A, Ozimek F, Stec K, Radzewicz C, Trawiński R S and Ciuryło R 2012 *J. Chem. Phys.* **136** 024201
[6] Long D A, Cygan A, van Zee R D, Okumura M, Miller CE, Lisak D and Hodges J T 2012 *Chem. Phys. Lett.* **536** 1–8
[7] Cygan A *et al.* 2015 *Opt. Express* **23** 14472
[8] Cygan A, Wójtewicz S, Zaborowski M, Wcisło P, Guo R, Ciuryło R and Lisak D 2016 *Meas. Sci. Technol.* **27** 045501
[9] Cygan A *et al.* 2016 *J. Chem. Phys.* **144** 214202–1–11
[10] Śliwczyński L *et al.* 2013 *Metrologia* **50** 133
[11] Morzyński P *et al.* 2015 *Sci. Rep.* **5** 17495
[12] Azoubib J *et al.* 2003 *Metrologia* **40** S245
[13] Jiang J *et al.* 2015 *Metrologia* **52** 384
[14] Nelkin M and Ghatak A 1964 *Phys. Rev.* **135** A4
[15] Wcisło P, Cygan A, Lisak D and Ciuryło R 2013 *Phys. Rev. A* **88** 012517
[16] Lance B, Blanquet G, Walrand J and Bouanich J-P 1997 *J. Mol. Spectrosc.* **185** 262–71
[17] Hinde R 2008 *J. Chem. Phys.* **128** 154308
[18] Song B, Wang X and Liu Z 2013 *Mol. Phys.* **111** 49–59
[19] Komasa J, Piszczatowski K, Łach G, Przybytek M, Jeziorski B and Pachucki K 2011 *J. Chem. Theory Comput.* **7** (10) 3105-15
[20] Ciuryło R, Shapiro D A, Drummond J R and May A D 2002 *Phys. Rev. A* **65** 012502
[21] Wcisło P, Tran H, Kassi S, Campargue A, Thibault F and Ciuryło R 2014 *J. Chem. Phys.* **141** 074301
[22] Salumbides E J, Koelemeij J C J, Komasa J, Pachucki K, Eikema K S E and Ubachs W 2013 *Phys. Rev. D* **87** 112008

IOP Conf. Series: Journal of Physics: Conf. Series **810** (2017) 012043 doi:10.1088/1742-6596/810/1/012043

The problem of numerical precision in spectral line shapes calculations

P Ablewski, P Wcisło and R Ciuryło

Institute of Physics, Faculty of Physics, Astronomy and Informatics, Nicolaus Copernicus University in Toruń, Grudziadzka 5, 87-100, Toruń

E-mail: piotra@fizyka.umk.pl

Abstract. Spectral line shapes can be described by the transport-relaxation equation (TRE). When *ab initio* collisional operator is incorporated, the TRE needs to be solved numericaly. We report a pure numerical problems encountered during tests of the iterative approach to solving the TRE. As a reference we have used Voigt profile, which can be easily calculated analytically with error function, as well as numerically by solving TRE with simple collisional operator. Our studies lead us to the conclusions about impact of numerical precion and matrix operators dimensions on the accuracy of the calculations.

1. Introduction

Modern high resolution spectroscopy provides experimental data with extremely high singnal-to-noise ratio [1, 2]. The theoretical analysis of these data should come with an accuracy at comparable level. The realistic theory must take into account many physical aspects and in ideal case be based on interaction potentials obtained with *ab initio* calculations [4, 3]. It is possible to find exact analytical formula describing spectral line shape. However, in general this is complicated or even impossible to solve analytically. In that case there is a need to find a stable and reliable numerical method providing a solution with the smallest posibble errors in a finite number of steps. The standard approach [5, 6] converts the physical problem to the set of linear equations which can be easly solved for high and middle pressures, where collsional width is larger or comparable to Doppler one. It falls down in cases of low pressures [7], where the Doppler effect is dominant. Iterative method of solving TRE [7] enables calculations for a wide range of pressures, including zero-pressure limit.

2. Iterative approach to stationary solution of transport-relaxation equation

According to [6], it is convinient to write TRE in as

$$1 = -i(\omega - \omega_0 - \vec{k} \cdot \vec{v})h(\omega, \vec{v}) - \hat{S}^f h(\omega, \vec{v}), \tag{1}$$

where ω_0 - ω is detuning from the resonant frequency, $\vec{k}$ is the wave wector, $\vec{v}$ is the absorber velocity and $\hat{S}^f$ is the collisional operator describing changes of both internal and translational state under collisions. The function $h(\omega, \vec{v})$ can be used to calculate the line-shape profile as

$$I(\omega) = \frac{1}{\pi} Re \int d^3\vec{v}\ f_m(\vec{v})h(\omega, \vec{v}), \tag{2}$$

where $f_m(\vec{v})$ is the Maxwellian velocity distribution. In the case of low pressures, where the usual solution of TRE fails, there is a need to modify Eq. (1) to ensure that the iterative process coverges. This can be done by introduction of nonphysical parameter Γ_{num} to Eq. (1)

$$1 = [\Gamma_{num} - i(\omega - \omega_0 - \vec{k}\cdot\vec{v})h(\omega,\vec{v})] - (\Gamma_{num} + \hat{S}^f)h(\omega,\vec{v}). \quad (3)$$

After some mathematical operations and decomposition of operators in finite-dimensional Burnet basis [8, 9], iterative form of the solution of TRE can be written as [7]

$$\mathbf{c}^{n+1}(\omega) = \mathbf{a}(\omega) + \mathbf{A}(\omega)\cdot\mathbf{B}\cdot\mathbf{c}^n(\omega), \quad (4)$$

where $\mathbf{c}^n$ is a vector of solutions in n-th iterative step, starting from $\mathbf{c}^0[0] = 1$ and $\mathbf{c}^0[\neq 0] = 0$, $\mathbf{A}$ is a complex matrix depending on detuning from resonance and Doppler shift, $\mathbf{B}$ is a matrix describing collisions and relaxation process (matrix form of collisional operator decomposed in Burnett basis) and $\mathbf{a}$ is a first column of $\mathbf{A}$ matrix. The dimensions of matrices are $(N_{max}\cdot L_{max})\times(N_{max}\cdot L_{max})$ corresponding to the dimensions of Burnet function basis used for decompositon of the operators. These functions are enumerated by a pair of indexes n, l such that $n = 0,\cdots,N_{max}$ and $l = 0,\cdots,L_{max}$. Further explanation of decomposition and forms of matrix operators can be found in [7].

3. Results

We have performed tests of the iterative method of solving transport-relaxation equation for Voigt profile, which has a simple analytical formula given by Faddeeva function [10]. In this case, the collisional operator $\hat{S}^f$ acts as a scalar quantity equal to $-\Gamma$, where Γ is a collisional half width at half maximum (HWHM) of the spectral line. Doppler effect is treated independently of collisional operator and taken into account in $\mathbf{A}(\omega)$.

In our calculations the ratio between collisional and Doppler line width was equal to $\frac{\Gamma}{\omega_D} = 1$. We have defined the iterative coverence as $|c^n[0] - c^{n-1}[0]|$ for different dimensions of operators matrices. We have used a single numerical precision `binary32` defined by IEEE 754-2008 standard [11].

As shown in Fig. 1, numerical accuracy of method depends on matrices' dimensions. For small dimensions of matrices, the convergence criterion leads to the proper solution. Further iterations do not destroy the agreement with analytical form of the used profile. For the optimal base dimension in single precision calculations, cnovergence criterion does not lead to the minimum of the difference between analytical and numerical solution. Interesting fact is that this difference stabilizes at the same level during further iterations. In case of too large base, range between minimum and maximum value in $\mathbf{A}$ matrix can not be covered by representation of floating point values in the used arithmetic. This causes the destabilization of iterative process and leads to a wrong solution. It means that it is not possible to arbitrarily extend matrixces' dimensions using the same numerical precision. This effect is caused by accumulation of numerical erros and finite representation of significand in floating point arithmetics.

4. Conclusions

We have performed simple test to check why extending of the dimensions of matrix operators does not lead to the better convergence of the calculations. The main problem is numerical precision which is not able to handle range of the values of perturbative corrections. There is only one remedy for that problem - use of larger precision, which could encode value with more significant digits. In this paper we have formed the research problem such that iterative approach to solving TRE for spectral line shapes calculations needs to put more attention on numerical precision and values of nonphysical parameters.

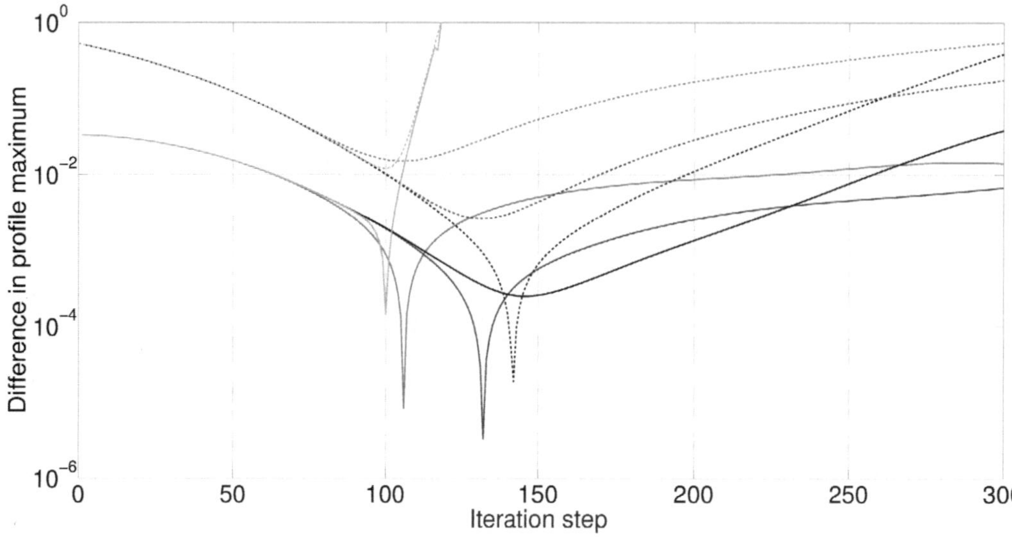

Figure 1. Evolution of the iterative process as a function of the number of steps. Solid lines indicate iterative process covergence defined as $|c^n[0] - c^{n-1}[0]|$ and dashed lines show difference between analytical solution and value obtained by iterative method in n-th step. Colors refer to $L_{max} = N_{max}$ dimensions as following: red - 4, blue - 6, black - 7, yellow - 8.

Acknowledgments
This work is supported by the National Science Centre, Poland, Project No. 2015/19/D/ST2/02195. The research is part of the program of the National Laboratory FAMO in Toruń, Poland. It is also supported by the COST Action CM1405 MOLIM.

References
[1] Cygan A, Lisak D, Wójtewicz Sz, Domysławska J, Hodges J T, Trawiński R S and Ciuryło R 2012 *Phys. Rev. A* **85** 022508
[2] Lin H, Reed Z D, Sirinneau V T and Hodges J T 2015 *J. Quant. Spectrosc. Radiat. Transfer* **161** 11
[3] May D A, Liu W K, McCourt F R W, Ciuryło R, Sanchez-Fortún Stoker J, Shapiro D and Wehr R 2013 *Can. J. Phys.* **91** 879
[4] Blackmore R, Green S and Monchik L 1989 *J. Chem. Phys.* **91** 3846
[5] Blackmore R 1987 *J. Chem. Phys.* **87** 791
[6] Shapiro D A, Ciuryło R, Drummond J R and May A D 2001 *Phys. Rev. A* **65** 012501
[7] Wcisło P, Cygan A, Lisak D and Ciuryło R 2013 *Phys. Rev. A* **88** 012517
[8] Ciuryło R, Shapiro D A, Drummond J R and May A D 2001 *Phys. Rev. A* **65** 012502
[9] Lindenfield M J and Shizgal B 1979 *Chem. Phys.* **41** 81
[10] Gradshteyn I S and Ryzhik I M, 1965 *Table of Integrals, Series and Products* Academic Press. New York
[11] IEEE Computer Society 2008 *IEEE Standard for Floating-Point Arithmetic: IEEE Std 754-2008*

Computer simulation and experimental studies of the N I $3p\,^2S^o_{1/2} - 3d\,^2P_{1/2}$ spectral line Stark broadening

W Olchawa, A Bartecka and A Bacławski

University of Opole, Institute of Physics, ul. Oleska 48, 45-052 Opole, Poland

E-mail: bartecka@uni.opole.pl

Abstract. A computer simulation method, dedicated to lines, which energy levels are influenced only by the quadratic Stark effect, is introduced and tested in this paper. The simulated profiles of the well isolated N I line $3p^2S^o_{1/2}-3d^2P_{1/2}$ are compared with experimental ones.

1. Introduction

The computer simulation method (CSM) [1] is successfully used for calculation of the Stark broadening of the plasma spectral lines mainly for hydrogen, hydrogen-like and helium lines, e.g. [2]. Profiles obtained in this way, are characterized by high reliability. Whereas in the case of the line profiles of non-hydrogen atoms, the so-called $j(x)$ profile [3] is commonly applied in computation. The impact approximation for electrons, the quasi-static approximation for ions and the separation of the strong and the weak collisions are basic assumptions lying on the basis of the $j(x)$ model. The CSM do not require such simplifications. In this paper the CSM for non-hydrogen lines is presented and delivered by this method results are compared with other ones. The required in semi-empirical calculation atomic data (line strengths) have been taken from NIST [4].

2. Simulation method

The model of the plasma is essentially the same as in commonly used by computer simulation method (CSM) [1]. Ions and electrons, the charged components of plasma, are treated as classical particles, which move along straight-line trajectories independently from each other and from the emitter. The shape of simulated volume is a sphere, with the electrically neutral emitter resting at its center. The Schrödinger equation describes an evolution of a quantum state of the emitter caused by varying in time electric fields produced by ions and electrons. The starting point of this consideration is the relation between the spectral line profile and the average of the dipole autocorrelation function $C(t)$, which can be written in the following way:

$$I\left(\Delta\omega\right) = \lim_{t_f\to\infty} \pi^{-1} \int_0^{t_f} C\left(t\right) e^{i\Delta\omega t} dt, \qquad C\left(t\right) = \mathrm{Tr}\left\{\vec{d}_{if} \cdot U^{+}_{ff'}\left(t,0\right) \vec{d}_{f'i'}\, U_{i'i''}\left(t,0\right) \rho_{i''i}\right\}_{\mathrm{av}}, \quad (1)$$

where $\vec{d}$ is the dipole operator for the emitter, while i, i' and f, f' indicate the sublevels of the initial (E_i) and final (E_f) states of the unperturbed atom, respectively. The frequency separation from the line center is given by $\Delta\omega = \omega - (E_i - E_f)/\hbar$, whereas $U(t,0)$ is the operator of the time development of the emitter in the presence of the electric field produced by electrons and ions. The averaging $\{\cdots\}$ is taken over all initial simulated field strengths and possible time histories. The time-evolution operators $U_{ii'}(t,0)$ and $U^{+}_{ff'}(t,0)$ (corresponding to the initial and final states, respectively) satisfy the following Schrödinger equation:

$$i\hbar\dot{U}(t,0) = [H_0 + V(t)]\,U(t,0), \tag{2}$$

where H_0 is the Hamiltonian of the isolated emitter and $V(t) = -\vec{d}\cdot\vec{F}(t)$ is the emitter-plasma interaction potential. The Strang symmetrical splitting formula [5], was applied to solve the Schrödinger equation of evolution, leads to unitary evolution of time:

$$U(t+\Delta t, t) \cong \exp(id_xF_x\Delta t/2\hbar)\exp(id_yF_y\Delta t/2\hbar)\exp(id_zF_z\Delta t/2\hbar)\times \exp(-iH_0\Delta t/\hbar)\exp(id_zF_z\Delta t/2\hbar)\exp(id_yF_y\Delta t/2\hbar)\exp(id_xF_x\Delta t/2\hbar). \tag{3}$$

3. Comparison of the simulation and experimental results

Line shape measurements were performed using a wall-stabilized arc, operated at atmospheric pressure. The radiation of the plasma, emitted from homogeneous plasma layers in end-on direction, was measured using a grating-spectrometer with a two-dimensional CCD detector. In this way the spectrum corresponding to the N I $3p\,^2S^o_{1/2} - 3d\,^2P_{1/2}$ spectral line have been measured. In Figure 1 the comparison of the simulated profile with the pure experimental one is shown.

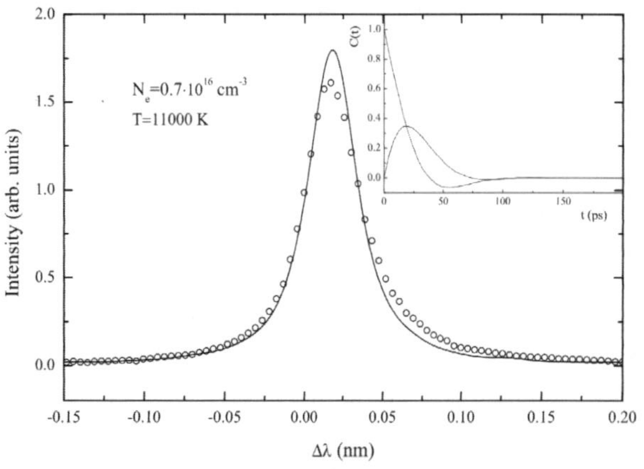

Figure 1. The simulated profile (solid line) of the spectral line N I $3p\,^2S^o_{1/2} - 3d\,^2P_{1/2}$ is presented and compared with experimental result (circles). Doppler and instrumental broadening are included. As can be seen, the experimental profile is more asymmetric than the simulated one. Real and imaginary parts of the autocorrelation function are shown in the inset.

Experimental values of Stark broadening parameters have been obtained from the best fit procedures applying the Griem-like profile $j(x)$ [3], convoluted with the corresponding Doppler and apparatus profile. Profile $j(x)$ has three parameters: electron impact width w_e, electron impact shift d_e and ion asymmetry parameter A. We may regard all three parameters of the spectral line as free, or restrict the number of free parameters to two, using relation between w_e and A: $A = (w_{i0}/w_{e0})^{3/4}(N_e/10^{16})^{1/4}$, where the constant w_{i0} is related to the quadratic Stark splitting in the normal electric field F_0. This new approach to determination of the asymmetry parameter is described in detail in [6]. The results of the fitting procedure with two different w_{i0} parameters are shown in Figure 2.

In Figure 3 the electron impact width obtained from the fitting procedure for the simulated and measured profiles are compared. The results of the calculation performed according to the impact and quasi-static theory are also shown.

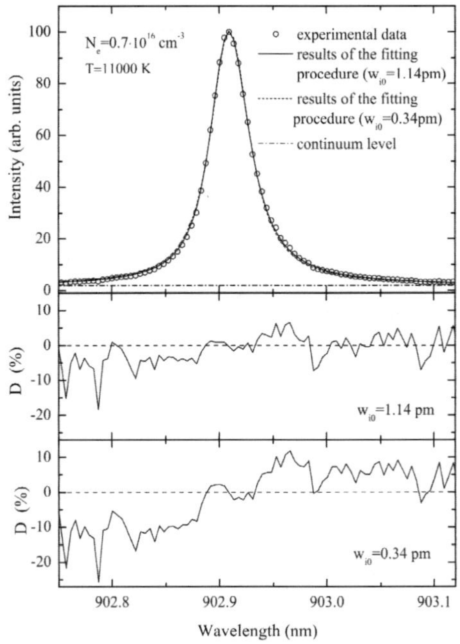

Figure 2. An example of final fitting procedure using Griem-like profile with constrained A ion broadening parameter. In the case of the investigated multiplet w_{i0} is experimentally evaluated to 1.14 pm, whereas its calculated, according to the Griem theory, value is 0.34 pm. The results of the fitting procedure using two different w_{i0} parameters are compared. The difference between the measured spectral intensity distribution and the fitted spectrum, expressed by the quantity $D = (I_{expt} - I_{fit})/I_{expt}$ for two parameters w_{i0} is presented. As can be seen the selection of the w_{i0} value doesn't have nearly any impact on the width and shift of the investigated profile, however is important for ion asymmetry parameter determination.

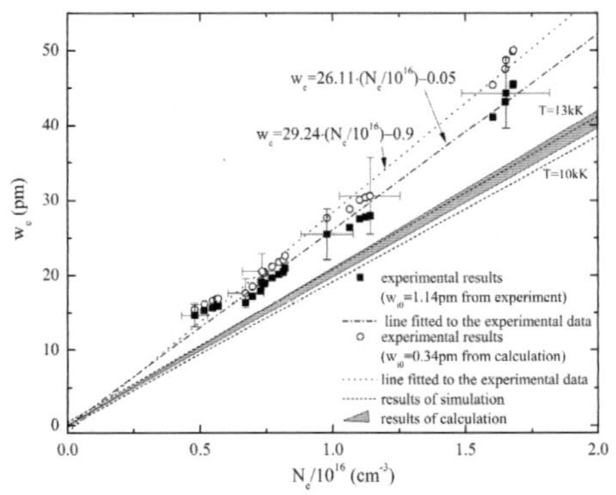

Figure 3. Measured (squares and circles), simulated (dashed lines) and calculated according to Griem theory (grey area) electron impact widths (w_e) of the studied N I line are presented. The w_e as a function of the electron density is shown. The simulation method gives result similar to the calculated according to Griem theory result and similar to the experimental ones.

Acknowledgments
Calculations were carried out in Wroclaw Centre for Networking and Supercomputing (http://www.wcss.wroc.pl)

References
[1] Seidel J and Stamm R 1982 *J. Quant. Spectr. Rad. Transf.* **27** 499
[2] Gigosos M and Cardeñoso V 1996 *J. Phys. B* **29** 4795
[3] Griem H R 1974 *Spectral line broadening by plasmas* (Academic Press, New York)
[4] *NIST Atomic Spectra database* http://physics.nist.gov/PhysRefData/ASD/index.html
[5] Strang G 1968 *SIAM J. Numer. Anal.* **5** 506
[6] Bartecka A *et al* 2013 *Eur. Phys. J ST* **222** 2381

EUV induced photoionized plasma, a medium for spectral investigation

A Bartnik, P Wachulak, H Fiedorowicz, W Skrzeczanowski, T Fok, R Jarocki, L Węgrzyński
Institute of Optoelectronics, Military University of Technology, Warsaw, Poland

Email: andrzej.bartnik@wat.edu.pl

Abstract. In this work a method of low temperature photoionized plasma creation is presented. Photoionized plasmas are produced by irradiation of gases with extreme ultraviolet (EUV) radiation pulses of high intensity. The gases are injected into a vacuum chamber synchronously with the EUV pulses. Such configuration allows for spectral investigation in the EUV range. Examples of emission and absorption spectra are shown.

1. Introduction

Photoionized plasmas produced by irradiation of gases with intense soft X-ray (SXR) or extreme ultraviolet (EUV) pulses are interesting objects for spectral investigations. This kind of plasmas are produced for experimental simulation of similar plasmas formed in Space, especially close to X-ray radiating compact stars. These experiments are called laboratory astrophysics and are conducted using high energy density facilities delivering X-ray pulses of extremely high energy [1,2]. Photoionized plasmas can be also produced using small scale laboratory SXR/EUV sources. In this case interaction of the ionizing radiation with a small portion of gas results in creation of a low temperature plasma. In contrast to low temperature plasmas produced using standard generators employed in technological processes, electron density of photoionized plasmas can be a few orders of magnitude higher, reaching 10^{18} cm^{-3} [3]. Apart from that such plasmas are produced locally in a vacuum chamber, hence, can be investigated using different methods requiring at least low vacuum environment, including EUV/SXR spectroscopy. Parameters of low temperature photoionized plasmas created in laboratory can be easy controlled and adjusted for specific conditions that can be encountered in Space. It concerns for example hydrogen plasmas with parameters similar to photospheres of white dwarfs. The corresponding experiments allowing for investigation of hydrogen Balmer emission and absorption line profiles were performed at the Z Pulsed Power Facility at Sandia National Laboratories [4]. Plasmas of similar parameters can be also produced using the above mentioned low energy EUV/SXR sources.

In this paper some results concerning experimental investigations of low temperature, photoionized plasmas are presented. Photoionized plasmas were created by EUV irradiation of small portions of gases injected into the interaction region synchronously with the radiation pulse. A laser plasma EUV source based on a 10 J/ 10 ns Nd:YAG laser system and equipped with a multifoil [5] or ellipsoidal [6] EUV collector was employed for the gas irradiation. Measurements of emission and absorption spectra for photoionized plasmas created this way were performed. Examples of emission spectra acquired in UV/VIS and EUV ranges are presented. Absorption spectra measured in the EUV range for photoionized plasma created in neon are also shown.

2. Emission spectra

The EUV emission and absorption spectra were recorded using a grazing incidence, flat-field spectrometer (McPherson, Model 251), equipped with a 450 lines/mm toroidal grating. The spectral range of the spectrometer was $\lambda = 10 - 95$ nm and the resolution $\lambda/\Delta\lambda \approx 500$. Optical axis of the spectrometer was perpendicular to the optical axis of the EUV collector thus the emission spectra are well separated from the spectra of laser-plasma radiation. It is very important because the laser-plasma emission is approximately 10^5 times higher comparing to the photoionized plasma emission. The spectra, integrated over 50 pulses, were recorded using a CCD detector, cooled down to a temperature of -65°C.

The second instrument was an Echelle Spectra Analyzer ESA 4000, equipped with the ICCD Kodak KAF 1001 camera. The spectrometer system allowed for simultaneous measurements of complex spectra within the wide ultraviolet and visible (UV/VIS) spectral range, $\lambda = 200 \div 780$ nm. Spectral resolution of the spectrograph was $\lambda/\Delta\lambda \approx 20000$. Similarly to the EUV spectrograph, optical axis of the UV/VIS spectrograph was perpendicular to the optical axis of the EUV collector.

In Fig.1 examples of emission spectra obtained for photoionized plasmas created in molecular gases are presented. In case of the EUV measurements, the spectral lines originate exclusively from atomic ions. In case of the UV/VIS range mostly molecular spectra were recorded.

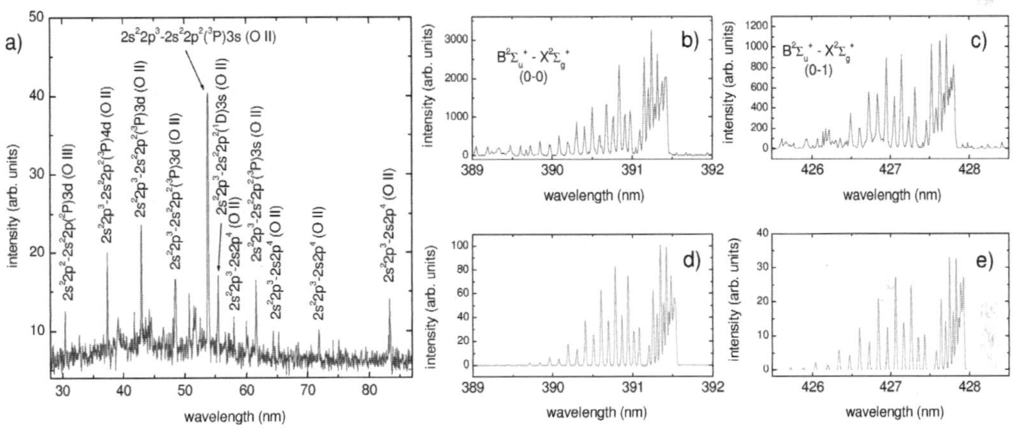

Figure 1. Emission spectra obtained from photoionized plasmas created in molecular gases: a) O II and O III EUV spectrum; b,c) N_2^+ spectra; d,e) N_2^+ spectra simulated using the LIFBASE [7] code

The EUV spectrum, measured for photoionized plasmas created in oxygen, contains mainly O II lines, however a single O III line was also recorded. Similar spectra were obtained for different gases, either atomic or molecular [5,6,8]. In most cases mainly singly charged ions were detected. Numerical simulations of EUV spectra performed for Ne, He and F plasmas allowed for estimation of electron temperatures [6,9,10] to the value of the order of 1 eV. Similar values were obtained employing Boltzmann plots constructed for S or Kr plasmas [10,11].

The UV/VIS spectrum measured for photoionized plasmas created in nitrogen, contains exclusively line series corresponding to excited N_2 molecules or N_2^+molecular ions. Numerical simulation of the N_2^+spectrum performed using the LIFBASE [7] code, for vibrational temperature of 1500 K and rotational temperature 130 K, gave spectral distribution similar to the measured one (Fig.1 b-e).

3. Absorption spectra

The absorption spectra were measured employing a continuous part of Xe laser plasma radiation for backlighting of photoionized plasmas. The most interesting spectra were obtained for He [6] and Ne (Fig.2) plasmas. In both cases the pronounced absorption lines were recorded for neutral atoms present in the photoionized plasmas. In case of not ionized gases absorption lines were hardly visible.

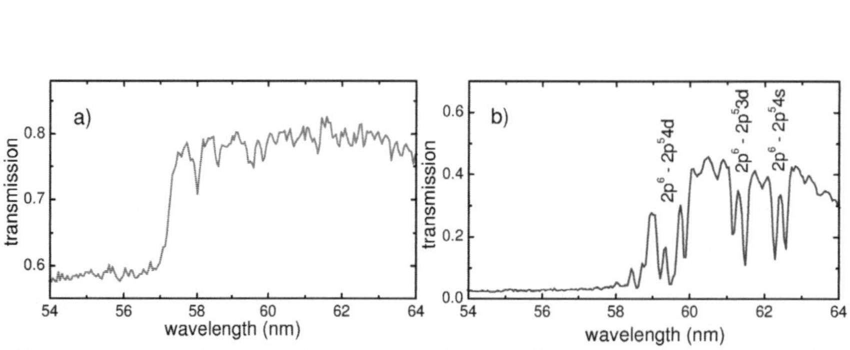

Figure 2. Absorption spectra obtained for a neon gas (a) and a Ne photoionized plasma (b)

4. Summary

A possibility of creation and spectral measurements of low temperature photoionized plasmas was presented. Photoionized plasmas were created by irradiation of gases, injected into the vacuum chamber, using a laser plasma EUV source. Examples of emission spectra acquired in EUV and UV/VIS ranges for plasmas created in molecular gases are shown. Numerical simulation of N_2^+ spectrum allowed for estimation of vibrational and rotational temperatures. Significant differences between absorption spectra obtained for Ne gas and low temperature Ne plasma are presented.

Acknowledgments

This work was supported by Grant No. UMO-2013/09/430 B/ST2/01625 of the National Science Centre, Poland

References

[1] S. Fujioka, H. Takabe, N. Yamamoto, D. Salzmann, F. Wang, H.Nishimura, Y. Li, Q. Dong, S. Wang, Y. Zhang, Y. Rhee, Y. Lee, J. Han, M. Tanabe, T. Fujiwara, Y. Nakabayashi, G. Zhao, J. Zhang, and K.Mima, Nature Phys. 5, 821–825 (2009)

[2] D. H. Cohen, J. J. MacFarlane, J. E. Bailey, and D. A. Liedahl, Rev. Sci. Instrum. 74, 1962 (2003)

[3] A. Bartnik, P. Wachulak, T. Fok, Ł. Wegrzynski, H. Fiedorowicz, T. Pisarczyk, T. Chodukowski, Z. Kalinowska, R. Dudzak, J. Dostal, E. Krousky, J. Skala, J. Ullschmied, J. Hrebicek, T. Medrik, Physics of Plasmas 22, 4, 043302 (2015)

[4] R.E. Falcon, G.A. Rochau, J.E. Bailey, J.L. Ellis, A.L. Carlson, T.A. Gomez, M.H. Montgomery, D.E. Winget, E.Y. Chen, M.R. Gomez, T.J. Nash. High Energy Density Physics 9, 82-90 (2013)

[5] A. Bartnik, P. Wachulak, H. Fiedorowicz, T. Fok, R. Jarocki, M. Szczurek, „Extreme ultraviolet-induced photoionized plasmas", Physica Scripta T161, 014061 (2014)

[6] A. Bartnik, P. Wachulak, H. Fiedorowicz, T. Fok, R. Jarocki, M. Szczurek, Physics of Plasmas 20, 11, 113302 (2013)

[7] J. Luque and D.R. Crosley, "LIFBASE: Database and spectral simulation (version 1.5)", SRI International Report MP 99-009 (1999)

[8] A. Bartnik, H. Fiedorowicz and P. Wachulak, Physics of Plasmas 21, 073303 (2014)

[9] A. Bartnik, R. Fedosejevs, P. Wachulak, H. Fiedorowicz, C. Serbanescu, E.G. Saiz, D. Riley, S. Toleikis, D. Neely, Laser and Particle Beams 31, 2, 195-201 (2013)

[10] A. Bartnik, P. Wachulak, H. Fiedorowicz, W. Skrzeczanowski, R. Jarocki, T. Fok, Ł. Wegrzynski, Journal of Instrumentation 11, Issue 3, C03009 (2016)

[11] A.Bartnik, P. Wachulak, H. Fiedorowicz, W. Skrzeczanowski, Physics of Plasmas 23, 043512 (2016)

A very large aperture spectrometer for low light optical emission spectroscopy

R Baude[1], A Escarguel[1], P David[1]

[1]Aix Marseille University, CNRS, PIIM UMR 7345, Marseille, France

E-mail: romain.baude@univ-amu.fr

Abstract. In low light experiments such as low temperature plasma or astronomical spectroscopy usual tools such as Czerny Turner based spectrometer (CTS) may have a limited efficiency due to the fact that a large entrance slit implies a large broadening of the instruments profile. To overcome the brightness issue of spectrometer, the so-called Bowen Chamber spectrometer (BCS) was designed to accept a large aperture slit, without significant deterioration of the spectrum. The BSC characteristics are presented and a close comparison with a CTS spectrometer is detailed. A simple geometrical model have been developed to still improve the BSC specifications.

1. Introduction

The Optical emission spectroscopy diagnostics have been widely used for many decades [1]. A simple experimental set up, composed of a spectrometer, can provide in-situ information regarding a light source from the electron, ion, neutral and metastable temperature and density [2]. Space and time resolved spectroscopy can be experimentally established readily with more care [3].

However, in low light experiments such as low temperature and low pressure plasma or astronomical spectroscopy usual tools such as Czerny Turner based spectrometer (CTS) may have a limited efficiency. The brightness limitation of the spectrometer is due, in the first place, to the fact that a large entrance slit aperture implies a large broadening of the spectrum. As a consequence most spectrometers are used with a maximum aperture of a few tens of micrometers to minimize the spectrum degradation, limiting in the same time the quantity of light entering the spectrometer.

Nevertheless, in the early seventies the CERCO (the French center for optical research and computation) designed the so-called Bowen Chamber spectrometer (BCS) under the supervision of Prof. A. Lesage from the "Observatoire de Meudon". The spectrometer was built to obtain plasma spectrum in low light shock wave experiments where the plasma is observed during only a few millisecond. The data were used to establish the transition rate coefficient and the Stark broadening of atomic spectrum for metallic atoms and ions [4].

This paper propose to reevaluate the specificity of the Bowen chamber spectrometer in low light experiments or for fast imaging spectroscopy. The authors do not clam the paternity of the design but wants to fill in the lack of information regarding the BCS in the literature.

2. The CERCO's Bowen Chamber spectrometer

The BCS was designed to collect a maximum of light, therefor it is characterized by a large numerical aperture of 0.294. The spectral dispersion and the magnification were experimentally measured at respectively 2.344 nm/mm and 0.14. The BCS is composed of three stages as seen in Figure 1: the collimating optics, the dispersive system and the Bowen Chamber.

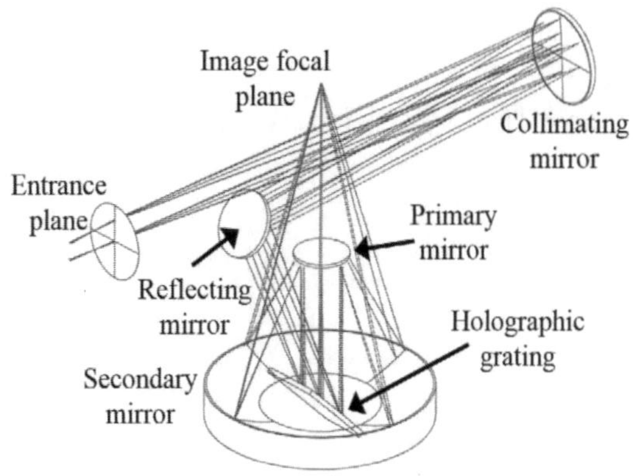

Figure 1. Schematic view of spectrometer showing the path of light throughout the BCS.

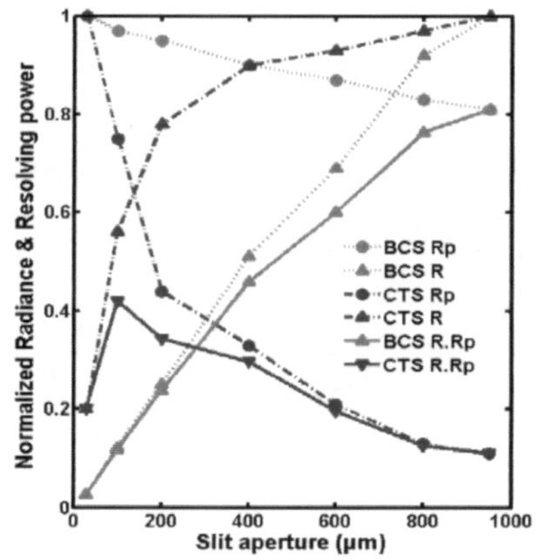

Figure 2. Comparison of the BCS and the CTS instrument profile for different slit aperture.

The collimating stage is composed of a micro-controlled entrance slit (ES) 60 mm wide. The slit opens symmetrically up to 1000 µm. Behind the ES a field lens gathers the light from the ES and direct it toward the collimating mirror. The collimating mirror has a focal length of 1300 mm. To reduce the size of the spectrometer a flat reflecting mirror guides the light toward the dispersive system. The second part of the spectrometer is composed of a holographic reflection grating of dimension 130x220 mm^2. The grating used during the experiments has 2230 g/mm and is optimized for radiations in the range 500 to 600 nm. The last part is the so-called Bowen chamber, composed of two spherical mirrors. The primary mirror is convex and gathers the light dispersed by the grating and reflects it toward the secondary mirror. This mirror is concave and images the light at the focal plane of the Bowen chamber.

3. BCS versus CTS

To evaluate the BCS brightness and instrument profile (IP) we use a classical Czerny-Turner spectrometer of focal length 500 mm, f/#=8.0 with an holographic grating blazed à 500 nm and dimension 64x64 mm^2. A He-Ne laser source at 632.8 nm has been used to provide a Dirac spectrum. The enlargement of the measured laser spectrum is assumed to be induced only by the IP of the spectrometer. Therefore, the determination of the full width at half maximum (FWHM) gives a direct access to the IP. The injection scheme and the detectors are strictly set in the same condition for both spectrometers. Experiments have established a resolving power of 0.15 nm and 0.18 nm respectively for the BCS and the CTS for an entrance slit of 30 µm. In the same condition the radiance is measured at 936.5 counts (cc) and 2560 cc respectively for the BCS and the CTS.

On Figure 2 is shown the evolution of the He-Ne spectrum normalized radiance and resolving power as a function of the entrance slit aperture from 30 to 950 µm. We see that the radiance profile of the CTS increases according to a saturation profile. On the other hand the BCS exhibits a linear radiance profile, from 30 µm up to 950 µm, where approximately doubling the ES aperture double the radiance. The resolving power profile point out the fact that the CTS quickly enlarges the He-Ne spectrum. For an ES aperture of 950 µm the spectrum is 90% larger than for an ES of 30 µm. The BCS presents a linear reduction of the resolving power with a maximum of 20% enlargement for an ES of 950 µm. As described in [5], the product of the normalized radiance and resolving power (R.Rp) indicates, when R.Rp reaches a maximum, the optimal slit aperture of a spectrometer. Figure 2 clearly shows that the CTS has an optimal slit aperture around 100 µm. Over this limit the gain in radiance does not justify the loss in resolving power. However the BCS R.Rp product reveals no maximum showing without ambiguity that the optimal ES aperture is wider than 950 µm.

Finally the absolute brightness of the two spectrometers is compared by measuring the light intensity at the image focal plane as a function of the light emitted by a calibrated black body source. The spectrometers are set to limit the IP enlargement to a few per cent. Figure 3 indicates, on a log-log scale, that the BCS (red down triangle) is 8 to 10 times brighter than the CTS (blue up triangle).

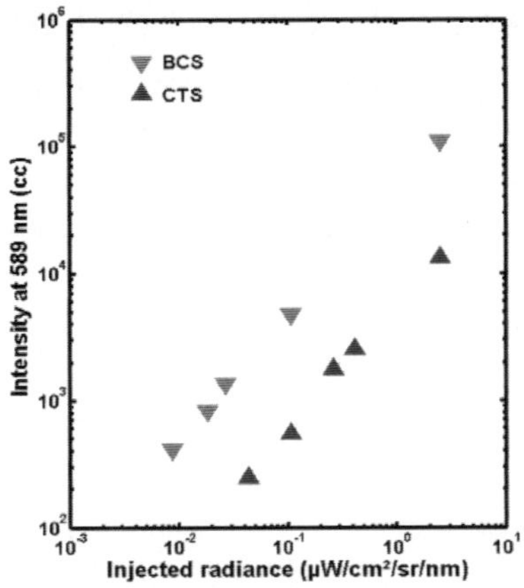

Figure 3. Absolut brightness comparison between the BCS and the CTS with a calibrated black body source. The spectrometers are set to optimise the spectral resolution.

4. Modelling and optimisation

Despite the brightness of the BCS attentive studies of the collimating stage have pointed out that a large part of the incoming light is lost during the collimating process. To optimize the light injection into the BCS a 3D geometrical ray tracing like model has been developed. Giving the optical characteristics of the different mirrors, the model gives information regarding the mirror position and dimensions. The model is able to quantify the light transmitted from the entrance slit to the image focal plane and gives insight regarding spherical and astigmatism aberration. This model is being used to design a new collimating stage for the actual Bowen chamber spectrometer.

5. Conclusion

The Bowen chamber spectrometer shows remarkable brightness due to its ability to work with a very large slit aperture up to 1000 µm. A 3D ray tracing model is in development to optimize the BCS and to guide us in the prototyping of a Bowen chamber based spectrometer with a reasonable size grating in order to provide a new powerful tool to spectroscopists.

Acknowledgment

This work has been carried out thanks to the support of the A*MIDEX project (n° ANR-11-IDEX-0001-02) funded by the "Investissements d'Avenir" French Government program, managed by the French National Research Agency (ANR)

References

[1] T. Cao, H. Zhang, B. Yan, W. Lu, and Y. Cheng, RSC Adv 4, 15131 (2014)
[2] J.B. Boffard et al., Plasma Sources Sci. Technol. 18, 035017 (2009)
[3] T. Okamoto, A. Takahashi, and I. Yamaguchi, Appl. Spectrosc. 47, 1198 (1993)
[4] A. Lesage et al., Phys. Rev. A, 16 (1977)
[5] P. Bousquet et P. Jacquinot, Spectroscopie Instrumentale, Dunot (1969)

Relativistic theory of excitation and ionization of Rydberg atomic systems in a Black-body radiation field

V V Buyadzhi, P A Zaichko, M Y Gurskaya, A A Kuznetsova, E L Ponomarenko, and V B Ternovsky

Odessa State Environmental University, L'vovskaya str.15, Odessa-16, 65016, Ukraine

E-mail: vbuyad@mail.ru

Abstract. The combined relativistic energy approach and relativistic many-body perturbation theory with the zeroth model density functional approximation are used for computing the thermal Blackbody radiation ionization characteristics of the Rydberg atoms, in particular, the sodium and caesium in Rydberg states with n=40-100. The comparison of the calculated ionization rate values with available theoretical and experimental data is carried out.

1. Introduction

This work goes on our research on the thermal Blackbody radiation (BBR) ionization characteristics of the Rydberg atoms. A great progress in experimental laser physics and appearance of the so called tunable lasers allow to get the highly excited Rydberg states of atoms. In fact this is a beginning of a new epoch in the atomic physics with external electromagnetic field. It has stimulated a great number of papers on the ad and dc Stark effect [1-12]. The experiments with Rydberg atoms had very soon resulted in the discovery of an important ionization mechanism, provided by unique features of the Rydberg atoms. Relatively new topic of the modern theory is connected with consistent treating the Rydberg atoms in a BBR field [1]. The account for the ac Stark shift, fast redistribution of the levels' population and photoionization provided by the environmental BBR became of a great importance for successfully handling atoms in their Rydberg states. The most popular theoretical approaches to computing ionization parameters of the Rydberg atom in the BBR are based on the different versions of the model potential (MP), quasiclassical methods. It should be mentioned the simple Simons-Fues MP approximation to compute the thermal ionization rate for Rydberg atoms [1]. In fact, using the MP approach is very close to the quantum defect method and other semi-empirical methods, which were also widely used in the past few years for calculating atom–field interaction amplitudes in the lowest orders of the perturbation theory. The significant advantage of the Simons-Fues MP method in comparison with other models is the possibility of presenting analytically (in terms of the hypergeometric functions) the quantitative characteristics for arbitrarily high orders, related to both bound–bound and bound–free transitions. Naturally, the standard methods of the theoretical atomic physics, including the Hartree-Fock and Dirac-Fock ones, should be used in order to determine a thermal ionization characteristics of neutral and Rydberg atoms [2]. A correct treating the heavy Rydberg atoms parameters in an external electromagnetic field requires using strictly relativistic models. Here we apply an energy approach [11-15] and relativistic perturbation theory (PT) with the model density functional Dirac-Kohn-Sham zeroth approximation [16-21] to computing the thermal BBR ionization characteristics of the Rydberg atoms, in particular, sodium (as a test) and caesium.

2. Ionization of the Rydberg atoms in the Blackbody radiation

Qualitative picture of the BBR Rydberg atoms ionization is in principle easily understandable. Even for temperatures of order T=10^4 K, the frequency of a greater part of the BBR photons ω does not exceed 0.1 a.u. One could use a single-electron approximation for calculating the ionization cross section σ_{nl} *(ω)*. The latter appears in a product with the Planck's distribution for the thermal photon number density:

$$\rho(w,T) = \frac{\omega^2}{\pi^2 c^3[\exp(\omega / kT) - 1]}, \tag{1}$$

where k=3.1668×10^{-6} a.u., K^{-1} is the Boltzmann constant, c = 137.036 a.u. is the speed of light. Ionization rate of the bound state nl results in the integral over the Blackbody radiation frequencies:

$$P_{nl}(T) = c\int_{|E_{nl}|}^{\infty} \sigma_{nl}(\omega)\rho(\omega,T)d\omega \cdot \tag{2}$$

The ionization cross-section from a bound state with a principal quantum number n and orbital quantum number l by photons with frequency ω is as follows:

$$\sigma_{nl}(\omega) = \frac{4\pi^2\omega}{3c(2l+1)}[lM^2_{nl\to El-1} + (l+1)M^2_{nl\to El+1}], \tag{3}$$

where the radial matrix element of the ionization transition from the bound state with the radial wave function $R_{nl}(r)$ to continuum state with the wavefunction $R_{El}(r)$ normalized to the delta function of energy. The corresponding radial matrix element looks as: $M_{nl\to El'} \sim \int_0^{\infty} R_{El'}(r)r^3 R_{nl}(r)dr\cdot$ We apply a generalized energy approach [11-15] to computing the Rydberg atoms ionization parameters. The radiation decay probability (cross-section) is connected with the imaginary part of electron energy shift. The latter is presented as: ΔE = ReΔE + i $\Gamma/2$, where Γ is a level width, and decay probability P = Γ. The imaginary part of the shift ΔE is defined in the PT second order as (in atomic units):

$$\mathrm{Im}\Delta E = (1/4\pi) \sum_{\alpha>n>f} V^{|\omega_{\alpha n}|}_{\alpha n\alpha n}, \tag{4}$$

where (α>n>f) for electron and (α<n<f) for vacancy. The matrix element is determined as follows:

$$V^{|\omega|}_{ijkl} = \iint dr_1 dr_2 \Psi^*_i(r_1)\Psi^*_j(r_2)\frac{\sin|\omega|r_{12}}{r_{12}}(1-\alpha_1\alpha_2)\Psi^*_k(r_2)\Psi^*_l(r_1) \tag{5}$$

Their detailed description of the matrix elements and procedure for their computing is presented in Refs. [12,13,15]. The relativistic wave functions are calculated by solution of the relativistic Dirac equation with the model Dirac-Kohn-Sham zeroth approximation plus correlation potential [16-21]. All calculations are performed on the basis of the numeral code Superatom-ISAN (version 93).

3. Results

In Table 1 we list the data of the ionization rate computing for the Rydberg sodium atom in the states (17,18D,18P) at temperatures of 300 K and 500 K: Th5 – our (relativistic MP theory) data [3], Th1-theory (nonrelativistic Simons-Fues MP) by Glukhov-Ovsyannikova [9], Th2- theory of Lehman [8], Th3- quasiclassical model by Dyachkov-Pankratov [10], Th4- theory by Beterov et al [1] and E1 – experimental data [4,5]. In whole there is physically reasonable agreement between the theoretical and experimental data. The accuracy of theoretical data is provided by a correctness of the corresponding relativistic wave functions and accounting for the exchange-correlation effects. In Table 2 we firstly list our data of ionization rate (s^{-1}) for the Cs Rydberg states induced by BBR radiation (T = 300 K) .

Table 1. Theoretical and experimental values of the ionization rate (10^3 s^{-1}) of sodium Rydberg states: E1- Kleppner et al Burkhardt etal; Th1 – theory by Glukhov-Ovsyannikova; Th2- theory by Lehman, Th3- theory by Dyachkov-Pankratov; Th4- theory by Beterov et al; Th5 – this work.

T	nL	E1	Th1	Th2	Th3	Th4	Th5
300	17D	10^3	$1.08\cdot10^3$	$0.95\cdot10^3$	$0.9\cdot10^3$	$1.147\cdot10^3$	$1.02\cdot10^3$
500	18P	-	$4.18\cdot10^3$	-	-	-	$5.54\cdot10^3$
500	18D	-	$4.07\cdot10^3$	-	-	-	$5.46\cdot10^3$

Table 2. Ionization rate (s^{-1}) for Cs Rydberg states (n= 40-100), induced by BBR (our data; see text)

Atom	40	50	70	100
Cs S	151	114	67.9	31.5
Cs P	452	330	170	81.1
Cs D	369	261	142	67.9

References

[1] Beterov I I, Tretyakov D V, Ryabtsev I I, Entin V M, Ekers A, Bezuglov N N 2009 *New J. Phys.* **11** 013052

[2] Li W, Noel M W, Robinson M P, Tanner P J, Gallagher T F 2004 *Phys. Rev. A.* **70** 042713

[3] Svinarenko A A, Khetselius O Yu, Buyadzhi V V, Florko T A, Zaichko P A, Ponomarenko E L 2014 *J. of Phys.: Conf. Ser.* **548** 012048

[4] Spencer W P, Vaidyanathan A, Kleppner D, Ducas T 1982 *Phys.Rev.A.* **26** 1490.

[5] Burkhardt C, Corey R, Garver W, Leventhal J, Allergini M, Moi L 1986 *Phys. Rev. A.* **34** 80.

[6] Malinovskaya S V, Glushkov A V, Khetselius O Yu, Loboda A V, Lopatkin Yu, Nikola L, Svinarenko A A, Perelygina T 2011 *Int.Journ.Quant.Chem.* **111** 288

[7] Khetselius O Yu, Florko T A, Svinarenko A A, Tkach T B 2013 *Phys. Scr.* **T153** 014037

[8] Lehman G W 1983 *J. Phys. B: At. Mol. Phys.* **16** 2145.

[9] Glukhov I and Ovsiannikov V 2009 *J.Phys.B:At.Mol.Phys.* **42** 075001.

[10] D'yachkov L G and Pankratov P M 1994 *J. Phys. B: At. Mol. Opt. Phys.* **27** 461.

[11] Glushkov A V and Ivanov L N 1992 *Phys. Lett. A* **170** 33; 1993 *J. of Phys. B* **26** L379

[12] Glushkov A V, Ambrosov S V, Ignatenko A V, Korchevsky D A 2004 *Int. Journ. Quant. Chem.* **99** 936; Glushkov A V 1988 *Sov. Journ. Struct. Chem.* **29**(4) 3

[13] Glushkov A V 2012 *Advances in the Theory of Quantum Systems in Chemistry and Physics. Ser. Progress in Theor. Chemistry and Physics*, vol. 28, ed. by K Nishikawa, J Maruani, E Brandas, G Delgado-Barrio, P Piecuch (Berlin: Springer) p 131

[14] Glushkov A V 2012 *J.of Phys.: Conf. Ser.* **397** 012011; 2014 *J.of Phys.: Conf. Ser.* **548** 012020

[15] Glushkov A V, Malinovskaya S V, Chernyakova Y G, Svinarenko A A 2004 *Int. Journ. Quant. Chem.* **99** 889; Khetselius O 2012 *J. of Phys.: Conf. Ser.* **397** 012012

[16] Glushkov A V, Khetselius O Yu, Svinarenko A A 2013 *Phys. Scr.* **T153** 014029

[17] Malinovskaya S V, Glushkov A V, Khetselius O Yu, Svinarenko A A, Mischenko E V, Florko T A 2009 *Int.Journ.Quant.Chem.* **109** 3325; Glushkov A V, Malinovskaya S V, Khetselius O Yu, Loboda A V, Sukharev D E, Lovett L 2009 *Int.Journ.Quant.Chem.* **109** 1717

[18] Khetselius O Yu 2009 *Phys.Scr.* **T135** 014023; 2009 *Int. J. Quant. Chem.* **109** 3330

[19] Glushkov A V, Malinovskaya S V, Gurnitskaya E P, Khetselius O Yu, Dubrovskaya Yu V 2006 *J. of Phys.: Conf. Ser.* **35** 425; Glushkov A V, Khetselius O Yu, Lopatkin Y M, Florko T A, Kovalenko O A, Mansarliysky V F 2014 *J. of Phys.: Conf. Ser.* **548** 012026

[20] Serga I N, Dubrovskaya Yu V, Kvasikova A S, Shakhman A N, Sukharev D E 2012 *J. of Phys.: Conf. Ser.* **397** 012013; Malinovskaya S V, Dubrovskaya Yu V, Vitavetskaya L A 2005 *AIP Conf. Proc.* **796** 201; Svinarenko A A 2014 *J.of Phys.: Conf. Ser.* **548** 012039

[21] Glushkov A V, Malinovskaya S 1988 *Rus.J.Phys.Chem.* **62**(1) 100; Glushkov A V, Khetselius O Yu, Malinovskaya S V 2008 *Europ.Phys.Journ.ST* **160** 195; 2008 *Mol.Phys.* **106** 1257

IOP Conf. Series: Journal of Physics: Conf. Series **810** (2017) 012048 doi:10.1088/1742-6596/810/1/012048

Line shapes in the presence of nonlinear wave collapse in a plasma

I Hannachi[a,b], R Stamm[a], J Rosato[a], Y Marandet[a]

[a]Aix-Marseille University and CNRS, PIIM, 13397 Marseille, France
[b]University of Batna, Department of Physics, Batna, Algeria

E-mail: ibtissem.hannachi@univ-amu.fr

Abstract. In a plasma submitted to an external source of energy, such as a beam of energetic charged particles, Langmuir, ion sound and electromagnetic wave may couple nonlinearly and change the structural and radiative properties of the plasma. Wave packets localize in many regions of the plasma and evolve cyclically toward very large values of the electric field. How the dipole autocorrelation functions (Fourier transform of the line shape) are modified by such conditions may be investigated with the help of a renewal model for the electric field. We compare the results of our model to the thermal line shape of Lyman α in low density plasmas.

1. Introduction

In a plasma submitted to a beam of energetic charged particles, perturbations of the plasma parameters may appear and develop. In the presence of a beam in the plasma, one most often observes the development of waves driven by kinetic instabilities. In the linear regime, an unmagnetized plasma supports three weakly damped wave modes: Langmuir, ion sound and electromagnetic waves. In cases of a strong perturbation, the density fluctuations associated with ion sound waves allow nonlinear interactions between the three basic wave modes. This coupling between waves can change the structural and radiative properties of the plasma. Langmuir wave packets concentrate in regions of low densities, and evolve to shorter scales and higher intensities. Computer simulations reveal the existence of a wave packet cycle with a collapse ultimately arrested by dissipation, and a nucleation mechanism allowing the creation of new wave packet cycles [1]. The picture gained from 3D simulations is the following: numerous localized wave packets coexist amid a uniform background plasma. We propose a model for calculating the effect of a sequence of collapsing wave packets on the dipole autocorrelation function, i.e. the line shape emitted by an atom in such conditions. A basic element of our model is an envelope soliton representing the oscillating electric field felt by an emitter in the vicinity of a collapsing wave packet. Our model relies on a renewal process assuming that each emitter is submitted to a sequence of collapsing waves, characterized by given statistical properties. Using the Lyman α (L_α) line of hydrogen as a benchmark, we compare this effect to the thermal Stark effect.

2. A model for the electric field in a plasma affected by wave collapse

The first model proposed for describing wave collapses is provided by the Zakharov equations [2], and their adiabatic limit, the nonlinear Schrödinger equation (NSE). In one dimension the NSE admits

IOP Conf. Series: Journal of Physics: Conf. Series **810** (2017) 012048 doi:10.1088/1742-6596/810/1/012048

stable soliton solution, with a localized and oscillating electric field modulated by a Gaussian or Lorentzian envelope [3]. The basic piece of our model is a soliton with a Lorentzian envelope for an electric field oscillating at a high-frequency close to the plasma frequency $\omega_p = \sqrt{N_e e^2 / m\varepsilon_0}$, with N_e the electronic density of the background plasma, e and m the electron charge and mass, and ε_0 the permittivity of free space. Using 3D computer simulations [1], solitons lose their stability, and the plasma exhibits numerous wave packets, each one undergoing a cycle starting with a nucleation mechanism, after which the wave packet narrows and strongly increases its intensity (wave collapse), then dissipates its energy mostly by wave-particle interactions, and finally reforms by renucleation [1]. The characteristic time for this cycle may be estimated from simulations [4], and is expressed as $\tau \approx 40\langle W\rangle^{-1}\omega_p^{-1}$, where W is the ratio of wave energy density to the plasma energy density, and the brackets denote an average value. The ratio W is given by:

$$W = \frac{\varepsilon_0 E_L^2}{4N_e k_B T_e}, \qquad (1)$$

where k_B is the Boltzmann constant, T_e the electron temperature, and E_L is the magnitude of the wave collapse electric field. We will consider in the following values of W around unity, as observed in plasmas in interaction with an energetic beam of charged particles [5], or in laser plasmas [6]. For such values of W, the magnitude of the electric field E_L reaches values which can be several hundred times larger than the plasma microfield $E_0 = e/(4\pi\varepsilon_0 r_0^2)$, with r_0 the mean interparticle distance. An atom in such a plasma is submitted to a sequence of collapsing wave packet that we propose to model as a renewal process for envelope solitons.

3. Stochastic model for the dipole autocorrelation function

A sequence of oscillating solitons may be viewed as a stochastic renewal process, with for each new electric field a random direction and phase in each time interval. We need for this process a probability density function (PDF) for the magnitude of the electric field, and a waiting time distribution for sampling the sequence of solitons. Gaussian PDF of the electric field magnitude have been measured and will be used here [5]. A Markovian choice for the waiting time distribution (WTD) is an exponential law $\nu\exp(-\nu t)$, with ν the jumping frequency chosen as the inverse of the characteristic time τ for the wave collapse cycle. In order to evaluate the effect of such an electric field $\vec{E}(t)$ on a hydrogen emitter, we numerically integrate the Schrödinger equation for the evolution operator of an atom submitted to a perturbing potential $-\vec{D}.\vec{E}(t)$, with $\vec{D}$ the atomic dipole operator. We can then write the expression of the dipole autocorrelation function (DAF) $C(t)$:

$$C(t) = Tr\langle\vec{D}\,U^{+}(t)\vec{D}U(t)\rho\rangle, \qquad (2)$$

where the trace is over the atomic states, ρ is the density matrix, and the angle brackets imply an average over all the realizations of the electric field. Our simulation calculations have been performed with a time step small compared to the inverse plasma frequency, and involve 5000 time histories. We present in the following calculations for the hydrogen L_α DAF, neglecting fine structure in order to speed up the numerical evaluation.

4. Results: Lyman α dipole autocorrelation function in the presence of wave collapse

We compute the DAF for the single effect of a sequence of solitons, and compare to the DAF of a thermal plasma, here obtained with an impact approximation. We consider a hydrogen plasma with a density $N=N_e=N_i=10^{19}$ m^{-3}, and two temperatures $T=T_e=T_i=4\text{x}10^4$ K and $T=10^5$ K. In a previous study, we showed that a Lorentzian time width equal to 20% of each soliton duration is compatible with a computer simulations of wave collapse, and also minimize the damping effect of the DAF or the broadening effect of the line shape [7]. Here we examine the effect of E_L, the magnitude of the wave

collapse electric field, or in an equivalent manner of the ratio W. We also use the relation between the cycle time τ and W to determine the value of the jumping frequency.
For the case T=4x10^4 K, we plot on figure 1(a) the wave collapse DAF for three values E_L=75, 90 and 105 E_0, and compare them to the impact limit representing thermal Stark effect. For E_L=105 E_0, corresponding to W=1.33 , and $\nu=\omega_p/30$, the DAF is similar to that of the impact limit, indicating that wave collapse damping is important and equivalent to Stark broadening. The effect is still significant for E_L=90 E_0 (W=0.98, $\nu=\omega_p/40$), and becomes a small correction to Stark effect only for electric field values smaller than E_L=75 E_0 (W=0.68, $\nu=\omega_p/60$). For the case T=10^5 K (figure 1(b)), the DAF with E_L=105 E_0 (W=1.09, $\nu=\omega_p/37$) shows more damping than Stark effect. Wave collapse damping is still significant for E_L=125 E_0 (W=0.76, $\nu=\omega_p/53$), and becomes a small correction to Stark effect only for electric field values smaller than E_L=100 E_0 (W=0.48, $\nu=\omega_p/83$).

5. Conclusion
Nonlinear wave collapse may appear in plasmas submitted to a strong external source of energy. We have presented the main characteristics of this phenomenon, and have proposed a renewal stochastic process modelling its effect on an emitting atomic dipole in the plasma. The field used in our model is a sequence of Lorentzian envelope solitons oscillating at the plasma frequency. We use this field for solving numerically the Schrödinger equation of an emitting hydrogen atom in a plasma with a density of 10^{19} m^{-3}. Comparisons of wave collapse DAF with those of thermal Stark broadening show that this nonlinear wave collapse can be a significant contribution to line broadening in low density plasmas.

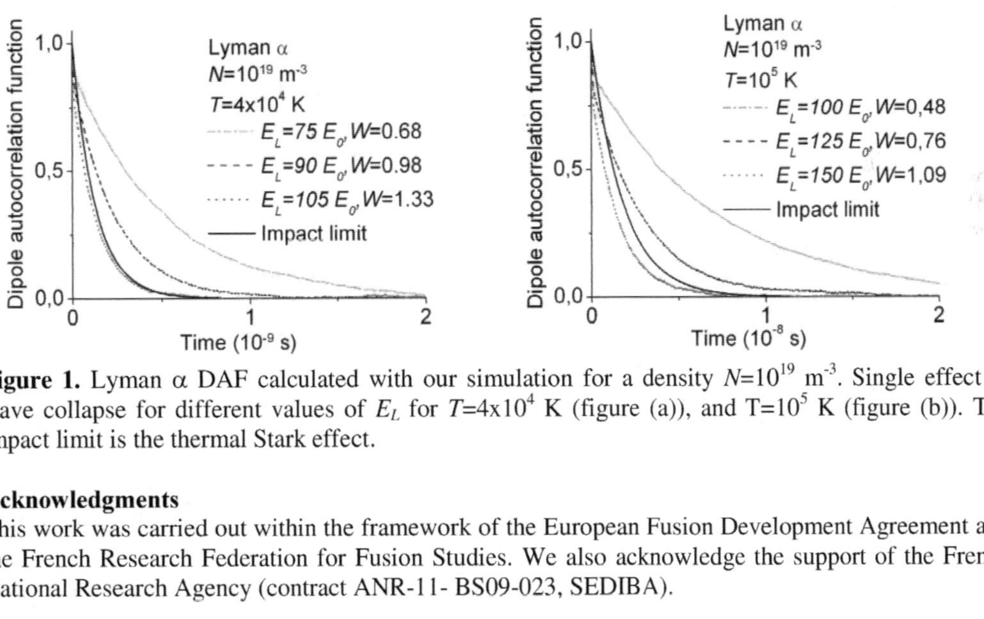

Figure 1. Lyman α DAF calculated with our simulation for a density N=10^{19} m^{-3}. Single effect of wave collapse for different values of E_L for T=4x10^4 K (figure (a)), and T=10^5 K (figure (b)). The impact limit is the thermal Stark effect.

Acknowledgments
This work was carried out within the framework of the European Fusion Development Agreement and the French Research Federation for Fusion Studies. We also acknowledge the support of the French National Research Agency (contract ANR-11- BS09-023, SEDIBA).

References
[1] Robinson P A 1997 *Rev. Mod. Phys.* **69**, 507.
[2] Zakharov V E 1972 *Sov. Phys. JETP* **35**, 908.
[3] Kuznetsov E A, Rubenchik A M, Zakharov V E 1986 *Phys. Rep.* **142** 103.
[4] Robinson P A, Newman D L *Phys. Fluids B* 1990 **2** 3120.
[5] Levron D, Benford G, Tzach D 1987 *Phys. Rev. Lett.* **58** 1336.
[6] Kruer W L 1988 *The Physics of Laser-Plasma Interactions* (Redwood City: Addison-Wesley).
[7] Hannachi I, Stamm R, Rosato J, Marandet Y 2016 *EPL* **114** 23002.

IOP Conf. Series: Journal of Physics: Conf. Series **810** (2017) 012049 doi:10.1088/1742-6596/810/1/012049

Modeling X-ray Line Shapes from Asphaltenes

H. Alqahtani*, J. Courtenay Lewis* and J. Shirokoff**

*Department of Physics and Physical Oceanography, **Department of Process Engineering, Memorial University of Newfoundland, St. John's, NL, Canada A1B 3X7

john.lewis@mun.ca

Abstract. Structural studies of asphaltene in crude oil have been performed by x-ray diffraction (XRD). The XRD spectra were taken with a Rigaku Dmax 2200V/PC, and built-in Jade^{TM} software was used for initial analysis. XRD was performed with Cu-K-α radiation operating at 40 KV and 40 mA, with a scan rate of 0.001° 2θ per second. The XRD data were first fitted with Pearson VII profiles and then with pseudo-Voigt profiles, and then modeled in Mathematica© using a generalized Fermi function (GFF). These fits also included several different backgrounds (linear, exponential, and Lorentzian). The results are discussed in terms of their accuracy with different combinations of background and line profile.

1. Introduction

Asphalt binder finds many applications such as in asphalt cement pavement for roads and highways, roofing materials, and heavy oil fuel upgrading. The latter application may involve asphaltene removal technology with solvents, or heavy oil catalysis into lighter fuels. So why study asphaltene? Recent history suggests that conventional light oil sources have decreased leaving the more abundant heavy oil sources to be developed. With this trend in mind, there has been a growing interest in further understanding the constituents in heavy oil such as asphaltene [1].

Asphaltene is a chemical constituent of crude oil reservoirs, bitumen and natural sources. Asphaltene chemistry, molecular size and structure are important variables. Crude oils can be separated into saturates, aromatics, resins and asphaltenes (SARA) in the laboratory. The asphaltene component has repeat units consisting of straight or branched chain molecules known as aliphatic or paraffinic; saturated rings with a high hydrogen to carbon ratio; unsaturated ring structures that are aromatic; and additional metals and heteroatoms as part of the ring structures. These repeat units have been previously studied by various analytical techniques including x-ray diffraction (XRD) originally developed for colloids in order to quantify molecular length scales [2-4]. XRD characterization is also useful in the case of asphaltene samples when one or more chemical phases (eg. asphaltene, wax, metal etc.) are present. This paper will focus on analyzing the X-ray line shapes of asphaltenes as measured by XRD for the purpose of further understanding chemical phases and related structure in these samples with three different combinations of backgrounds (linear, parabolic, 3rd order polynomial).

2. Experimental

The asphaltene samples were obtained from Canadian crude oil and asphalt by using solvent separation methods [5]. They were then coated onto glass slide holders by heating to 150°C for 10 minutes in a drying oven followed by cooling to ambient (25°C) temperature.

The X-ray diffraction (XRD) system used was a Rigaku Dmax 2200V-PC and Jade™ software (version 6.1) [4-5]. XRD was performed by using Cu-K-α radiation operating at 40 KV and 40 mA, scan rate of 0.001° 2θ per second and detector count time of 5 seconds/step. The XRD spectra obtained was then peak searched (screened out K-α-2 peaks) over the angular range of 5° to 110°. The full width at half-maximum (FWHM) and profile fits were obtained by using either Pearson VII and pseudo-Voigt function and three background types (linear, parabolic, 3rd order polynomial) over the approximate ranges 5° to 35° 2θ and 60° to 110° 2θ on the x-ray diffraction line spectra of interest.

The background radiation produced in the XRD spectra is an important aspect in studying asphaltene. Three background types (linear, parabolic, 3rd order polynomial; see table 1) were used in the case of the XRD analysis for generating profile fits with Pearson VII and pseudo-Voigt. The XRD spectra modeled in Mathematica© using a generalized Fermi function (GFF) also used three background types. The GFF modeled fits and plots were compared to Pearson VII and pseudo-Voigt. However, the calculations for the aromaticity (f_a), and crystallite parameters for interlayer distance between aromatic sheets (d_M) and interchain layer distance (d_γ) are as follows:

$$f_a = C_a/C_{total} = C/C(\text{graphene} + C\gamma) = \text{Area}_{\text{graphene}}/(\text{Area}_{\text{graphene}}+\text{Area}_\gamma) \quad (1)$$

$$d_M = \lambda/(2\sin\theta) \quad (2)$$

$$d_\gamma = 5\lambda/(8\sin\theta) \quad (3)$$

For each calculation they are all treated differently due to the types of background used and the different mathematical functions for each background.

3. Results and discussion

The results in table 1 highlight the effect of background type on accuracy of fit as indicated by the residual error of fit when using two different mathematical functions (Pearson VII, pseudo-Voigt) to fit asphalt binder samples. By inspection it is seen that the residual error of fit is lowest for parabolic, highest for 3rd order polynomial and low to intermediate for linear types of background. This trend in accuracy of fit is consistent for both types of mathematical functions used in profile fitting.

The asphaltene XRD patterns consists of four peaks: gamma peak (γ), (002) graphene, (100), and (110). The first two peaks in the XRD patterns and GF-mathematica© profile fit spectra (figure 1) are the gamma and graphene at about 18-20° 2θ while the peaks and planes for (100) and (110) are much broader at higher angles.

The XRD peak profile fitting function of Pearson VII and pseudo-Voigt have different peak shapes. Pseudo-Voigt is a combination of Gaussian and Lorentzian. Pearson VII is an exponential mixing of Gaussian and Lorentzian. The four most commonly used empirical peak shape functions (y) and the Generalized Fermi function are expressed mathematically in equations (4) to (8) [6]:

Table 1. The effect of background on fit.

Sample	Mathematical Function	Background	Residual Error of Fit %
AT1	Pseudo-Voigt	Linear/Parabolic/3rd Order Polynomial	8.8/5.4/25.3
AT3	Pseudo-Voigt	Linear/Parabolic/3rd Order Polynomial	8.6/5.4/26.4
AT5	Pseudo-Voigt	Linear/Parabolic/3rd Order Polynomial	7.8/5.4/22.3
AT1	Pearson VII	Linear/Parabolic/3rd Order Polynomial	10.0/6.1/23.3
AT3	Pearson VII	Linear/Parabolic/3rd Order Polynomial	9.6/5.8/24.0
AT5	Pearson VII	Linear/Parabolic/3rd Order Polynomial	8.2/5.7/20.5

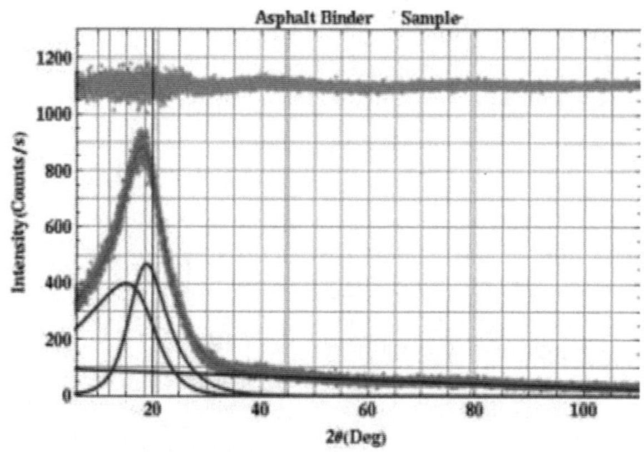

Figure 1. GF-mathematica© profile fit spectra with fixed background.

Gauss: $y(x) = G(x) = (C_G^{1/2})/(\sqrt{\pi}H)\ \exp(-C_G x^2)$ (4)

Lorentz: $y(x) = L(x) = (C_L^{1/2})/(\sqrt{\pi}H')\ (1+C_L x^2)^{-1}$ (5)

Pseudo-Voigt: $y(x) = \eta\ (C_G^{1/2})/(\sqrt{\pi}H)\ \exp(-C_G x^2) + (1-\eta)\ (C_L^{1/2})/(\sqrt{\pi}H')\ (1+C_L x^2)^{-1}$ (6)

Pearson: $y(x) = PVII(x) = \Gamma(\beta)/(\Gamma(\beta)-1/2)\ (C_P^{1/2})/(\sqrt{\pi}H)\ (1+C_P x^2)^{-\beta}$ (7)

Generalized Fermi: $y(x) = A/[\exp(-a(x-c)) + \exp(b(x-c))]$ (8)

Where: H and H' are the full widths at half maximum (FWHM), $x = (2\theta_i-2\theta_k)/H_k$, where x is the XRD Bragg angle of the ith point in the diffraction pattern with its origin in the position of the kth peak divided by the peak's FWHM; $2\theta_i$, is the Bragg angle of the ith point of the diffraction pattern; $2\theta_k$ is the Bragg angle of the kth Bragg reflection [6].

A Gauss function is taller at the top with little to no side tails at the bottom relative to the Lorentz that has a shorter sharp top with long tails at the bottom. Both functions are symmetric since G(x)

= G(-x) and L(x) = L(-x). These distributions can vary in XRD results. In pseudo-Voigt the ratio of η and η-1 varies for Gaussian to Lorentzian behavior, while in Pearson VII exponent (β) varies [6].

The determination of x-ray background intensity can be an issue in line profile analysis. When line profiles have long tails (at high 2θ angles in our data) there is always overlap of profiles leading to apparent background intensity [7]. These conditions can create errors in structural parameters obtained from line profile analysis. Usually a linear background under the profile is assumed from the height and slope as calculated by a fixed straight line from the extremities of the profile over a specified measuring range. For relative determinations this is usually not a serious problem. However, in the case of comparing profiles using Jade™ software to modeling in Mathematica© the determination of x-ray background intensity can differ thus leading to mixed results in aromaticity and structure (crystallite parameters) as indicated by the values of residual error of fit in table 1. The lowest to highest residual error of fit was parabolic<linear<3rd order polynomial using the methods defined in this research paper.

Table 2: Aromaticity (f_a), crystallite parameters (interlayer distance between the aromatic sheets (d_M), and interchain layer distance (d_γ)) for Pearson VII (P), pseudo-Voigt (V), and Generalized Fermi Function (GF).

Sample AT5	f_a			d_M(Å)			d_γ(Å)		
Background	**P**	**V**	**GF**	**P**	**V**	**GF**	**P**	**V**	**GF**
Linear	0.4	0.6	0.7	4	4	5	5	5	5
Parabolic	0.5	0.7	0.7	4	4	5	5	5	5
3rd Order	0.5	0.6	0.7	4	4	5	5	5	5

The XRD results modeled in GFF (GF) provided a smooth shape (Figure 3), however analysis of aromaticity and crystallite size parameters (Table 2) using the XRD data were sometimes mixed due to assymetry in the GFF data and differences in calculating background intensity (Table 1).

4. Conclusions

XRD pattern profile fits by Pearson VII and pseudo-Voigt using Jade™ software compared to generalized Fermi function data modeled in Mathematica© showed mixed correlation. The results are explained in terms of the calculation of x-ray background intensities, peak shape function, profile fitting, and assymetry of XRD data.

References

[1] Akbarzadeh K et al 2007 *Oilfield Review* **19** 2 22-43
[2] Yen TF 1992 *Fuel Sci. Technol. Int.* **10** 723-33
[3] Eyssautier JL et al 2011 *J. Phys. Chem. B* **115** 6827-37
[4] Gebresellasie K, Shirokoff J and Lewis JC 2012, *21st Int. Conf. Spectral Line Shapes, J. Phys. Conf. Ser.* ***375*** *012069*
[5] Hesp SAM, Illiuta S and Shirokoff J 2007 *Energy & Fuels* **21** 1112-21
[6] Pecharsky VK and Zavilij PY 2005 *Fundamentals of Powder Diffraction and Structural Characterization of Materials,* Springer, New York
[7] Delhez R et al 1982 *Fresenius Z. Anal. Chem.* **312** 1-16

X-ray emission in interaction of highly charged xenon ions with Be foil

Ł Jabłoński[1], **D Banaś**[1], **J Braziewicz**[1], **J Czub**[1], **P Jagodziński**[2], **A Kubala-Kukuś**[1], **D Sobota**[1], **I Stabrawa**[1], **M Pajek**[1]

[1] Institute of Physics, Jan Kochanowski University, 25-406 Kielce, Poland
[2] Department of Physics, University of Technology, 25-314 Kielce, Poland

E-mail: ljablonski@ujk.edu.pl

Abstract. In this work we report on the measurements of X-rays emitted in interaction of ~100 keV Xe^{35+} ions with metallic Be foil. Measured X-ray spectrum was interpreted as a series of satellite and hypersatellite nf-3d transitions with n≥4. The observed X-ray transitions from very high Rydberg states indicate the formation of so-called "hollow atoms" in the process of neutralization of highly charged ions on the metallic surface.

1. Introduction

The interaction of highly charged ions (HCI) with surfaces is dominated by a high coulombic potential energy of the projectile, as compared to its kinetic energy [1]. This conditions provide unique opportunity for formation the so-called "hollow atoms"[2] in the process of neutralization of HCI at surfaces in which, first, the electrons are captured into high Rydberg states while inner shells remain empty. Such highly excited exotic „hollow atoms" quickly deexcite in a cascade of Auger and radiative transitions. Consequently, the emitted Auger electrons and X-rays carry information about the structure and relaxation processes for these exotic, highly excited states.

We report here the results of experiment in which the "hollow atoms" are created in interaction of highly charged xenon ions with the metallic surface. The measured spectrum of emitted X-rays is interpreted using available atomic structure calculations for dominating dipole-allowed transitions and the predictions of the so-called Classical Over-barrier Model describing the neutralization of slow HCI colliding with a surface.

2. Experiments

In this experiment the X-rays emitted in interaction of ~100 keV Xe^{35+} ions with metallic Be foil were measured. The highly charged xenon ions were produced in the EBIT trap of the EBIS-A facility [3-5] by successive electron impact ionization of xenon atoms. The electron beam energy and current were 8.2 keV and 80 μA, respectively. The highly charged xenon ions created in the EBIT were extracted from the trap in a pulse mode and after selecting Xe^{35+} charge state in the analyzing dipole magnet the ion beam was focused on a Be foil mounted on a manipulator. The X-rays were measured with a 25 μm thick silicon drift detector (XFlash 5010, Bruker) having energy resolution of about

127 eV for Mn Kα line. In these experiments the whole system was kept at UHV conditions, at 10^{-10} mbar level.

3. Results and discussion

The measured spectrum of X-rays emitted in interaction of slow Xe^{35+} ions with Be foil is shown in figure 1. This spectrum can be interpreted as a result of electronic dipole-allowed radiative transitions from high Rydberg states, populated in a process of neutralization of HCI at the surface, into n=3 vacancy states in Xe^{35+} leading to emission of M-X-rays.

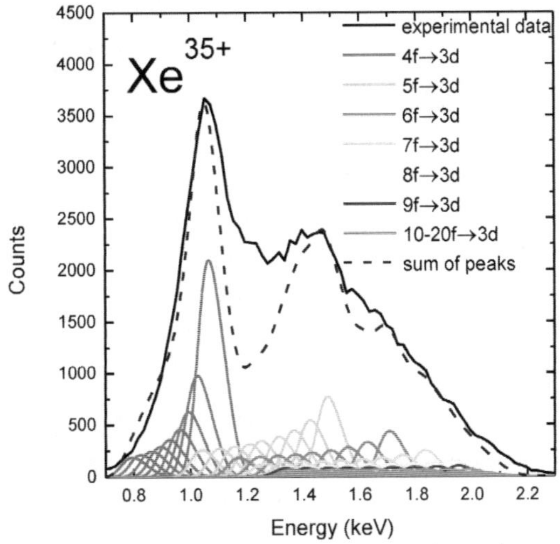

Figure 1. Measured spectrum of X-rays emitted in interaction of ~3 keV×q Xe^{35+} ions with Be foil compared with theoretical predictions.

Xe^{35+} ions have vacancies in the 3d subshell (see figure 2). Consequently, the X-ray spectrum is expected to be dominated by nf-3d transitions assuming the electric dipole selection rules and hydrogenic oscillator strength values. For any of these transitions we expect nine hypersatellite lines corresponding to nine 3d vacancies which can be filled by electrons. The energies of the transitions were calculated using the electron binding energies taken from [6]. The relative intensities of hypersatellits were assumed to be scaled as $1/(M-1)^{\alpha}$, were M is a number of electrons in the 3d subshell in the final state and α is an adjustable parameter that describes the relative intensities of hypersatellites taking into account its possible dependence on the principal quantum number of high Rydberg stated considered, or more precisely a change of amount of electrons in a higher state caused by the capture of electrons from higher shells and Auger processes. In case of analysed spectrum, the estimated value of α varies from 1.1 (transitions from n=4 shell) to 0 (transitions from the n≥10 shells). The FWHM width of hypersatellite lines was estimated to be about 106 eV and it takes into account the Gaussian experimental broadening for analysed energy range (about 70 eV) and the effective line broadening caused by its satellite structure reflecting the different number of electrons present in the high n-states while the X-rays are emitted.

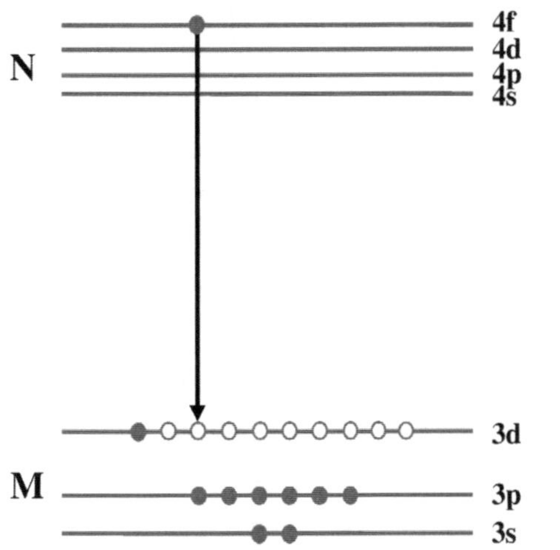

Figure 2. The diagram showing the electron configuration and 4f-3d transition in highly charged Xe^{35+} ion.

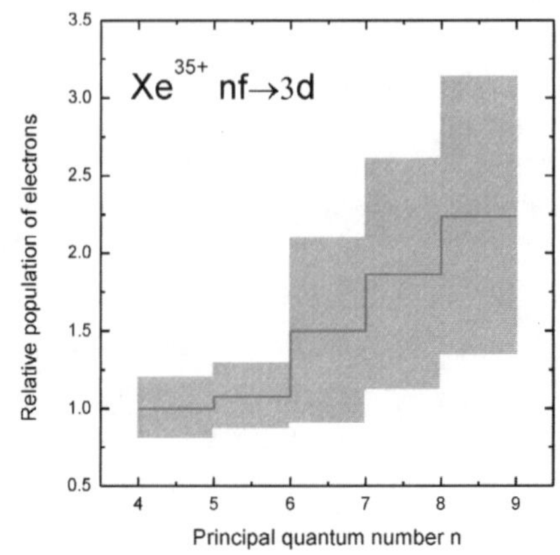

Figure 3. Dependence of the measured relative intensities of nf-3d X-ray transitions with respect to 4f-3d. The gray rectangles indicate the error bars.

The calculations based on Classical Over-barrier Model [7][8] suggests that in collision electrons from beryllium conduction band are resonantly captured into high Rydberg states (up to n=23). This hypothesis is consistent with analysis of the measured spectrum, where X-ray transitions from high n-states (up to n=20) to n=3 state are observed. The relative intensities of the strongest spectral lines in each series of hypersatellites were chosen to best describe the experimental spectrum. The results presented in figure 3 where normalised to the relative intensities predicted by hydrogen-like atoms model [9-10]. Because the total intensity of the line is proportional to the population of higher state, this graph carry information on a number of electrons at the higher shells (relative to the amount of electrons on the shell n=4). Analysis of both graphs (see figures 2 and 3) leads to a picture showing that after neutralization of HCI at a surface many electrons are in high Rydberg states, while the inner shells remains less populated or almost empty. This indicates in fact a formation of the "hollow atom" in the process of interaction of slow Xe^{35+} ions with Be foil.

4. Acknowledgments

The equipment was purchased thanks to financial support of the European Regional Development Fund in the framework of the Polish Innovative Economy Operational Program (Contact No. WNP-POIG.02.02.00-26-023/08).

References

[1] H. Winter, F. Aumayr 1999 *J. Phys B* **32** R39
[2] J-P Briand 1990 *Phys. Rev. Lett.* **65** 159
[3] G Zschornack et al. 2008 *Rev. Sci. Instrum.* **79** 02A703
[4] D. Banaś et al. 2010 *J. Instrum* **5** C09005
[5] D. Banaś et al. 2015 *Nucl. Instr. Meth. Phys. Res. B* **354** 125
[6] B. Rudek et al. 2012 *Nature Photonics* **6** 858
[7] J. Burgdörfer, F. Meyer 1993 *Phys. Scripta* **T46** 225
[8] J. Burgdörfer et al. 1996 *Aust. J. Phys.* **49** 527
[9] O. Jitrik, C. F. Bunge 2004 *J. Phys. Chem. Ref. Data* **33** 1059
[10] W.L. Wiese, J.R. Fuhr 2009 *J. Phys. Chem. Ref. Data* **38** 565

IOP Conf. Series: Journal of Physics: Conf. Series **810** (2017) 012051 doi:10.1088/1742-6596/810/1/012051

Reduction of the ionization energy for 1s-electrons in dense aluminum plasmas

C Lin[1], H Reinholz[1,2] and G Röpke[1]

[1] Universität Rostock, Institut für Physik, 18051 Rostock, Germany

[2] The University of Western Australia, School of Physics, WA 6009 Crawley, Australia

E-mail: chengliang.lin@uni-rostock.de

Abstract. The properties of a bound multi-electron system immersed in a plasma environment are strongly modified by the surrounding plasma. In particular, the modification of the ionization energy is described by the electronic self-energy within the framework of the quantum statistical theory. We present the energy shift of the eigenstates and the lowering of the continuum edge of free electrons in a plasma. The reduction of the ionization potential is determined by their difference. This ionization potential depression for the 1s-levels in dense aluminum plasmas is calculated. Comparisons with other theories and the experimental data are shown for aluminum plasma at solid density 2.7 g/cm^3.

1. Introduction

The controversy in the interpretation of two recent measurements on dense aluminum plasma [1,2] renews the interest on a better theoretical description of the ionization potential depression (IPD) in dense plasmas. Neither the generally accepted Stewart-Pyatt (SP) model [3] nor the Ecker-Kröll (EK) model [4] is able to explain both experiments. In this communication, a quantum-statistical approach [5,6] is used to calculate the reduction of the ionization energy for the 1s-electron in dense aluminum plasma. We consider atoms or ions as $(N+1)$-particle clusters, consisting of a nucleus with Z protons and N bound electrons ($N \leq Z$), as separate species. An effective two-particle description is introduced within this chemical model.

2. Quantum-statistical approach

To investigate the reduction of the ionization energy, two different effects should be mentioned. Firstly, the continuum edge of free electrons in plasmas is modified due to screening. Secondly, the energy levels of the bound electrons are shifted because of the plasma microfield. Within the Green's function method, both effects are described by the change of self-energies in plasma.

In the following, we only discuss the electronic contribution to the self energies. The self-energy of free electrons in plasma is given in the Montroll-Ward approximation [6] as

$$\Delta_{\rm MW} = -\frac{e^2}{8\pi\epsilon_0}\sqrt{\frac{n_e e^2}{\epsilon_0 k_{\rm B} T_e}} + \frac{m_e e^4}{8\epsilon_0^2 (2\pi\hbar)^2}\left\{\frac{\sqrt{2\pi^2}\, e^4\, n_e \lambda_e}{8\,(4\pi\epsilon_0)^2\,(k_{\rm B}T_e)^2} - \frac{n_e\lambda_e^3}{8\sqrt{2}} + \frac{e^2\, n_e\lambda_e^2}{4\,(4\pi\epsilon_0)\, k_{\rm B}T_e}\right\} \tag{1}$$

with the electron density n_e and the electron temperature T_e. $\lambda_e = \sqrt{2\pi\hbar^2/(m_e k_{\rm B} T_e)}$ denotes the thermal wavelength.

The correction of the eigenenergy of the electron in level $|\nu\rangle$ due to the plasma microfield is determined by the self-energy Σ_ν, which can be generally splitted into an electronic and an ionic part $\Sigma_\nu = \Sigma_\nu^{\rm el} + \Sigma_\nu^{\rm ion}$ because of their different mobilities. The electronic contribution $\Sigma_\nu^{\rm el}$ [5] is expressed in the dynamically screened Born approximation as

$$\Sigma_\nu^{\rm el} = -\frac{1}{e^2}\int \frac{d^3\mathbf{q}}{(2\pi\hbar)^3} V(\mathbf{q}) \sum_{\nu'} |M_{\nu\nu'}(\mathbf{q})|^2 \int_{-\infty}^{\infty} \frac{d\omega}{\pi}\left[1+n_{\rm B}(\omega)\right] \frac{{\rm Im}\,\epsilon^{-1}(\mathbf{q},\omega+i0)}{E_\nu - E_{\nu'} - (\hbar\omega + i0)}, \tag{2}$$

with the interaction potential $V(\mathbf{q})$ between an electron in eigenstate E_ν and the plasma electrons. $n_{\rm B}(\omega)$ and $\epsilon(\mathbf{q},\omega+i0)$ are the Bose-Einstein distribution and the dielectric function of the plasmas, respectively. The summation over virtual states $|\nu'\rangle$ should include both bound and scattering states. In other words, three different contributions need to be considered: the elastic scattering channel, the bound-bound transition, and the bound-free transition.

We are interested in the level shift of bound electrons, which is obtained from the real part

$$\Delta_\nu^{\rm el} = {\rm Re}\ \Sigma_\nu^{\rm el} = -\frac{2}{\Omega^2}\mathcal{P}\left\{\sum_{\mathbf{p},\mathbf{q},\nu'} V^2(\mathbf{q})\, f_{\rm e}(E_{\mathbf{p}}) \cdot \frac{|M_{\nu\nu'}(\mathbf{q})|^2}{E_\nu - E_{\nu'} + E_{\mathbf{p}} - E_{\mathbf{p}+\mathbf{q}}}\right\}, \tag{3}$$

see [7], with the Fermi-Dirac distribution $f_{\rm e}(E_{\mathbf{p}})$ which can be replaced by the Maxwell-Boltzmann distribution in the non-degenerate limit. The essentially important quantity is the transition matrix element $M_{\nu\nu'}(\mathbf{q})$. In the Green's function technique, a generalized matrix element $M_{\nu\nu'}(\mathbf{q})$ for a $(N+1)$-particle bound state should be considered [6]. Within the mean-field approximation, the independent-particle approximation is usually performed. An effective two-particle description can be introduced by decomposing the bound cluster into two parts: an involved electron in subshell $|\nu\rangle = |n_\nu l_\nu\rangle$ that is defined as the "transition electron" and is assumed to move in an effective mean field induced by the remaining N-particle ion core $|\beta\rangle$. The effective charge, seen by the transition electron in the subshell $|\nu\rangle$, is given by [8]

$$Z_\nu = Z - \sum_{\nu'}^{\nu_{\rm max}} \sigma_{\nu\nu'} \cdot (g_{\nu'} - \delta_{\nu\nu'})\,, \tag{4}$$

where $\nu_{\rm max}$ denotes the highest occupied subshell. The constant $\sigma_{\nu\nu'}$ represents the screening of an electron in the subshell $|\nu\rangle$ by the electron in another subshell $|\nu'\rangle$ [8]. This phenomenological approach is now generalized to take into account the spatial distribution of bound electrons through expressing the effective charge via a charge distribution

$$Z_\nu(\mathbf{q}) = Z - \sum_{\nu'}^{\nu_{\rm max}} \sigma_{\nu\nu'} \cdot (g_{\nu'} - \delta_{\nu\nu'}) \cdot F_{\nu'\nu'}(\mathbf{q}), \tag{5}$$

with the diagonal element of the atomic form factor $F_{\nu\nu'}(\mathbf{q}) = \int d^3\mathbf{r}\, e^{i\mathbf{q}\cdot\mathbf{r}}\, \psi^*_{\nu'}(\mathbf{r})\psi_\nu(\mathbf{r})$. It is expressed via the hydrogenic wave function $\psi_\nu(\mathbf{r}) = R_{n_\nu l_\nu}(r) Y_{l_\nu m_\nu}(\Omega_{\mathbf{r}})$. The total transition matrix element can be derived from the general vertex function [6] and is given by

$$M_{\nu\nu'}(\mathbf{q}) = S_\beta(q) \cdot \left\{ Z_\nu(\mathbf{q})\, \delta_{\nu\nu'} - F_{\nu\nu'}(\mathbf{q}) \right\} \tag{6}$$

with a screening function $S_\beta(q)$ resulting from the ion core $|\beta\rangle$, which is expressed as

$$S_\beta(q) = 1 - \frac{1}{Z}\sum_{\nu'}^{\nu_{\rm max}} F_{\nu_1\nu_1}(\mathbf{q}) \cdot (g_{\nu'} - \delta_{\nu\nu'})\,. \tag{7}$$

With the aformentioned effective two-particle description (6) of the matrix element, the expression (3) for the level shift is determined for the multi-electron bound system. Then the IPD of individual energy levels can be calculated by

$$\Delta^\nu_{\rm IPD} = \Delta_\nu^{\rm el} - \Delta_{\rm MW}\,. \tag{8}$$

3. Results and Conclusions

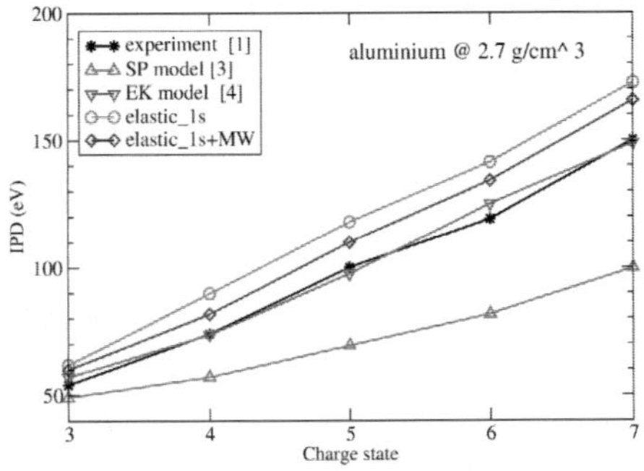

Figure 1. The IPD (8) for $1s$-electron of aluminum $\Delta^{1s}_{\rm IPD}$. The plasma parameters (electron density and electron temperature) for the considered charge states are taken from Ref. [9]. The results of the EK-model [4] are calculated with a modified prefactor $C_{\rm EK} = 1$.

Exemplarily, the IPD of the $1s$-electron in a laser induced Al-plasma [1] in comparison to different theoretical models is shown in Figure 1. In experiment, the local plasma conditions for each charge state are different as shown in [9]. Within our calculation, only the elastic scattering channel ($|\nu'\rangle = |\nu\rangle$) is considered. Because of the phase occupation by other bound electrons and the strong coupling between the $1s$-electron and the nucleus, the bound-bound and bound-free transitions only give small contributions to the IPD. Furthermore, the IPDs are evaluated only for the ground state of each charge state. Our results are larger than the experimental data and the results estimated by the SP and EK model, since no ionic correlation effect is considered and only the electronic Debye screening in the interaction potential $V(\mathbf{q})$ is used. The ionic correlation has strong influence on the IPD and should be reflected in the boundary condition of the confined atoms in plasmas. In the SP model, the IPD is determined by the Wigner-Seitz radius $R_{\rm WS} = \sqrt[3]{3/(4\pi n_{\rm ion})}$ and the Debye length including both ions and electrons. However, in the EK model, the characteristic length determining the IPD is calculated by the radius $R_{\rm EK} = \sqrt[3]{3/(4\pi(n_{\rm ion} + n_e))}$ where both the electron density n_e and ion density $n_{\rm ion}$ are considered. Comparing the two models with the experimental data, it shows the necessity of a deeper understanding of the concept "IPD".

We have presented an effective two-particle description within the framework of the thermodynamic Green's function technique to investigate the level shift of a bound system in dense plasmas. The inclusion of an ionic microfield and the application of our approach to K_α spectral lines [10] will be discussed in a forthcoming paper. Furthermore, the dependence of the IPD on the atomic configurations, e.g. excited states, also deserves further discussion.

4. Acknowledgement

This work is supported by the German Research Foundation DFG within SFB 652.

References

[1] Ciricosta O, Vinko S M, Chung H K, *et al.* 2012 *Phys. Rev. Lett.* **109** 065002
[2] Hoarty D J, Allan P, James S F, *et al.* 2013 *Phys. Rev. Lett.* **110** 265003
[3] Stewart J C and Pyatt K D 1966 *Astrophys. J.* **144** 1203
[4] Ecker G and Kröll W 1963 *Phys. Fluids* **6** 62
[5] Günter S, Hitzschke H and Röpke G 1991 *Phys. Rev. A* **44** 6834
[6] Kraeft W-D, Kremp D, Ebeling W and Röpke G 1986 *Quantum Statistics of Charged Particle Systems* (Akademie-Verlag Berlin)
[7] Omar B, Günter S, Wierling A and Röpke G 2006 *Phys. Rev. E* **73** 056405
[8] Faussurier G, Blancard C and Renaudin P 2008 *High Energy Dens. Phys.* **4** 114
[9] Son S-K, Thiele R, Jurek Z, Ziaja B and Santra R 2014 *Phys. Rev. X* **4** 031004
[10] Chen Y, Sengebusch A, Reinholz H and Röpke G 2014 *Contrib. Plasma Phys.* **54** 772

AlGaInN diode-laser technology for optical clocks and atom interferometry

S P Najda[1], P Perlin[1,2], T Suski[2], L Marona[2], S Stanczyk[1,2], M Leszczyński[1,2],PWisniewski[1,2] , R Czernecki[1,2], G Targowski[1], C Carson[3], D Stothard[3], L J McKnight[3]

[1]TopGaN Ltd., ul. Sokolowska 29/37, 01-142 Warsaw, Poland.
[2]Institute of High Pressure Physics PAS, ul. Sokolowska 29/37, 01-142 Warsaw, Poland.
[3]Fraunhofer Centre for Applied Photonics, Fraunhofer UK, Technology and Innovation Centre, 99 George Street, G1 1RD, Glasgow, U.K.

E-mail: s.najda@topganlasers.com

Abstract. Optical clocks have demonstrated an improvement in temporal accuracy of several orders of magnitude over existing time standards based on caesium.Such systems hold great promise in many industrial sectors including financial time stamping, GPS-free navigation and network synchronisation. Such systems require a number of sophisticated lasers in a compact and reliable format for use outside of a laboratory environment, suitable for commercialisation and user transportation. Of particular interest, is emerging AlGaInN laser diode technology that has the potential to provide practical solutions for next generation optical clock technology.

1.Introduction

The AlGaInN material system allows for single transverse mode laser diodes to be fabricated with optical powers up to 100mW and over a very wide range of wavelengths from U.V., ~380nm, to the visible, ~530nm.By tuning the indium content and thickness of the GaInN quantum well, we have developed a range of AlGaInN diode-lasers [1] targeted to meet the wavelength and power requirements suitable for optical clocks and atom interferometry systems.Hence, there is considerable interest in using AlGaInN laser technology for the next generation optical clocks and atom interferometry systems. This includes the $[5s^2S_{1/2}\text{-}5p^2P_{1/2}]$ cooling transition in strontium$^+$ ion optical clocks at 422 nm, the $[5s_2{}^1S_0\text{-}5p^1P_1]$ cooling transition in neutral strontium clocks at 461 nm and the $[5s^2s_{1/2} - 6p^2P_{3/2}]$ transition in rubidium at 420 nm. These transitions are a few among an extensive list required.In addition to the requirements outlined above, narrow linewidths (<5 MHz) are required for the identified transitions and few kHz levels for other transitions. To meet these requirements we are pursuing both extended cavity laser (ECDL) geometries in the short term and distributed feedback architectures in the long term.

2.AlGaInN laser diode results

The LIV and the beam profile characteristics for a AlGaInN LD structure packaged in a TO5.6mm package are shown in figure 1a). The device has an optical power of ~80mW, threshold current of ~65mA, a threshold voltage of ~5 V, a lasing wavelength of 410nm, and a characteristic temperature T_0 of ~120 K. A single transverse mode optical beam profile is observed in both the slow and fast axis (see fig.1 b).

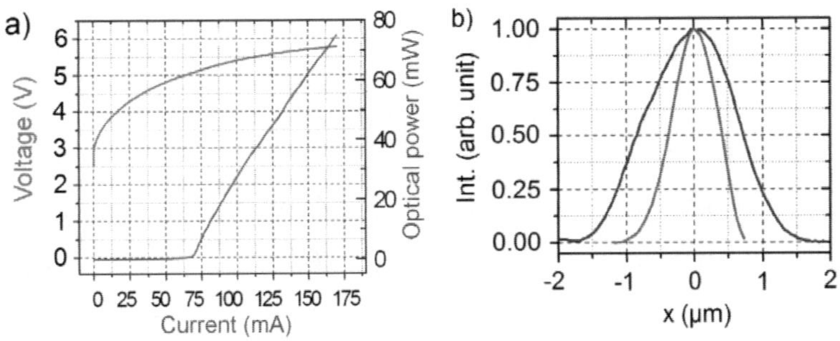

Fig.1 410nm laser diode characteristics, a) LIV , b) Near-Field (slow axis – red line, fast axis – green line).

The spectral output of a ~422nm laser is measured as a function of increasing drive current. For 200mA (14mW) operation, a dominant single longitudinal mode at 421.6nm, with multiple small side modes is observed. At a higher drive current of 300mA (36mW) the dominant single longitudinal mode jumps to a spectrally wide (~1-2nm) mode comb as is more typical of a Fabry-Perot LD device (see fig.3).

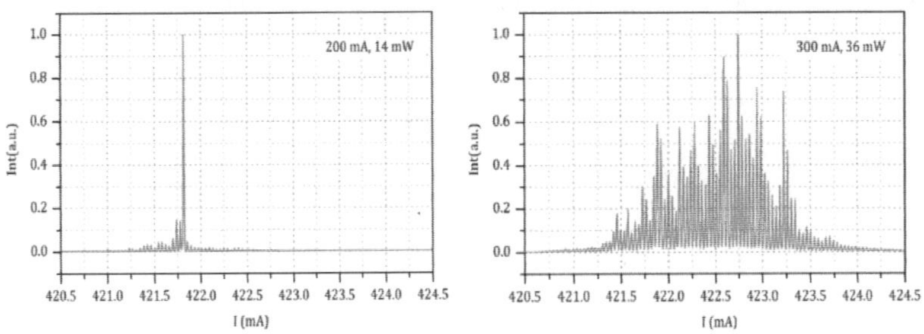

Fig 3 The evolving spectra of a LD's vs increasing drive current mA/optical power mW (cw) at 20^0C

Similar single longitudinal mode characteristics has also been observed in the spectral output of other AlGaInN laser diodes and was explained by surface roughness inadvertently introduced during growth[2] and that the single mode is stabalised by longitudinal mode competition caused by optical gain saturation. Similarly we observe a surface topology of the order of ~10nm in height and a periodicity of 100nm, eventhough the epitaxy growth is done on very low defectivity ($<5x10^4cm^{-2}$) GaN substrates with a flatness of <0.1nm [3], the surface topology features appear inadvertantly in the last epitaxy layer of growth.

3.ECDL results

A packaging system has been developed optimised for ECDL narrow-linewidth operation. An achromatic 4 mm focal length lens was used to produce a near-collimated beam from the diode. A 1800 lines/mm reflective diffraction grating was used to provide feedback to the diode in Littrow configuration.

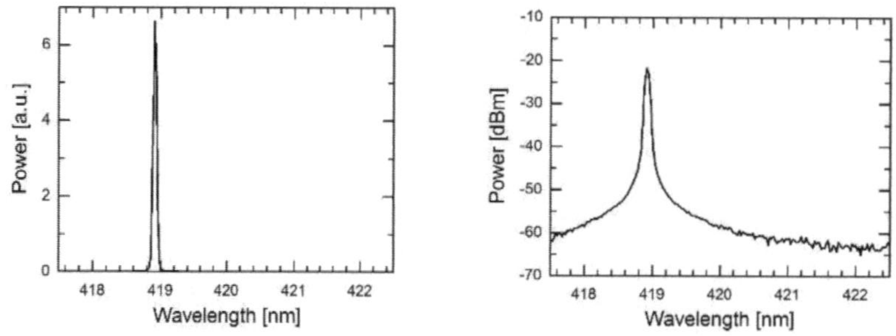

Figure 4 Spectral output of ECDL at 40mW, 134mA operation

The spectral output of the ECDL is shown in Figure 4 with a spectral resolution <0.05 nm (85 GHz). Single frequency operation with 40 mW was achieved with no side peaks anda 30dB drop to the background. A scanning Fabry-Perot measurement shown in Figure 5further confirms single-frequency operation and demonstrates a linewidth of less than 10 MHz limited by finesse of cavity (< 0.006 pm). No mode-hopping was observed for extended lengths of measurement time.

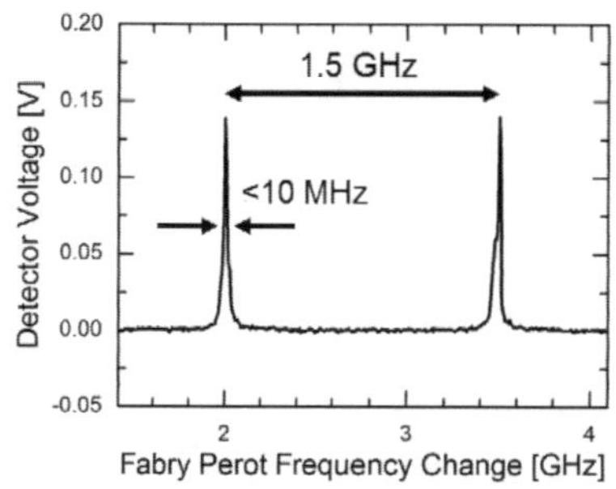

Figure 5 Scanning Fabry-Perot measurement

Conclusions

We have developed an ECDL system using GaN laser diodes to achieve single frequency operation with a linewidth <10MHz and side mode suppression of more than 30dB. This architecture is suitable for a range of requirements in optical clocks and atom interferometers.

References:

[1] Najda S P,Perlin P,Suski T,Marona L,Boćkowski M,Leszczyński M, Wisniewski P,Czernecki R, Kucharski R, Targowski G, Smalc-Koziorowska J, Stanczyk S,Watson S, Kelly A E:Proc. SPIE 9363, Gallium Nitride Materials and Devices X, 93631A (13 March 2015); doi: 10.1117/12.2076268

[2] Meyer T, Braun H, Schwarz U, Tautz S, Schillgalies M, Lutgen S and Strauss U, p.6833, Vol.16, No.10, Optics Express 2008.

[3] Najda S P,Perlin P,Suski T,Marona L,Bockowski M, Leszczyński M, Wisniewski P, Czernecki R, Kucharski R, Targowski G,Watson S, Kelly A E. Proc. SPIE 9748, Gallium Nitride Materials and Devices XI, 9748 (17 March 2016), ISBN: 9781628419832

Lateral grating DFB AlGaInN laser diodes for optical communications and atomic clocks.

S P Najda[1], T Slight[3], P Perlin[1,2], O Odedina[4], T Suski[2], L Marona[2], S Stanczyk[1,2], M Leszczyński[1,2], P Wisniewski[1,2] , R Czernecki[1,2], G Targowski[1], and A E Kelly[4]

1 TopGaN Ltd., ul. Sokolowska 29/37, 01-142 Warsaw, Poland.
2 Institute of High Pressure Physics PAS, ul. Sokolowska 29/37, 01-142 Warsaw, Poland.
3CST Global, Hamilton International Technology Park, 4 Stanley Blvd, Blantyre, Glasgow, Lanarkshire G72 0BN, United Kingdom.
4 School of Engineering, University of Glasgow, Glasgow G12 8LT, U.K.
E-mail: s.najda@topganlasers.com

Abstract. AlGaInN laser diode technology is of considerable interest for telecom applications and next generation atomic optical clocks based on Sr (by using 422nm & 461nm) and Rb at 420.2nm.Very narrow linewidths (<1MHz) are required for such applications. We report lateral gratings on AlGaInN ridge waveguide laser diodes to achieve a single wavelength device with a good side mode suppression ratio (SMSR) that is suitable for atomic clock and telecom applications.

1.Introduction

Laser diodes fabricated from the AlGaInN material system is an emerging technology for next generation telecommunication systems and optical clocksbased on Sr (meeting the 88Sr+ cooling transition [5s2S1/2 -5p2P1/2] by using 422nm & 461nm) and a blue cooling transition for Rubidium (Rb) [4p65s2S1/2 - 4p66p2P3/2] at 420.2nm.A requirement for both of these applications is the need for a single longitudinal optical mode, with a very narrow linewidth (<1MHz), that is stable over a wide range of operating conditions. An AlGaInN laser diode with an etched DFB grating structure to achieve single longitudinal mode performance can be considered. However, there are significant challenges to fabricate a conventional buried DFB grating GaN laser diode.As an alternative, we report preliminary work on lateral grating AlGaInN laser diodes to achieve continuous wave single longitudinal optical mode performance that is suitable for atomic clock and telecom applications.

2.AlGaInN laser diode results

We fabricate AlGaInN ridge waveguide laser diode structures for single transverse mode operation and package the processed laser diode chips into TO5.6mm packages [1].The LIV and the beam profile characteristics for a 2μm ridge waveguide LD structure are shown in figure 1a). The device has an optical power of ~80mW, threshold current of ~65mA, a threshold voltage of ~5 V, a lasing wavelength of 410nm, and a characteristic temperature T_0 of ~120 K. A single transverse mode optical beam profile is observed in both the slow and fast axis (see fig.1 b).

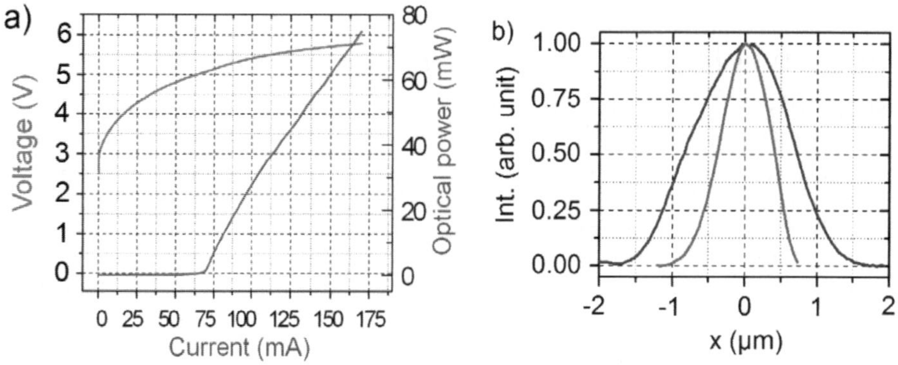

Fig.1 AlGaInN 410nm laser diode characteristics, a) LIV , b) Near-Field (slow axis – red line, fast axis – green line).

High resolution spectral measurements of the AlGaInN LD's reveal the fine mode structure with a characteristic dominant single longitudinal mode in all of these devices more reminiscent of a DFB type of laser device with etched grating, providing optical feedback for mode selection, rather than a more standard 'mode comb' Fabry-Perot device with no etch grating (see fig.3)[1].

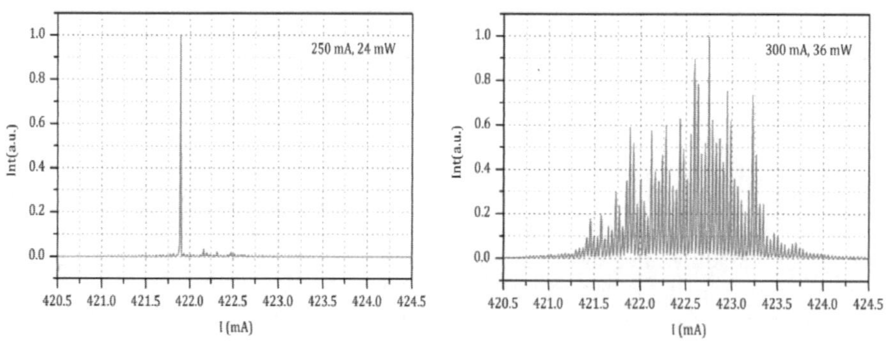

Fig 2 The evolving spectra of a LD's vs increasingcurrent mA/optical power mW (cw) at 20^{0}C

The above results for a Fabry-Perot laser diode (with no etched grating) show that single longitudinal mode operation is achievable; however the mode is unstable and hence is unsuitable for most applications. Therefore we consider sidewall etched as one of the simplest way of manufacturing gratings on ridge waveguide laser devices (figure 3). They have the advantage of simplified fabrication compared to more conventional DFB structures with no need for complex overgrowth steps and have been demonstrated in a range of other material systems and wavelengths [2,3]. However at wavelengths around 400 nm the fabrication of gratings becomes technologically challenging with a first order grating requiring a feature size of around 40 nm, dimensions not practical with our current fabrication technology. In this study we report on preliminary etch and modelling studies of AlGaInN laser diode structures in order to understand the issues of side-wall etching versus laser diode single longitudinal mode performance in order to obtain a practical solution.

IOP Conf. Series: Journal of Physics: Conf. Series **810** (2017) 012053 doi:10.1088/1742-6596/810/1/012053

3.Fabrication

Device fabrication was carried out on ~410 nm GaN laser grade material. The fabrication for both third order and high order gratings follows the same general process. Grating and ridge patterns were defined in ZEP 520 resist using electron beam lithography (EBL). The EBL tool used was a Vistec VB6 UHR having a write resolution of 1.0nm, enabling the accurate targeting of the emission wavelength. Reactive ion etching was used to transfer the pattern into a 100 nm thick SiO_2 hard mask. Inductively coupled plasma (ICP) etching on an STS Multiplex tool was then used to form the grating and ridge. A Cl_2/N_2 based ICP etch process with 300 W platen and 600 W coil power produced the vertical and smooth etch profile required for good grating performance.

Figure 3 SEM micrographs of the (a) 3rd order grating and (b) the 39th order grating.

4. Conclusions

We observe single longitudinal optical mode performance from a ~420nm AlGaInN laser diode at moderate optical powers (~20mW) and small SMSR. To stabilise the optical mode we investigate side-wall lateral gratings as a way to develop a practical AlGaInN laser diode. We have performed modelling and etch studies of 3rd and 39th order grating to understand the issues with side-wall gratings in AlGaInN ridge waveguide laser structures. Future work will focus on achieving cw operation via an optimised grating design.

Acknowledgements: This work was supported by the European Union under the project E9776.

REFERENCES

[1] Najda S P, Perlin P, Suski T, Marona L, Leszczyński M, Wisniewski P, Czernecki R, Kucharski R, Targowski G,Watson M A, White H, Watson Sand Kelly A E.*Opt. Eng.***55**(2), 026112 (2016). doi: 10.1117/1.OE.55.2.026112

[2] Slight T J, Odedina O, Meredith W, Docherty K E, andKelly A E, SPIE OPTO 2016 (pp. 97481A-97481A). International Society for Optics and Photonics.

[3] Zhong Y,Zhu X,Song G,Huang Y,Chen L, Semicond. Sci. Technol. 19(8), 971–974 (2004).

IOP Conf. Series: Journal of Physics: Conf. Series **810** (2017) 012054 doi:10.1088/1742-6596/810/1/012054

Self-absorbed Al lines in laser-induced plasma

D. M. Surmick and C. G. Parigger

University of Tennessee Space Institute, 411 B.H. Goethert Parkway, Tullahoma, TN 37388, USA

E-mail: dsurmick@utsi.edu,cparigge@utsi.edu

Abstract. Self-reversed line profiles of Al $3s^23p \rightarrow 3s^24s$ transitions in a laser ablation plasma are considered following plasma initiation on the surface of an Al alloy target. Experimental self-absorption correction factors are tabulated by measuring plasma line profiles with and without its reflection using the mirror duplication method. At a delay of 60 ns, measured line profiles 0.5 mm from the Al surface show self-reversed line profiles. Corrected line widths are determined to be 1.1 ± 0.2 and 1.7 ± 0.2 nm for the 394.4 and 396.15 nm lines, respectively, following fitting with a Voigt line profile. N_e are determined to be 3.8 ± 0.4 and $10 \pm 1 \times 10^{18}$ cm^{-3}, respectively.

1. Introduction

Laser-induced plasma initiated on solid surfaces are complex dynamical systems characterized by temperatures of 10 eV, electron densities, N_e, greater than 10^{19} cm^{-3}, and the rapid propagation of a shockwave. Of particular interest are the shapes of Stark broadened spectroscopic lines shortly after plasma initiation. At atmospheric pressures and with nominal 10 ns pulsed lasers, plasma self-absorption affects may become prominent [1, 2]. The plasma expansion produces two separate regions: (i) a plasma kernel and (ii) a cooler outer region [3]. Self-absorption occurs as emissions from the plasma kernel emit through the outer region and absorb energy from the kernel emissions. Due to self-absorption, measured spectroscopic line profiles become distorted through line broadening, peak flattening and, in extreme cases, reversal of the line peak [3]. Such affects are most pronounced for ground state transitions.

Here, we present experimentally corrected, self-reversed aluminium line profiles measured from a laser-induced plasma. Corrected profiles are obtained through application of a wavelength dependent correction factor. The correction factor may be tabulated by measuring line profiles with and without a duplication of the plasma. Duplication may be achieved by reproducing the plasma behind the the original plasma or through the use of optics to re-image the plasma radiation back onto its focal volume. The correction factor is given as

$$K(\lambda) = \frac{\ln(y(\lambda))}{y(\lambda) - 1}, \qquad (1)$$

with

$$y(\lambda) = \frac{R_c(\lambda) - 1}{R(\lambda) - 1}, \qquad (2)$$

where $R_c(\lambda)$ is the ratio of the spectral continuum and $R(\lambda)$ is the ratio of the spectral line profile, as measured with and without the plasma duplication. The above formulation is derived from the equation of radiation transport [4].

2. Experimental Details

We investigate the ground state $3s^2 3p \rightarrow 3s^2 4s$ Al transitions following ablation of an Al alloy target sample. The plasma was initiated using 1064 nm, 14 ns pulsed, Nd:YAG laser radiation. The energy per pulse of the laser was 120 mJ. The target sample was an Al alloy 6061 target that was 2 cm square. The target sample was placed into a vacuum chamber that was evacuated and filled with a 90% hydrogen, 10% nitrogen gas atmosphere. The pressure of the cell during ablation was 840 Torr. Plasma emissions were dispersed using a 0.64 meter Jobin-Yvon, HR640 spectrometer installed with a 1200 grooves/mm grating. The observed spectra were recorded using an Andor iStar ICCD. Further details on the experimental apparatus are presented in Reference [5]. The spatial resolution of the apparatus amounted to 0.1 mm and the spectral resolution was 0.15 nm. The plasma emissions were recorded with and without a doubling mirror in order to accomplish measurements of $R_c(\lambda)$ and $R(\lambda)$. Light from the plasma source was focused onto the slit with a 100 mm focal length, uncoated, UV fused silica, 50 mm diameter (2 inch) lens. The plasma emission was retro-reflected using a plane reflecting mirror. Light was focused onto the plane mirror using a copy of the lens used to image the plasma light onto the spectrometer slit. The two lenses were each placed at a distance equal to to 2× their focal length from the respective objectives. The plasma was initiated at a distance of 40 cm from both the spectrometer slit and the reflecting mirror.

3. Results

In order to determine the experimental correction factor, the ratio of spectra collected with and without the mirror and their associated wavelength dependent continua were determined by fitting the measured spectra. The observed spectra were fit using Voigt profiles. The line profiles were calculated using the so-called Faddeeva function, whose real part component is equivalent to the Voigt integral [6]. Calculation of the Faddeeva function was completed using the Algorithm 916 method [7], and fitting to the measured profiles is accomplished using the Trust-Region nonlinear curve fitting method [8]. Each of the two Al profiles were fit simultaneously as well as considering a linear base line offset, i.e., a sum of two Voigt profiles and a linear baseline offset was fit to the measured profiles.

The ratio of the continuum, $R_c(\lambda)$, is taken to be the ratio of the fitted offsets for cases of profiles measured with and without the reflecting mirror. The ratio of spectroscopic line profiles, $R(\lambda)$, was taken to be the ratio of the cumulative fits. The result of correcting the measured profiles is displayed in Figure 1.

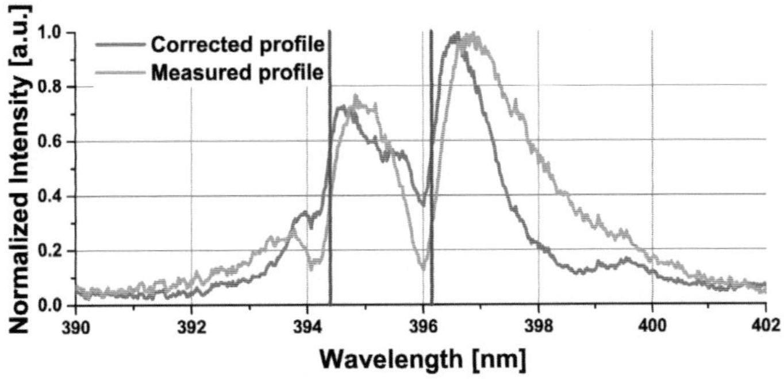

Figure 1. Measured *vs.* self-absorption corrected Al line profiles. The blue lines indicate the unperturbed Al 394.4 and 396.15 nm line centers.

The spectrum that is displayed is measured at a distance of 0.5 mm from the sample surface and time delay of 60 ns following plasma initiation. The temporal gate width of the measurement was 10 ns. Deep absorptions are present near the line centers, which are consistent with a self-reversed line profile. The positions of these absorptions corresponds directly to the position of the unperturbed line centers of the Al lines in question. These absorptions are greatly diminished in the corrected profile. The red shift due to the Stark effect are also diminished and, as well, the line widths, particularly for the Al 396.15 nm line, are greatly narrowed by application of the correction factor, both of which lead to a reduction in the N_e in absence of self-absorption line distortions. After the profiles have been corrected for self absorption, the profiles are fit once more in an effort to determine N_e. The fitted widths of the corrected profile shown in Figure 1 are 1.1 ± 0.2 and 1.7 ± 0.2 nm for the Al the 394.4 and 396.15 nm lines, respectively. The densities are determined by considering the N_e *vs.* line-width relationship outlined in Reference [9]. The result is a power law trend as displayed in Figure 2 for both Al lines.

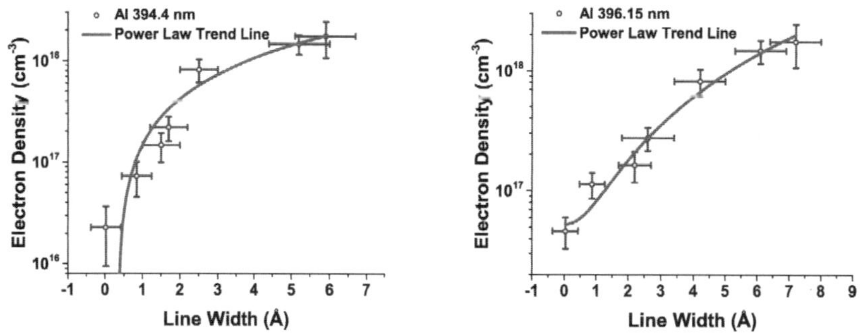

Figure 2. N_e power law trends for Al 394.4 (left) and 396.15 (right) nm line width diagnostics

The electron densities are determined to be 3.8 ± 0.4 and $10 \pm 1 \times 10^{18}$ cm^{-3} for the Al 394.4 and 396.15 nm lines, respectively. The large disparity between the density inferred from the two lines is likely due to the presence of a perturbing nitrogen ions at 395.5 and 399.5 nm. This causes the width of the Al 394.4 nm line to be slightly underestimated, thereby underestimating the determined N_e. The ability to correct self-reversed line profiles and infer N_e represents the opportunity to understand the plasma production process in laser-material interactions from spectroscopic observations.

Acknowledgments

This work is supported by the Center for Laser Applications at the University of Tennessee Space Institute.

References

[1] Parigger C G, Surmick D M, Gautam G and El Sherbini A M 2015 *Opt. Lett.* **40** 3346
[2] Surmick D M and Parigger C G *J. Phys. B: Atom. Mol. Opt. Phys.* **48** 115701
[3] Kunze H J 2009 *Introduction to Plasma Spectroscopy* (Springer-Verlag, Berlin)
[4] Moon H Y, Herrera K K, Omentto N, Smith B W and Winefordner JD 2009 *Spectrochim. Acta B* **64** 702
[5] Parigger C G, Woods A C, Witte M J, Swafford L D and Surmick D M 2014 *J. Vis. Exp.* **84**
[6] Abrarov S M and Quin B M 2011 *Appl. Math. Comp.* **218** 1894
[7] Zaghloul M and Ali A 2012 *ACM Trans. Math Soft.* **38** 1
[8] Byrd R H, Schnabel R B and Shultz G A 1987 *SIAM J. Numer. Anal.* **24** 1152
[9] Parigger C G, Hornkohl J O and Nemes L 2007 *Appl. Opt.* **46** 4026

IOP Conf. Series: Journal of Physics: Conf. Series **810** (2017) 012055 doi:10.1088/1742-6596/810/1/012055

Electron density and temperature diagnostics in laser-induced hydrogen plasma

G Gautam and C G Parigger

University of Tennessee Space Institute, Center for Laser Applications, 411 B.H. Goethert Parkway, Tullahoma, TN 37388-9700, USA

E-mail: `cparigge@tennessee.edu`

Abstract. Laser-induced optical breakdown is achieved by using Q-switched, Nd:YAG radiation focused into ultra-high-purity (UHP) hydrogen gas at a pressure of $1.08 \pm 0.03 \times 10^5$ Pa inside a cell. The plasma emission spectra are dispersed by a Czerny-Turner type spectrometer and detected with an intensified charge-coupled device (ICCD). Stark-broadened hydrogen Balmer series H_α and H_β line profiles are used as a spectroscopic tool for the determination of electron density and excitation temperature. Spatial variation of electron density and temperature at 0.40 μs are extracted from the recorded intensities of H_α and H_β lines. Temporal variations of electron density and excitation temperature are also presented for the time delay range of 0.15 μs to 1.4 μs.

1. Introduction

Hydrogen Balmer series lines have been explored in various theoretical and experimental investigations as well as tested experimentally to evaluate plasma parameters [1, 2]. Here, the Balmer series H_α and H_β lines are used for the electron density and excitation temperature diagnostics at various time delays. Figures 1 and 3 illustrate the electron density computed from the measured FWHM using Equation (1) [3]. These equations have been derived from previously collected data [4, 5],

$$N_{e,\alpha}[\mathrm{cm}^{-3}] = \left[\frac{\Delta\lambda[\mathrm{nm}]}{1.31}\right]^{1.56 \,\pm\, 0.07} \times 10^{17} \quad , \quad \mathrm{N}_{\mathrm{e},\beta}[\mathrm{cm}^{-3}] = \left[\frac{\Delta\lambda[\mathrm{nm}]}{4.50}\right]^{1.41 \,\pm\, 0.06} \times 10^{17}. \quad (1)$$

Electron excitation temperature is determined with the Boltzmann plot method [6] by utilizing the measured values of relative intensities of H_α and H_β lines. From these intensities, a straight line is plotted by applying the equation,

$$\ln\left[\frac{I_{ki}\lambda_{ik}}{g_k A_{ki}}\right] = -\left[\frac{E_k}{k_B T}\right] + \ln C, \quad (2)$$

where the following values are used [7]: Transition probabilities, $A_{ki} = 0.4410 \times 10^8\ \mathrm{s}^{-1}$ and $0.08419 \times 10^8\ \mathrm{s}^{-1}$, statistical weights, $g_k = 18$ and 32, for H_α and H_β lines, respectively. Temperature values are computed from straight-line slopes. Intensity profiles of H_α and H_β lines along the lateral direction, *i.e.*, along the slit height, are used to determine the spatial variation of temperature.

2. Experiment

Micro-plasma is generated inside a cell containing 99.999% purity hydrogen gas by using Nd:YAG laser, operated at the fundamental wavelength of 1064 nm, frequency 10 Hz generating 13 ns pulses with an energy of 120 mJ. Spatially and temporally resolved plasma spectra are recorded with a Czerny-Turner type spectrometer and ICCD. A 1200 grooves/mm holographic grating is used to disperse the plasma emission into various wavelengths.

The ICCD consists of a 1024×1024 array with 1024 horizontal for wavelength and 1024 vertical pixels for intensity. In this experiment, eight vertical pixels are combined to reduce noise and improve the signal-to-noise ratio. A total of 128 lateral spectral measurements are possible along the slit height with this pixel binning. Each pixel is a square with 13.6 μm sides, therefore, each lateral measurement has a resolution of 0.1088 mm. The spectra recorded in the different zones indicate the variations of electron density and temperature.

A gate width of 20 ns is used for the time delay range of 0.15 μs to 1.4 μs, and time steps of 0.25 μs are selected for the measurements of H_α and H_β emission spectra. Each collected spectrum corresponds to an average of 100 consecutive laser-plasma events.

3. Results

Spatial variations of an electron density, N_e, and temperature, T_e, for a time delay of 0.4 μs are displayed in Figures 1 and 2. The spatial distances indicated in Figs. 1 and 2 correspond to the slit height. The laser radiation propagates parallel to the slit, viz. from right to left.

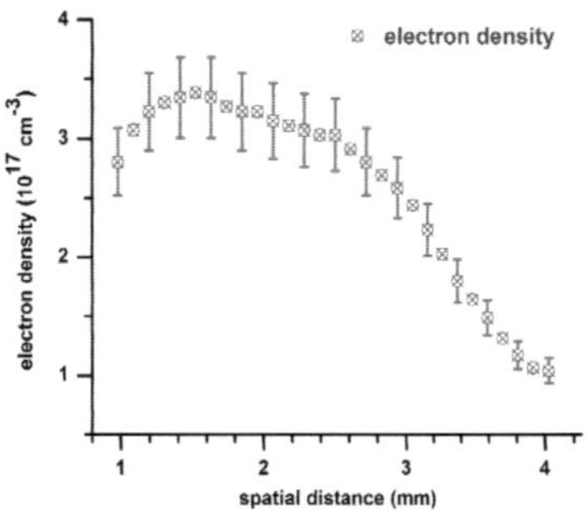

Figure 1. N_e variation from H_α, 0.4 μs time delay.

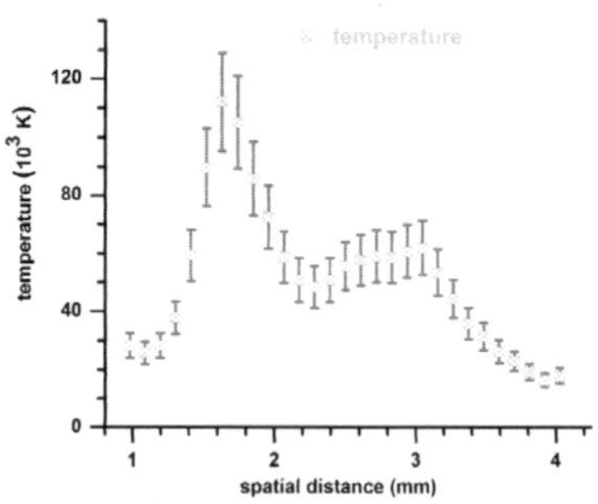

Figure 2. T_e variation from H_α, 0.4 μs time delay.

Full width at half maximum (FWHM) values are extracted from Lorentzian fits to the experimental data. Variation in FWHM values are found in the different zones of the plasma. This variation in FWHM values affects the electron densities due to approximately power of 3/2 dependence in Eq. (1).

Spatial variations of N_e indicate up to a factor of 3 difference and an average value of $2.6 \times 10^{17}\text{cm}^{-3}$ for a time delay of 0.4 μs. The results of the spatial variation of electron density inferred from the H_β line are communicated in a recent article [8].

The spatial variation of temperature shows up to a factor of 7 difference with an average value of 52×10^3 K. The estimated ±15% error bars used for temperature values are higher than ±10% error bars for electron density due to the increased uncertainties in determining temperatures.

Temporal variations of electron density and temperature determined from H_α and H_β lines are displayed in Figures 3 and 4. The electron densities agree with the line-of-sight values obtained from H_β lines [8] and are within the error bars of previous experiments in hydrogen gas [2, 4].

The measured plasma volume is represented by 29 values of the FWHM. At the time delay of 0.15 μs, incomplete line profiles of H_β cause increased error margins in the measured electron densities. The temporal electron density variations are in the range of $8 \times 10^{17}\text{cm}^{-3}$ to $0.5 \times 10^{17}\text{cm}^{-3}$ for time delays of 0.15 μs to 1.4 μs.

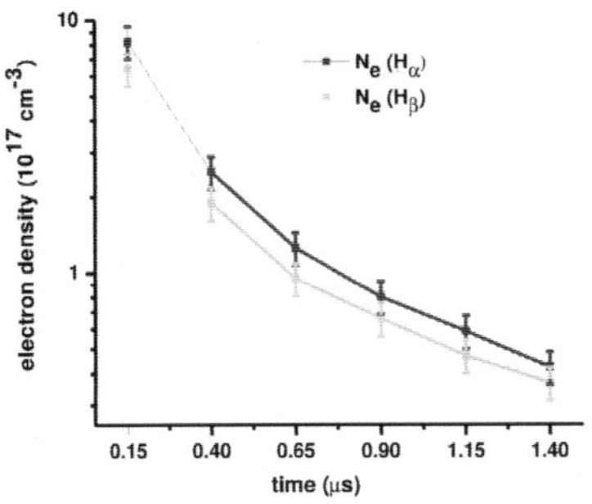

Figure 3. Temporal variation of electron density.

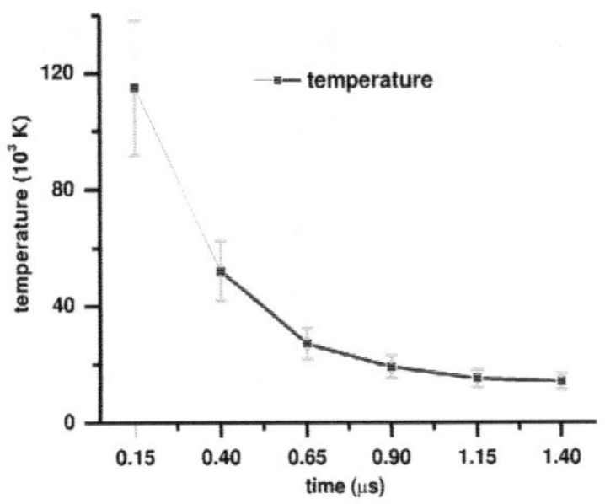

Figure 4. Temporal variation of excitation temperature.

The temperature changes decrease with time as indicated in Figure 4. The temperatures are in the range of 115×10^3 K to 14×10^3 K for time delays of 0.15 μs to 1.4 μs. These values are consistent with the previous hydrogen experiments [4]. Estimated error bars for electron density and excitation temperature are ±15% and ±20%, respectively. The error bars for the temporal variation are slightly larger than for the spatial variation due to the use of a 20 ns gate width in the measurements reported here.

4. Conclusions

Recently derived empirical formulae are implemented for the electron density diagnostics. These formulae are applicable for the analysis of electron density in the indicated time delay range. Furthermore, both H_α and H_β lines can be used for the electron density diagnostics. The spatial variation of electron density and excitation temperature at 0.4 μs show up to a factor of 3 and 7 difference, respectively. Application of Abel inversion methods to extract radial distribution of electron density and excitation temperature is of continued interest.

Acknowledgments

The authors thank for support in part by the Center for Laser Applications at the University of Tennessee Space Institute.

References

[1] Griem H R 1974 *Principles of Plasma Spectroscopy* (McGraw-Hill, New York)
[2] Parigger C G 2013 *Spectrochim. Acta Part B* **79-80** 4
[3] Surmick D M and Parigger C G 2014 *Int. Rev. At. Mol. Phys.* **5** 73
[4] Parigger C G, Plemmons D H and Oks E 2003 *Appl. Opt.* **42** 5992
[5] Parigger C G, Surmick D M, Gautam G and EL Sherbini A M 2015 *Opt. Lett.* **40** 3436
[6] Wiese W L 1991 *Spectrochim. Acta Part B* **46** 831
[7] Wiese W L and Fuhr J R 2009 *J. Phys. Ref. Data* **38** 565
[8] Gautam G and Parigger C G 2014 *Int. Rev. At. Mol. Phys.* **5** 115

IOP Conf. Series: Journal of Physics: Conf. Series **810** (2017) 012056 doi:10.1088/1742-6596/810/1/012056

Multicomponent line profile restoring by means of ill-posed inverse task solution

G Revalde[1,3], N Zorina[1,2], A Skudra[2]

[1]Ventspils University College, Inzenieru str. 101, Ventspils, Latvia;
[2] Institute of Atomic Physics and Spectroscopy, University of Latvia, Skunu str. 4, Riga, Latvia;
[3]Institute of Technical Physics, Faculty of Materials Science and Applied Chemistry, Riga, Technical university, P.Valdena 3, Riga, Latvia

E-mail: gitar@latnet.lv

Abstract. The investigation of the criteria of usage of the Tikhonov regularization method for multicomponent overlapping line profiles restoring was done by means of model task solution. The influence of the width and kind of the instrumental function, number of the components of the profile and distance between components are discussed.

1. Introduction

The procedure of restoration of the spectral line profiles from the measured ones is ill-posed task. The small experimental uncertainties can cause large deviations in the solution. In general, for obtaining the real shape of the spectral line, two main directions of solution exist: there are modelling [1] of the spectral line profiles and the real shape obtaining by means of ill-posed inverse task solution.

In this work the "experimental" profiles were modeled by means of a nonlinear multiparameter X^2 in the first stage. The values of the parameters were chosen so, that the "experimental" model profiles were maximally close to real spectral profiles, emitted from microsizes light sources[2]. After that the solution of the Fredholm integral equation of the first kind was performed to remove the instrumental function by means of Tikhonov regularization.[in detail in[3]]

2. Results and discussion

Below we can see the examples of our calculations. On the figures 1(a-b), we can see comparisons between "modeled -real" profiles and "calculated- real" profiles. "Modeled" profiles with one component were generated with following parameters: Doppler width (FWHM), $\delta\nu D=0.03cm^{-1}$, and Lorentz width, $\delta\nu_L=0.01cm^{-1}$, instrumental function, Gauss profile, $\delta\nu_{Instr.}=0.06cm^{-1}$. For modeling of profile with self absorption were used optical density in the line centre: $k_o l=4$ and homogeneity of the radiation source, n=13. The multicomponent profiles were generated with instrumental function, Gauss profile, $\delta\nu_{Instr.}=0.04cm^{-1}$ (figures 2(a),(b)) and $\delta\nu_{Instr.}=0.06cm^{-1}$(figure 2(c)). The distances between components $0.1cm^{-1}$ (figures 2(a), (c)) and $0.3cm^{-1}$ (figure 2(b)).

As we can see on the figures 1(a-b) and 2(a-c), there are very well coincidence between "modeled - real" profile and "calculated- real" profiles as for one component(figures 1(a-b)) as well for multicomponents profiles (figures 2(a-c)).

IOP Conf. Series: Journal of Physics: Conf. Series **810** (2017) 012056 doi:10.1088/1742-6596/810/1/012056

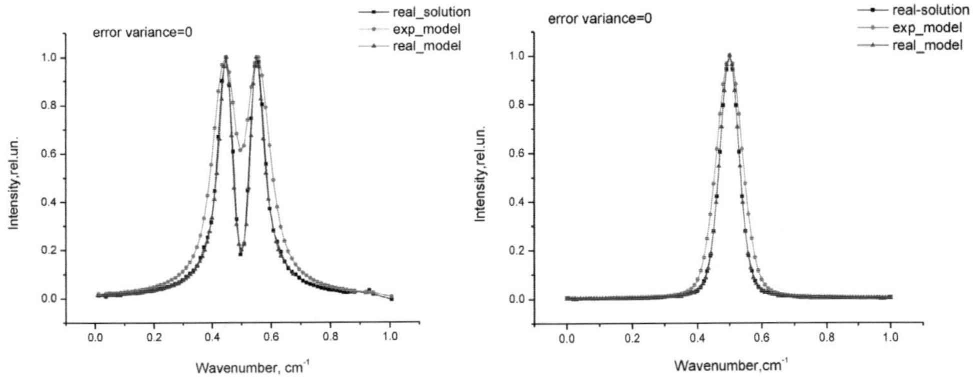

Figure 1. Comparison between "Real modeled" (modeled by means of non-linear multi-parameter modelling) one component profile and "Real solution" profile(obtained by means of Tikhonov regularization method). (a)initial profile was generated with self absorption; (b) initial profile was generated without self absorption.

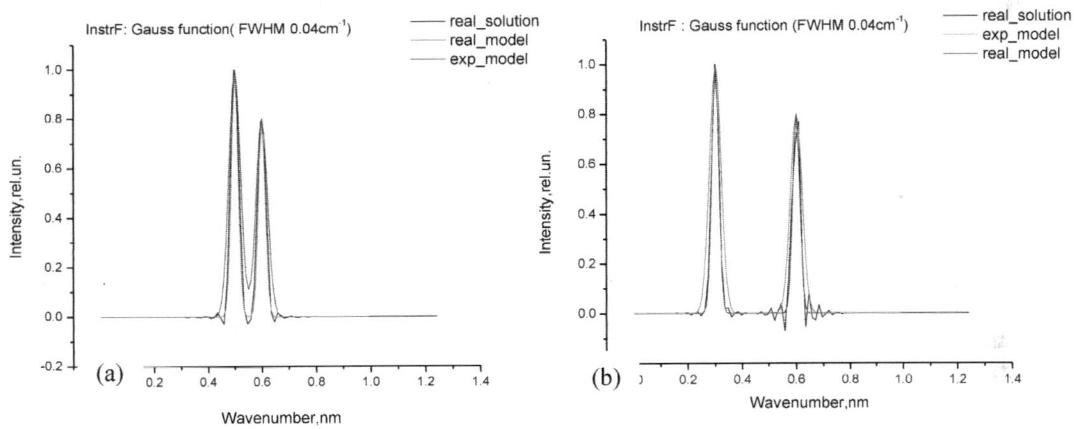

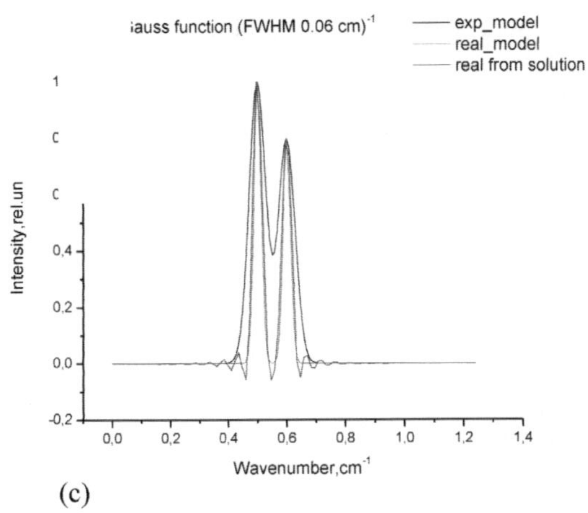

Figure 2. Comparison between "Real modeled" (modeled by means of non-linear multi-parameter modelling) multicomponent profile and "Real solution" profile(obtained by means of Tikhonov regularization method). Instrumental function is Gauss function.

On the figures 3(a-d) we can see results, obtained for cases when was used absolutely different instrumental function. The instrumental function was given as data massive (Airy function).

The mirrors refraction coefficients are: 0.6, 0.7, 08 and 0.9 accordingly. "Experimental" profile was generated without errors. As we can see, too broad instrumental function (refraction coefficient is 60%) leads to the errors in the solution.

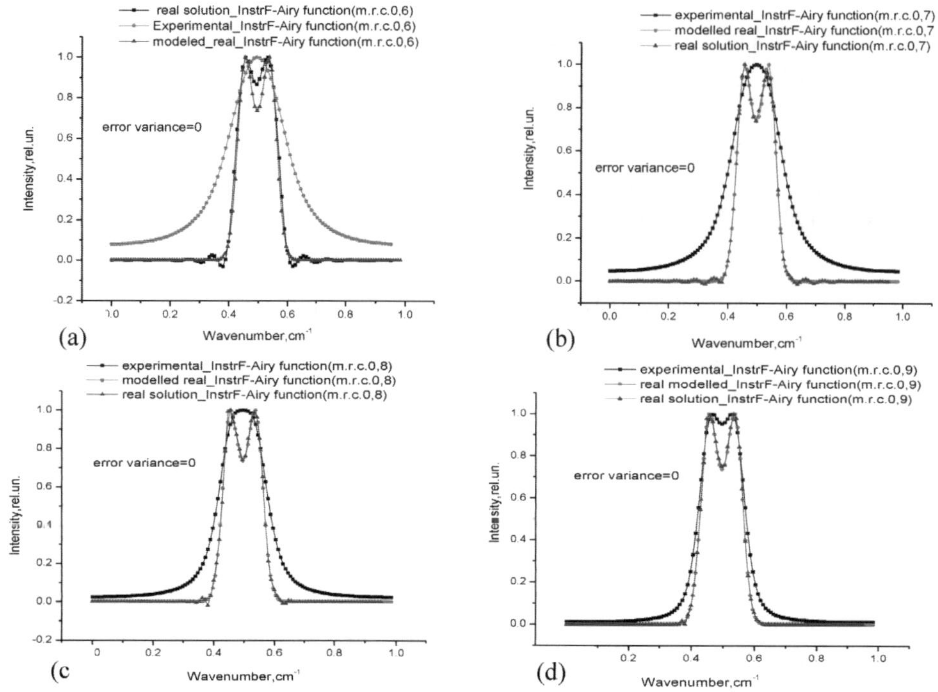

Figure 3. Comparison between "Real modeled" (modeled by means of non-linear multi-parameter modelling)profile and "Real solution" profile(obtained by means of Tikhonov regularization method). Mirrors refraction coefficients are: (a)= 60%,(b)= 70%, (c)= 80%,(d)= 90%. Instrumental function is Airy function.

3. Conclusion

As we can see FWHM of „calculated real" profiles and dip of self absorption coincide with „modelled real" profiles very well as for profiles with one component as well as for multicomponent profiles. This is true, in error limit, for both cases, when the instrumental function was used as formula and as data massive and independent on kind of instrumental function. All this work results show, that instrument function has not been neglected in case of low –pressure or cold plasma: when the instrument function is on the same order that experimental profile.

Acknowledgements

This research has been partly supported by the National Research Programme of Latvia"The next generation of information and communication technologies" (NexIT).

References

[1] A. Skudra,G. Revalde, Mathematical modeling of the spectral line profiles in the high-frequency discharge ,1999, *JQSRT* 61, N6, pp 717-728

[2] G. Revalde, N. Zorina, A. Skudra, Z. Gavare, Deconvolution of the line spectra of microsize light sources in the magnetic field, *Romanian Reports in Physics*, Vol. 66, No. 4, P. 1099–1109, 2014

[3] N. Zorina, G. Revalde, R. Disch, *Proceedings of SPIE Sixth International Conference on Advanced Optical Materials and Devices (AOMD-6)*, 24–27August (2008) Riga, Latvia, Vol. 7142 71420J-1

A spectroscopic diagnostic of the electron density in a corona discharge

J Rosato[1], N Bonifaci[2], Z Li[2,3] and R Stamm[1]

[1]Aix-Marseille Université, CNRS, PIIM UMR 7345, F-13397 Marseille Cedex 20, France
[2]Laboratoire G2Elab, CNRS and Grenoble University, 25 rue des Martyrs, 38042 Grenoble, France
[3]Guizhou Institute of Technology, Caiguan road 1, 550003, Guiyang, China

E-mail: joel.rosato@univ-amu.fr

Abstract. Spectral lines of hydrogen observed in a corona discharge are investigated. The corona discharge has been performed at the vicinity of a tip electrode under high voltage. The shape of the H-α and H-β lines is dominated by the Stark broadening due to the plasma microfield. Using a computer simulation method, we examine the sensitivity of the line shapes to the electron density. Our results indicate the possibility of a density diagnostic based on passive spectroscopy.

1. Introduction

Corona discharges of helium have been performed in an experiment devoted to the investigations of the dielectric properties of insulators in the context of electrical engineering. Here, we report on the analysis of hydrogen Balmer lines, which are strongly sensitive to the electron density and which have been observed in discharges at the room temperature. We show that a diagnostic of the electron density can be performed from an analysis of the width of these spectral lines. This work, which serves as a preliminary step, will be completed with further analyses of helium corona discharges that will be performed at low temperatures (a few K) with liquid helium.

2. Presentation of the experiment

The experimental setup consists of a point-plane electrode system placed inside a helium cryostat. The point electrode is negatively polarized by a stabilized high voltage DC power supply (Spellman RHSR/20PN60). The current-voltage characteristics were measured by using a Tektronix TDS540 oscilloscope and a Keithley 610C ammeter. The corona discharge in this geometry is axially symmetric and it appears as a luminous spherical region (ionization region) localized near the point electrode against the dark background. The measured intensity of the radiation is averaged over the observed lines and the exposure time. A liquid N_2 cooled 2D-CCDTKB-UV/AR detector is located directly in the exit plane of the spectrograph. The noise level of the CCD detector is determined only by the read-out noise as the dark current of the camera is less than 1 e/pixel/h at 153 K. The wavelength and intensity response of the detection system was calibrated by using low pressure helium and tungsten ribbon lamps. The line broadening due to the instrument response (1200g/mm grating) was estimated from the helium lines of a helium lamp as $\Delta\lambda_{ins} < 0.10$ nm.

3. An analysis of hydrogen lines observed at 300 K

As a preliminary step, we report here on an analysis of hydrogen lines that have been observed in helium discharges at the room temperature. The atomic hydrogen is present in traces and its emission is sufficiently strong so as to yield intense lines, especially in the H-α and H-β transitions. The discharges were carried out at pressures from 1 to 12 bar, with a current of 100 μA. In order to simplify the interpretation of spectra, we focus on the lowest pressure values (namely, between 1 and 2 bar) because the H-α and H-β lines overlap with nearby atomic lines and molecular bands at higher pressures. Furthermore, we focus on the H-β line because its spectral profile is cleaner than that of H-α.

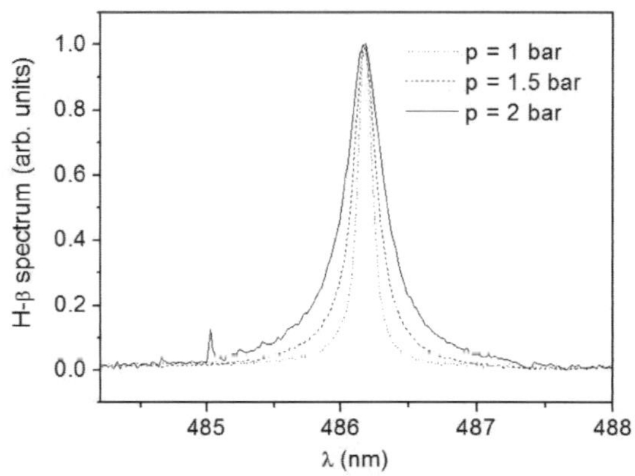

Figure 1. Plot of H-β observed in a helium corona discharge at room temperature, for different pressure values. The line width increases with the pressure. This trend is also observed on other lines.

Figure 1 shows a plot of the H-β line observed at 1, 1.5, and 2 bar. The line width increases with the pressure. An analysis using a van der Waals broadening model [1,2] indicates that this effect is not the dominant line broadening one. The instrumental broadening is also not dominant. We have examined the role of the Stark effect related to the plasma microfield. A computer simulation method [3] has been applied to the H-β line at the same pressure values as above. The He^+ ion microfield evolution is simulated from a quasiparticle model and the line broadening is calculated from a numerical integration of the time-dependent Schrödinger equation. The contribution of the electrons is evaluated using a collision operator. In our calculations, we have used the Griem-Kolb-Shen model [4] assuming an electron temperature of 10^4 K and leaving the electron density as an adjustable parameter. The ion temperature has been assumed equal to the atomic (300 K) temperature. Our calculations indicate that the Stark broadening is mainly due to the ions. Figure 2 shows an example of adjustment performed using the simulation method at p = 1 bar. A value of 10^{15} cm^{-3} has been obtained for the electron density. Our calculations at p = 1.5 bar and 2 bar yield higher values for the electron density (2×10^{15} cm^{-3} and 3×10^{15} cm^{-3}, respectively), which indicates an increase trend in terms of the pressure. This result is preliminary and will be completed with further analyses.

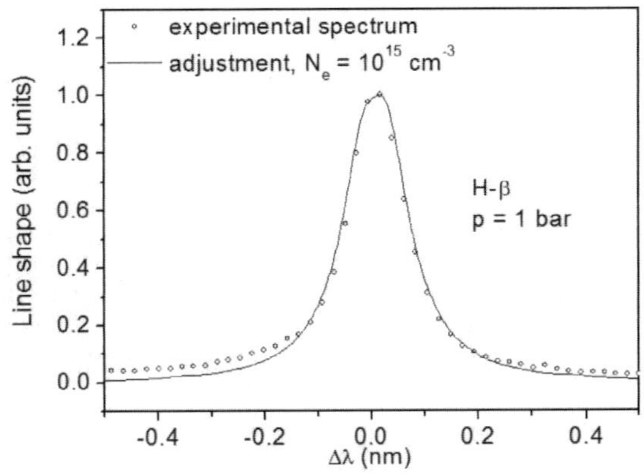

Figure 2. The plasma microfield yields an additional broadening, which can serve as a probe of the density. Here, the Stark broadening has been evaluated using a computer simulation method [3].

4. Conclusion

We have analyzed spectral profiles of hydrogen lines in helium corona discharges by means of a computer simulation method. An application to 1, 1.5, and 2 bar gas pressures indicates that the plasma microfield yields a visible Stark broadening, which can serve as a probe for the electron density. This result is a preliminary step in foregoing investigations of liquid corona discharges. New experiments, with liquid helium, are planned and will be analyzed using spectroscopic techniques.

References

[1] Peach G 1996 *Atomic, Molecular, and Optical Physics Handbook*, ed G W F Drake (Woodburry: AIP Press) p 669
[2] Allard N and Kielkopf J 1982 *Rev. Mod. Phys.* **54** 1103–82
[3] Rosato J, Marandet Y, Capes H, Ferri S, Mossé C, Godbert-Mouret L, Koubiti K and Stamm R 2009 *Phys. Rev. E* **79** 046408-1-7
[4] Griem H R, Kolb AC and Shen K Y 1959 *Phys. Rev.* **116** 4–16

IOP Conf. Series: Journal of Physics: Conf. Series **810** (2017) 012058 doi:10.1088/1742-6596/810/1/012058

Influence of Rydberg atom-atom collisional and (n-n')-mixing processes on optical properties of astrophysical and low-temperature laboratory plasmas

A A Mihajlov[1], V A Srećković[1], Lj M. Ignjatović[1], Z. Simić[2] and M S Dimitrijević[2,3,4]

[1] University of Belgrade,Institute of Physics, P. O. Box 57, 11001 Belgrade, Serbia
[2]Astronomical Observatory, Volgina 7, 11160 Belgrade 74, Serbia
[3]IHIS-Technoexperts, Bežanijska 23, 11080 Zemun, Serbia
[4]Observatoire de Paris, 92195 Meudon Cedex, France

E-mail: `mihajlov@ipb.ac.rs`

Abstract. Here, we will consider the influence of the (n-n')-mixing processes during atom Rydberg-atom collision processes on the intensity of chemi-ionization process. We will take into account H(1s) + H*(n) and He(1s^2) + He*(n,l) collisional systems, where the principal quantum number $n >> 1$. The corresponding calculations of the chemi-ionization rate coefficients are performed for the temperature region characteristic for the solar and DB white dwarf atmosphere.

1. Introduction and the theoretical remarks

The processes with participation of the Rydberg atoms are still important in the investigation of different stellar atmospheres as well as in the laboratory experiments [1, 2, 3, 4, 5, 6]. The main aim of this paper is the consideration of two kinds of atomic collision processes involving Rydberg atoms which simultaneously occur in the stellar atmospheres and the influence of one on the other. We will analyze the processes of chemi-ionization: in the hydrogen case

$$\mathrm{H}^*(n,l) + \mathrm{H}(1s) \to \mathrm{H}(1s) + \mathrm{H}^+ + \vec{e} \tag{1a}$$

$$\mathrm{H}^*(n,l) + \mathrm{H}(1s) \to \mathrm{H}_2^+ + \vec{e} \tag{1b}$$

and in the helium case

$$\mathrm{He}^*(n,l) + \mathrm{He}(1s^2) \to \mathrm{He}(1s^2) + \mathrm{He}^+(1s) + \vec{e} \tag{2a}$$

$$\mathrm{He}^*(n,l) + \mathrm{He}(1s^2) \to \mathrm{He}_2^+ + \vec{e} \tag{2b}$$

where the principal quantum number $n >> 1$, and the orbital quantum number l changes in the interval from 0 to $n-1$. Here, we also study the (n-n)-mixing processes:

$$\mathrm{H}^*(n,l) + \mathrm{H}(1s) \to \mathrm{H}(1s) + \mathrm{H}^*(n',l'), \tag{3}$$

$$\mathrm{He}^*(n,l) + \mathrm{He}(1s^2) \rightarrow \mathrm{He}(1s^2) + \mathrm{He}^*(n',l'), \tag{4}$$

where the final values of the quantum numbers n', $l' \neq n$, l. These processes have always been considered separately, although both processes are caused by the same dipole resonant mechanism [7, 8, 9]. Because of that the inclusion of (n-n')-mixing processes in the procedure of calculation of rate coefficients of the chemi-ionization processes is the main task of this paper. Thus, here will be considered the chemi-ionization processes (1a-1b) and (2a-2b) in the presence of (n-n')-mixing channel (3) and (4) from the astrophysical aspect and analyzed their real influence for the cases of different DB white-dwarf atmospheres and solar atmosphere.

The rate coefficients $K^a_{1,2}(n,l;T)$ and $K^b_{1,2}(n,l;T)$ of processes (1a,1b) or (2a,2b) are separately determined for given n, l and T. The total rate coefficient $K^{ab}_{1,2}(n,l;T)$ of processes (1a,1b) or (2a,2b) together, is obtained as $K^{ab}_{1,2}(n,l;T) = K^a_{1,2}(n,l;T) + K^b_{1,2}(n,l;T)$. Here, because of further applications, we determined the average total rate coefficient by

$$K_{1,2}(n,T) = \frac{1}{n^2} \cdot \sum_{l=0}^{n-1}(2l+1) \cdot K^{ab}_{1,2}(n,l;T). \tag{5}$$

The above mentioned partial rate coefficients $K^{ab}_{1,2}(n,l;T)$ and $K^b_{1,2}(n,l;T)$ are determined on the basis of standard expressions $K^{ab}_{1,2}(n,l;T) = \int_{E_{n;i}}^{\infty} \sigma^{ab}_{1,2}(n,l;E)\sqrt{\frac{2E}{m_{red}}} f_T(E)dE$, where E is impact energy, $\sigma^{ab}_{1,2}(n,l;E)$ is the corresponding cross section, m_{red} - reduced mass, and $f_T(E)$ is the Maxwell distribution function. Parameters $E_{n;i}$ $R_{n;i}$, cross sections $\sigma^{ab}_{1,2}(n,l;E)$ and all quantities needed for the calculations are described in details in [10, 11].

2. Results and discussion

In accordance with the aim of this work we consider here the model of solar atmosphere from [12] and the model for DB white dwarfs from Koester 2015 (private communication). In order to show the importance of the investigated chemi-ionization processes we compared their efficiencies with the efficiencies of the relevant concurrent processes, i.e. electron-Rydberg atom impact ionization $\mathrm{H}^*(n) + e \rightarrow \mathrm{H}^+ + e + e$, in the hydrogen case and $\mathrm{He}^*(n) + e \rightarrow \mathrm{He}^+ + e + e$, in the helium case. The main objective of this paper is to demonstrate the impact of these considered processes on the modeling of stellar atmospheres by analysing fluxes caused by atom-Rydberg atom and electron-excited atom impact ionization (using calculated rate coefficients). Fluxes generated in atom-Rydberg atom and electron-excited atom impact ionization are denoted by $I^{aa}_{1,2}(T;A^*) = K_{1,2}(n,T;A^*) \cdot N(\mathrm{A}^*) \cdot N(\mathrm{A})$ and $I^{ea}_{1,2}(T;A^*)\alpha^{ea}_{1,2}(n,T;A^*) \cdot N(\mathrm{A}^*) \cdot N(e)$ where $A^* = \mathrm{H}^*(n)$ or $\mathrm{He}^*(n)$, the rate coefficient $K_{1,2}(n,T;A^*)$ is given by Eq.(5) and the ionization rate coefficient $\alpha^{ea}_{1,2}(n,T;A^*)$ is determined on the basis of semiempirical expressions (see [11]). The relative importance of the chemi-ionization processes in comparison with electron-excited atom impact ionization can be characterized by parameters $F_{1,2}(n,T)$ defined as :

$$F_{1,2}(n,T;A^*) = I^{aa}_{1,2}(T;A^*)/I^{ea}_{1,2}(T;A^*) = \frac{K_{1,2}(n,T;A^*)}{\alpha^{ea}_{1,2}(n,T;A^*)}\frac{N(A)}{N(e)} \tag{6}$$

Illustrative calculations were carried out for photosphere of the Sun, for hydrogen case, and some DB white dwarfs with $\log g = 8$ and temperatures $T_{eff} = 12000$ K and 14000 K for the case of helium. On the one example (Fig. 1) we show the results of these calculations for DB white dwarf atmospheres with $T_{eff} = 14000$ K and $\log g = 8$. The parameters $F_{1,2}(n,T;\mathrm{He}^*)$ are shown as functions of $\log(\tau)$ for DB white dwarfs, where τ is the Rosseland optical depth. Fig. 1 shows that for the lower temperatures the chemi-ionization processes are still dominant over electron-excited atom ionization processes, for $3 \leq n \leq 5$ almost in the whole observed atmosphere. For

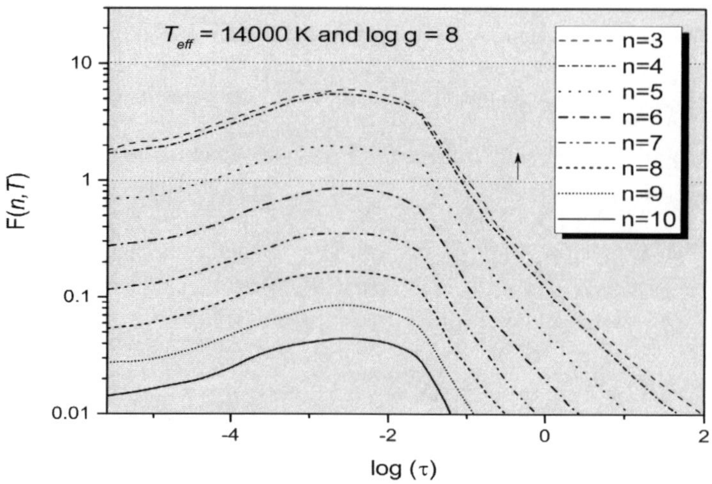

Figure 1. Parameter $F_i(n, T; \mathrm{He}^*(n))$ as a function of $\log(\tau)$, for principal quantum numbers $n = 3 - 10$, for model of DB white dwarf atmosphere with $T_{eff} = 14000$ K and $\log g = 8$.

$6 \leq n \leq 8$ and in the whole observed atmosphere, chemi-ionization processes are comparable with electron-excited atom ionization processes. It can bee noticed that if effective temperature T_{eff} is increasing the influence of chemi-ionization processes begins to decrease. Our analysis shows that the influence of processes (3) decreases with the increase of n, and for $n > 10$ it practically, does not exists.

From the presented material it is shown that the processes of (n-n')-mixing considerably influence on the rates of chemi-ionization processes, and in spite of the influence of (n-n')-mixing processes with $n' > n$ in the most parts of the photosphere of the Sun and most parts of white dwarf atmospheres remain dominant or at least comparable to the efficiency of the concurrent processes.

Acknowledgments

The authors are thankful to the Ministry of Education, Science and Technological Development of the Republic of Serbia for the support of this work within the projects 176002 and III4402.

References

[1] Barbier L, Cheret M 1987 *Journal of Physics B Atomic Molecular Physics* **20** 1229
[2] Barklem P S 2007 *A&A* **466** 327
[3] Gnedin Yu N, et al. 2009 *New Astronomy Reviews* **53** 259
[4] Mihajlov A A, Ignjatović L M, Srećković V A, Dimitrijević M S 2011 *ApJS* **193** 2
[5] Lim J, Lee H-g, Ahn J 2013 *Journal of Korean Physical Society* **63** 867
[6] Niederprüm T, Thomas O, Manthey T, Weber T M, Ott H 2015 *Phys.Rev.Lett.* **115** 013003
[7] Mihajlov A A, Ignjatović L M, Srećković V A, Djurić Z 2008 *JQSRT* **109** 853
[8] Mihajlov A A, Srećković V A, Ignjatović L M, Klyucharev A N 2012 *J.Clust.Sci.* **23** 47
[9] Srećković V A, Mihajlov A A, Ignjatović L M, Dimitrijević M S 2013 *A&A* **552** A33
[10] Mihajlov A A, Srećković V A, Ignjatović L M, et al. 2015 *J.Astrophys.Astron.* **36** 623
[11] Mihajlov A A, Srećković V A, Ignjatović L M, Dimitrijević M S 2016 *MNRAS* **458** 2215
[12] Fontenla J M, Balasubramaniam K S, Harder J 2007 *ApJ* **667** 1243

The inverse bremsstrahlung absorption coefficients and Gaunt factors in astrophysical plasmas

A A Mihajlov[1], V A Srećković[1], N M Sakan[1], Lj M Ignjatović[1], Z Simić[2] and M S Dimitrijević[2,3,4]

[1] University of Belgrade,Institute of Physics, P. O. Box 57, 11001 Belgrade, Serbia
[2]Astronomical Observatory, Volgina 7, 11160 Belgrade 74, Serbia
[3]IHIS-Technoexperts, Bežanijska 23, 11080 Zemun, Serbia
[4]Observatoire de Paris, 92195 Meudon Cedex, France

E-mail: mihajlov@ipb.ac.rs

Abstract. In this paper we present the method of determination of the electron-ion inverse Bremsstrahlung characteristics in order to cover the corresponding stellar atmosphere models. The used method is based on cut-off Coulomb model potential. It is shown that determination of these characteristics i.e. the absorption coefficients and Gaunt factors can be successfully performed in the region of electron densities from 10^{13}cm^3 to $3 \cdot 10^{19}$cm^3 and temperatures from 3000 K to 50000 K within the wavelength region 100 nm $\leq \lambda \leq$ 3000 nm.

1. Introduction

After examining the existing literature about inverse "Bremsstrahlung" process [1, 2, 3, 4, 5], it could be seen that the most of presented papers are devoted to the determination of the corresponding Gaunt factor. Namely, the exact relation for the differential cross section for the direct "Bremsstrahlung" process exist for a long time [6]. This automatically led to the possibility of the exact term for the inverse "Bremsstrahlung" cross section $\sigma_{i.b.}^{(ex)}(E, \varepsilon_{ph})$, for the considered inverse "Bremsstrahlung". However, the practical applicability of mentioned strict relation was complex, which suggested the search of a practical ways for describing inverse "Bremsstrahlung". Therefor, it has been taken the simple and widely used quasi classical (Kramer's) cross section for the direct and inverse "Bremsstrahlung" process. The idea was to present exact $\sigma_{i.b.}^{(ex)}(E, \varepsilon_{ph})$ in a form

$$\sigma_{i.b.}^{(ex)}(E, \varepsilon_{ph}) = \sigma_{i.b.}^{q.c.}(E, \varepsilon_{ph}) \cdot g_{i.b.}(E, \varepsilon_{ph}), \tag{1}$$

where $g_{i.b.}(E, \varepsilon_{ph})$ is the adequate defined Gaunt factor. Because of that, further step was to find a simple approximations for the Gaunt factor $g_{i.b.}(E, \varepsilon_{ph})$ in relation (1). The mentioned exact and approximated absorption coefficient are denoted with $k_{i.b.}^{(ex)}(\lambda, T; N_e)$ and $k_{i.b.}^{q.c.}(\lambda, T; Ne)$, where N_e is free electron density. Here it is taken into account that in our case (single charged ions) the ion density is equal to electron density N_e. Consequently, these coefficients are connected with the relation

$$k_{i.b.}^{(ex)}(\lambda, T; N_e) = k_{i.b.}^{q.c.}(\lambda, T; N_e) \cdot G_{i.b.}(\lambda, T), \tag{2}$$

where $G_{i.b.}(\lambda, T)$ is the required Gaunt factor. The determination of such averaged Gaunt factor as a function of λ and T was the object of investigation in the majority of previous papers devoted to the inverse "Bremsstrahlung" process (see e.g. [1, 2, 3, 4]).

In connection with the relations (1) and (2) it is needed to keep in mind the fact that exact terms for the inverse "Bremsstrahlung" process are determined in the case of scattering of the free electron onto the Coulomb potential. This means that mentioned exact terms are strictly applicable on the case of diluted plasma. It is clear that the same holds for all approximate relations which were obtained by now on the basis of these terms. From above mentioned it follows that all such relations become the unusable in the case of plasma of higher densities, e.g. non-ideal plasma. The aim of this work is to find such relations for $\sigma_{i.b.}^{(ex)}(E, \varepsilon_{ph})$ which could be applicable in the case of different stellar atmospheres (with higher densities), where the values of the electron density changes from $Ne \sim 10^{13}$ cm^{-3} to $Ne \sim 10^{20}$ cm^{-3}.

2. Theory

This work is continuation of our previous investigation devoted to the examination of the opacity of the weakly and strongly ionized parts of the solar photospheres and lower chromospheres and different helium rich white dwarfs atmospheres [7, 8, 9, 10, 11, 12, 13]. The aim of this work is determination of such absorption coefficient $k_{i.b.}^{(ex)}(\lambda, T; N_e)$ and Gaunt factor $G_{i.b.}(\lambda, T)$ which could be applicable on the case of plasma with higher densities. Because of that, we applied a model potential, specially adopted for the description of the electron scattering onto the ion inside plasma i.e. cut-off Coulomb potential, described by the relations $U_{cut}(r) = -\frac{e^2}{r} + \frac{e^2}{r_{cut}}$ if $0 < r \le r_{cut}$ and $U_{cut}(r) = 0$, if $r_{cut} < r$.

Here, we use the procedure of the determination of inverse Bremsstrahlung cross-section which is improved comparing to [12]. Using this improved procedure in this work were performed calculations of the corresponding Gaunt factor in the wide region of electron densities and temperature which can be used in connection with the different astrophysical and laboratory plasmas. Special calculations were made in order to determine free-free spectral absorption coefficients for the different stellar atmosphere models. Here we take $\kappa_{i.b.}^{(ex)}$ in the form

$$\kappa_{i.b.}^{(ex)}(\varepsilon_\lambda; N_e, T) = N_e^2 \cdot \int_0^\infty \sigma_{i.b.}^{(ex)}(E; E')v \cdot f_T(v) \cdot 4\pi v^2 dv \cdot \left(1 - \exp\left[-\frac{\hbar\omega}{kT}\right]\right), \qquad (3)$$

where $f_T(v)$ is the corresponding Maxwellian distribution function for given temperature T. On the other hand quasi classical Kramer's $k_{i.b.}^{q.c.}(\lambda, T; Ne)$ is given by the known expression

$$k_{i.b.}^{q.c.}(\lambda, T; Ne) = N_e^2 \cdot \frac{16\pi^{5/2}\sqrt{2}e^6}{3\sqrt{3}cm^{3/2}\varepsilon_{ph}^3} \frac{\hbar^2}{(kT)^{1/2}} \left(1 - \exp\left[-\frac{\hbar\omega}{kT}\right]\right), \qquad (4)$$

where $\varepsilon_{ph} = 2\pi\hbar c/\lambda$. According to this, averaged Gaunt factor $G_{i.b.}(\lambda, T)$ is determined here from Eq. (2), where $k_{i.b.}^{(ex)}(\lambda, T; Ne)$ and $k_{i.b.}^{q.c.}(\lambda, T; Ne)$ are given by Eqs. (3) and (4).

It is important that within this work the inverse "Bremsstrahlung" absorption coefficients $\kappa_{i.b.}^{(ex)}(\varepsilon_\lambda; N_e, T)$ are determined by Eq. (3), where the spectral absorption cross section $\sigma_{i.b.}^{(ex)}(E; E')$ is calculated using the improved procedure from [12], where all relevant quantities are determined strictly numerically without any additional approximation.

3. Results and discussion

The calculations of the Gaunt factor $G_{i.b.}(\lambda, T)$ and absorption coefficients $k_{i.b.}^{(ex)}(\lambda, T; Ne)$ were carried out for the electron densities in the range from $1 \cdot 10^{13}$cm^{-3} to $3 \cdot 10^{19}$cm^{-3} and temperatures from $3 \cdot 10^3$K to $1 \cdot 10^5$K, and the observed wavelength region 100nm $< \lambda <$ 3000nm

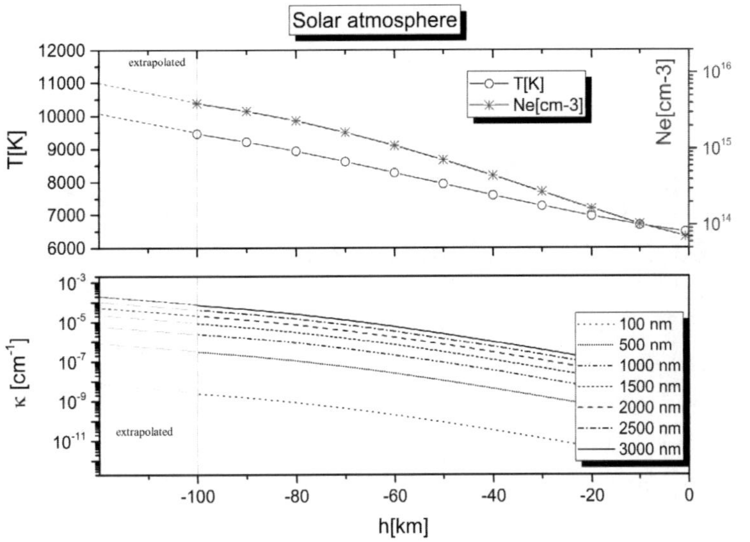

Figure 1. Upper panel: The behavior of T and N_e as a function of height within the considered part of the solar atmosphere model of [14]; lower panel: The absorption coefficient for inverse "Bremsstrahlung" processes within the considered part of the solar atmosphere model.

in order to cover the corresponding stellar atmosphere models. This is illustrated by Fig.1 which shows the behavior of the absorption coefficient for inverse "Bremsstrahlung" processes within the considered part of the solar atmosphere. Also the absorption coefficients for different DB white dwarf atmospheres were carried out. Here we highlight that the presented exact quantum mechanical method can be used to obtain the spectral coefficients for inverse "Bremsstrahlung" process for the broad class of plasma of higher non-ideality. We expect that the cut-off Coulomb potential model results are more accurate in comparison with other methods, which are so far used for cases of stellar plasma with higher non-ideality.

Acknowledgments
The authors are thankful to the Ministry of Education, Science and Technological Development of the Republic of Serbia for the support of this work within the projects 176002 and III4402.

References
[1] Berger J M 1956 *ApJ* **124** 550
[2] Karzas W J, Latter R 1961 *ApJS* **6** 167
[3] D'yachkov L G 1990 *J.Phys.B* **23** L429
[4] van Hoof P A M, Williams R J R, Volk K, et al. 2014 *MNRAS* **444** 420
[5] Armstrong G S J, Colgan J, Kilcrease D P, Magee N H 2014, *High Energy Density Physics* **10** 61
[6] Sommerfeld A 1953, *"Atombau und Spektrallinien"*, 2 Bnde (vol 2)
[7] Mihajlov A A, Sakan N M, Srećković V A, Vitel Y 2011 *J.Phys. A* **44** 095502
[8] Mihajlov A A, Sakan N M, Srećković V A, Vitel Y 2011 *Baltic Astronomy* **20** 604
[9] Mihajlov A, Ignjatović L, Srećković V A, Dimitrijević M, Metropoulos A. 2013 *MNRAS* **431** 589
[10] Ignjatović L M, Mihajlov A A, Srećković V A, Dimitrijević M S 2014 *MNRAS* **441** 1504
[11] Ignjatović L M, Mihajlov A A, Srećković V A, Dimitrijević M S 2014 *MNRAS* **439** 2342
[12] Mihajlov A A, Srećković V A, Sakan N M 2015 *Journal of Astrophysics and Astronomy* **36** 635
[13] Srećković V A, Mihajlov A A, Ignjatović L M, Dimitrijević M S 2014 *Adv.Sp.Res.* **54** 1264
[14] Fontenla J M, Balasubramaniam K S, Harder J 2007 *ApJ* **667** 1243

IOP Conf. Series: Journal of Physics: Conf. Series 81**0**(2017) 012060 doi:10.1088/1742-6596/810/1/01206

Does non-cyclic Berry phase play any role in the formation of pressure-broadened spectral line shapes?

R Ciuryło[1], **J Szudy**[1] **and W E Baylis**[2]

[1] Institute of Physics, Faculty of Physics, Astronomy and Informatics, Nicolaus Copernicus University, Grudziądzka 5, 87-100 Toruń, Poland

[2] Department of Physics, University of Windsor, Windsor, Ontario N9B 3P4, Canada

E-mail: szudy@fizyka.umk.pl

Abstract. The problem of the influence of contributions to the Berry phase coming from the motion of perturbers along open paths on the shapes of pressure-broadened spectral lines is examined by the use of the Born-Huang adiabatic approximation.

1. Introduction

In 1984 Berry [1] discovered that a state of a quantum system can acquire an extra phase of purely geometric origin when the system undergoes a cyclic adiabatic evolution, i.e. when it completes a closed path in parameter space. Recently, the concept of Berry's geometric phase was included in the formalisms of adiabatic theories of pressure broadening to derive expressions for the intensity distribution both in the core [2] and far wings [3] of spectral lines radiated by atoms placed in an external electric field rotating around a cone about a fixed axis. In the limit of slow rotation on a closed path such that the electric field has constant magnitude but returns to its original position, the atomic levels suffer energy shifts due to a Berry phase in addition to those due to the interaction of the radiating atom with the perturbers. Results of analyses reported in Refs. [2] and [3] seem to indicate the importance of the effect caused by Berry's phase on pressure-broadened spectral lines in cases when some parameters describing molecular system such as the external electric field traverse a closed path. Since the perturbers surrounding the radiating atom do not move along closed paths, the contributions to the Berry phase coming from the time dependence of the perturber coordinates $\mathbf{R}(t)$ were assumed to be negligible in these analyses. It was shown, however, by several researches that the geometric phase makes sense also for non-cyclic adiabatic evolution, corresponding to open paths in

IOP Conf. Series: Journal of Physics: Conf. Series **810** (2017) 012060 doi:10.1088/1742-6596/810/1/012060

parameter space [4-7]. Therefore, in the present work we address ourselves to the following problem: Should the phase contribution coming from open paths in **R** space, i.e. those resulting from non-cyclic perturber motions, give rise to observable effects on intensity distribution in pressure-broadened spectral lines? We were stimulated by two papers by Zygelman [8, 9], in which he demonstrated the appearance of Berry's phase in the context of atomic collision problems and suggested that this effect should be seen in spectral line shapes.

2. Berry vector gauge potentials

Let $\phi_\alpha(\mathbf{r},\mathbf{R})$ and $\phi_\beta(\mathbf{r},\mathbf{R})$ denote the electronic eigenfunctions of a diatomic quasimolecule consisting of a radiating atom in the initial (α) and the final (β) state, respectively, which undergoes a collision with one perturber at time t at the distance **R**= **R**(t). The Berry phase $\gamma_\alpha(t)$ (or $\gamma_\beta(t)$), can be written as the line integral along the curve from $\mathbf{R}(0)$ at $t = 0$ to $\mathbf{R}(t)$:

$$\gamma_\alpha(t) = \int_{R(0)}^{R(t)} \mathbf{A}_\alpha(\mathbf{R})d\mathbf{R}\,, \qquad (1)$$

where

$$\mathbf{A}_\alpha(\mathbf{R}) = i\langle\phi_\alpha(\mathbf{r},\mathbf{R})|\nabla_R|\,\phi_\alpha(\mathbf{r},\mathbf{R})\rangle \qquad (2)$$

is called now the Berry vector gauge potential for the state α. Both $\phi_\alpha(\mathbf{r},\mathbf{R})$ and $\phi_\beta(\mathbf{r},\mathbf{R})$ depend on the electron coordinates **r** and parametrically on **R.** For non-cyclic evolutions the Berry phases are traditionally assumed to be zero on the basis of the Born-Fock gauge transformation. However, it was pointed out in many papers that this choice of gauge is not always possible globally. This feature enables us to consider the possible influence of contributions to the geometric phase coming from open ended paths of the perturbers on the intensity distribution of pressure broadened spectral lines.

3. Appearance of Berry vector potentials in pressure-broadened spectral line shapes

In the adiabatic approximation the total wave function $\Psi_{\alpha,i}(\mathbf{r},\mathbf{R})$ for the initial electronic state α of the radiator is written as a product:

$$\Psi_{\alpha,i}(\mathbf{r},\mathbf{R}) = \phi_\alpha(\mathbf{r},\mathbf{R})\,\chi_{\alpha,i}(\mathbf{R}), \qquad (3)$$

where $\chi_{\alpha,i}(\mathbf{R})$ is the "perturber" wave function that describes the translational motion of a single perturber in the force field of the radiator being in the α- state. The index i (or f) denotes all quantum numbers for the initial (or final) continuum states with energies E_i (or E_f) of the relative motion of the single perturber. The intensity distribution in the $\alpha \to \beta$ line is obtained by calculating first the total electric dipole transition matrix element

$$T_{\alpha\beta}(\omega_{if}) = \langle \chi_{\alpha,i}(\mathbf{R}) | \ \mathbf{D}_{\alpha\beta}(\mathbf{R}) | \chi_{\beta,f}(\mathbf{R}) \rangle, \tag{4}$$

where

$$\mathbf{D}_{\alpha\beta}(\mathbf{R}) = \int d^3\mathbf{r}\ \phi_\alpha^*(\mathbf{r},\mathbf{R})\ \boldsymbol{\mu}\ \phi_\beta(\mathbf{r},\mathbf{R}) \tag{5}$$

is the electronic dipole transition moment of the radiating atom which depends on **R**, and **μ** is the dipole moment operator.

As was indicated by Mead and Truhlar [10], the calculation of the wave function $\chi_{\alpha,i}(\mathbf{R})$ (or $\chi_{\beta,f}(\mathbf{R})$) in the framework of the conventional Born-Oppenheimer (BO) approximation does not allow for any of the effects connected with the Berry vector potentials since no coupling between electronic and perturber motions is taken into account. For the same reason such effects do not appear in the conventional quantum-mechanical treatments of pressure broadening of spectral lines based on the BO approximation [11-14]. On the other hand, they showed that the Born-Huang version [15-17] of the adiabatic approximation gives rise to a natural manifestation of geometric phases.

Following Moody et al. [18] in the Born-Huang approximation the equation for the perturber wave function can be written in gauge-covariant form:

$$\left\{ -\frac{\hbar^2}{2m}\left(\nabla_R - i\,\mathbf{A}_\alpha(\mathbf{R})\right)^2 + W_\alpha(R) - E_i \right\}\chi_{\alpha,i}(\mathbf{R}) = 0, \tag{6}$$

where $\mathbf{A}_\alpha(\mathbf{R})$ is the Berry vector potential for the α state given by Eq. (2). Here $W_\alpha(R)$ (or $W_\beta(R)$) denotes the interatomic potential acting between a perturber and the radiator in the α (or β) state. Hence, the appearance of the Berry phase effect on the spectral line shape may be demonstrated by substituting the solutions of Eq. (6), $\chi_{\alpha,i}(\mathbf{R})$ and $\chi_{\beta,f}(\mathbf{R})$, into Eq. (4). An important feature is that $\mathbf{A}_\alpha(\mathbf{R})$ and $\mathbf{A}_\beta(\mathbf{R})$ transform as the one-dimensional unitary group U(1) when we change our choice of phases for the electronic wave functions $\phi_\alpha(\mathbf{r},\mathbf{R})$ and $\phi_\beta(\mathbf{r},\mathbf{R})$. These functions are defined here in a coordinate system with quantization axis fixed along the radius **R** between the radiating and perturbing atoms. In this molecule-fixed frame the molecular terms $W_\alpha(R)$ and $W_\beta(R)$ are characterized by their quantum number Λ, the projection of the electronic orbital angular momentum **L** along the interatomic axis **R**.

4. Line shapes not affected by Berry phases

Using spherical coordinates Zygelman [8] has shown that for the Σ states (Λ = 0) of a diatomic molecule, the only non-vanishing term for the Berry vector potential is the radial component. In such a case the curl of this vector vanishes at all **R,** and the Berry vector potential can be eliminated by the Born-Fock transformation. This means that if both $W_\alpha(R)$ and $W_\beta(R)$ belong to the Σ states, the

Berry phase effects cannot occur on spectral line profiles associated with the $\Sigma \rightarrow \Sigma$ molecular transitions.

Let us note that the Franck-Condon principle is equivalent to the assumption that the perturbation of electronic wave functions $\phi_\alpha(\mathbf{r}, \mathbf{R})$ and $\phi_\beta(\mathbf{r}, \mathbf{R})$ is so weak that their parametric dependence on **R** can be ignored so that the electronic transition moment in Eq. (5) $\mathbf{D}_{\alpha\beta}(\mathbf{R})$ can be replaced by a constant. In such a case one obtains from Eq. (2) that the Berry vector potentials $\mathbf{A}_\alpha(\mathbf{R})$ and $\mathbf{A}_\beta(\mathbf{R})$ become approximately equal to zero. Hence, one can conclude that for strong allowed transitions the effects of Berry's phase play – if any – a rather marginal role in pressure broadening of corresponding spectral lines.

5. Berry phases and shapes of forbidden spectral lines

The situation looks different, however, in cases where the perturbers disturb the electronic wave functions $\phi_\alpha(\mathbf{r},\mathbf{R})$ and $\phi_\beta(\mathbf{r}, \mathbf{R})$ so strongly that the assumption of a constant transition moment ceases to be valid. An example for which this assumption fails completely is that of collision-induced radiation at wavelengths corresponding to dipole-forbidden transitions, for which the unperturbed dipole transition moment for an isolated radiator vanishes: $\mathbf{D}_{\alpha\beta}(\mathbf{R}(\infty)) = 0$. The collision-induced spectra associated with forbidden lines may appear because the transition moment $\mathbf{D}_{\alpha\beta}(\mathbf{R})$, which is extremely small for isolated atoms, can become as large as for an allowed transition when the perturber is very close to the radiator. In such cases the parametric dependence of $\phi_\alpha(\mathbf{r},\mathbf{R})$ and $\phi_\beta(\mathbf{r}, \mathbf{R})$ on the coordinates of perturbers plays a crucial role and may create Berry vector potentials $\mathbf{A}_\alpha(\mathbf{R})$ and $\mathbf{A}_\beta(\mathbf{R})$ that do not vanish. As a consequence the line shape calculated on the basis of the matrix elements $T_{\alpha\beta}(\omega_{if})$ determined from Eq. (4) for the wave functions $\chi_{\alpha,i}(\mathbf{R})$ or $\chi_{\beta,f}(\mathbf{R})$ by the use of the Born-Huang approximation, Eq. (6), should differ from that evaluated on the basis of the conventional BO approximation. This means that the effect caused on spectral line shapes by the Berry vector potential can be expressed through the difference of intensity distributions corresponding to the Born-Huang and Born-Oppenheimer approximation.

In searching of these effects one can consider the application of the Born-Huang approximation to calculate the shapes of collision-induced spectra in the alkali-rare-gas and oxygen-rare-gas systems, which in the past were the subject of numerous experiments. Comparison of the experimental data with profiles calculated on the basis of the conventional BO approximation shows that in some cases, e.g. for the 5S – 6S and 5S – 4D transitions in Rb-Ar as well as 4S – 3D transition in K-He systems [19], complete disagreement occurs. One can thus speculate whether the inclusion of the Berry vector potential into the line shape calculations by the use of the Born-Huang wave functions instead of conventional BO ones would improve the agreement for these systems. Obviously, in order to obtain reliable results, the reliable potential curves $W_\alpha(R)$ and $W_\beta(R)$ must be used in these calculations.

6. Conclusions

We have examined the possible implications of the Berry vector gauge potential for the pressure broadening of spectral lines, and found that this effect is negligible for strong allowed transitions. In general, the effect due to Berry potential does not play any role for spectral features associated with Σ – Σ transitions. On the other hand, the Berry potential effect is expected to have some importance for collision-induced spectra associated with dipole forbidden transitions.

References

[1] Berry M V 1984 *Proc. Roy. Soc. London A* 392 45

[2] Difallah M, Meftah M T, Bekkouche A, and Bedida N 2012 *J. Quant. Spectr. Radiat. Transfer* **113** 299

[3] Ciuryło R, Szudy J, and Baylis W E 2015 *Phys. Rev. A* **92** 032503

[4] Pati A K 1995 Phys. Rev. A **52** 2576 and 1999 *Phys. Rev. A* **60** 121

[5] Kamleitner J D, Cresser J D, and Sanders B C 2004 *Phys. Rev. A* **70** 044103

[6] Chruściński D and Jamiołkowski A 2004 *Geometric Phases in Classical and Quantum Mechanics,* Birkhäuser, Boston-Basel-Berlin

[7] Kult D, Åberg J, and Sjöquist E 2006 *Phys. Rev. A* **74** 022 106

[8] Zygelman B 1987 *Phys. Lett. A* **125** 476

[9] Zygelman B 1990 *Phys. Rev. Lett.* **64** 256

[10] Mead C A and Truhlar D G 1979 *J. Chem. Phys.* **70** 2284

[11] Jabłoński A 1945 *Phys. Rev.* **68** 78

[12] Szudy J and Baylis W E 1975 *J. Quant. Spectr. Radiat. Transfer* **15** 641

[13] Peach G 1981 *Adv. Phys.* **30** 367

[14] Allard N F and Kielkopf J F 1982 *Rev. Mod. Phys.* **54** 1103

[15] Born M and Huang K 1956 *Dynamic Theory of Crystal Lattices,* Oxford Un. Press. New York.

[16] Kołos W and Wolniewicz L 1963 *Rev. Mod. Phys.* **35** 473

[17] Hirschfelder J O and Meath W J 1967 *Adv. Chem. Phys.* **12** 3

[18] Moody J, Shapere A and Wilczek F 1986 *Phys. Rev. Lett.* **56** 893

[19] Dubourg I, Ferray M, Visticot J P and Sayer B 1986 *J. Phys. B* **19** 1165; Dubourg I and Sayer B 1986 *J. Phys. B* **19** 2291

IOP Conf. Series: Journal of Physics: Conf. Series **810** (2017) 012061 doi:10.1088/1742-6596/810/1/012061

Multispectrum-fitting of phenomenological collisional line-shape models to a speed-dependent Blackmore profile for spectroscopic analysis and databases

P Wcisło[1], D Lisak[1], R Ciuryło[1], A S Pine[2]

[1] Institute of Physics, Faculty of Physics, Astronomy and Informatics, Nicolaus Copernicus University, Grudziadzka 5, 87-100 Toruń, Poland

[2] Alpine Technologies, 14401 Poplar Hill Road, Germantown, Maryland 20874 USA

E-mail: rciurylo@fizyka.umk.pl

Abstract. A variety of phenomenological line-shape models are compared with a speed-dependent Blackmore profile describing an O_2 spectral line. Multispectrum fitting techniques and quality of the fit parameter were used to quantify this comparison. It allowed to indicate which models are adequate for analysis of spectra measured with a given signal to noise ratio. This research has an impact on interpretation of spectral analysis results, improvement of spectroscopic databases, atmospheric remote sensing, trace gas metrology, and Doppler thermometry.

1. Introduction

High-quality spectroscopic line-shape data can exhibit deviations from the fitted profiles due to velocity-changing and speed-dependent collisional effects. To quantify these effects, we compare several phenomenological line-shape models [1, 2, 3, 4, 5, 6, 7, 8, 9] with a speed-dependent Blackmore profile [10, 11, 12] describing the self-broadened O_2 B-band R7 Q8 line measured earlier with a very high signal-to-noise ratio $S/N > 10^5$ [13, 14] at one relatively low pressure 7 Torr. In this reference Blackmore profile ($SD_hB_{12}P$), the speed-dependence of the self-broadening and shifting is given by a hypergeometric function characterized by an isotropic long-range attractive r^{-5} interaction potential, and the velocity-changing collision operator is calculated from a corresponding short-range r^{-12} repulsion.

2. Phenomenological line-shape models

We first compare the reference profile with the ordinary Voigt profile (VP), which can be calculated using a variety of fast algorithms [15, 16, 17]. Next we take into account the speed dependence of collisional broadening using hypergeometric [4] or quadratic [18] functions in the speed-dependent Voigt profile ($SD_{h/q}VP$). Boone at al. [19] have provided a fast algorithm expressing SD_qVP in terms of VP. We then include Dicke narrowing arising from pure velocity-changing collisions and correlations between velocity-changing and dephasing collisions [3] and dispersion asymmetry due to possible line-mixing and/or collision-time effects [20]. Velocity-changing collisions are described in terms of soft [1] or hard [2] collisions leading to the correlated asymmetric Galary profile (CAGP) or correlated asymmetric Nelkin-Ghatak profile (CANGP). The CAGP can be efficiently evaluated with numerical algorithm given in Ref. [21]. On the other hand the CANGP can be expressed in terms of the VP.

Both Dicke narrowing and speed-dependent effects are included using several models describing velocity-changing collisions. It starts with approximated correlated asymmetric speed-dependent Galatry profile (aCASD$_{h/q}$GP) [7, 8, 22] and correlated asymmetric speed-dependent Nelkin-Ghatak profile (CASD$_{h/q}$NGP) [5, 23]. In case of quadratic speed dependence and soft collision model the transport/relaxation equation can be solved exactly leading to correlated asymmetric speed-dependent Galatry profile (CASD$_q$GP) [24]. Ngo et al. [9] provided fast algorithm evaluating CASD$_q$NGP expressed in terms of VP.

A more flexible description of velocity-changing collisions is provided by linear combination of soft and hard collision model [3] leading to approximated correlated asymmetric speed-dependent Rautian-Sobelmann profile (aCASD$_{h/q}$RSP) [8, 25, 26] or exact correlated asymmetric speed-dependent Rautian-Sobelmann profile (CASD$_q$RSP) [27]. An alternative approach is provided by linearized Kielson-Storer model leading to correlated asymmetric speed-dependent linearized Kielson-Storer profile (CASD$_{h/q}$LKSP) [6].

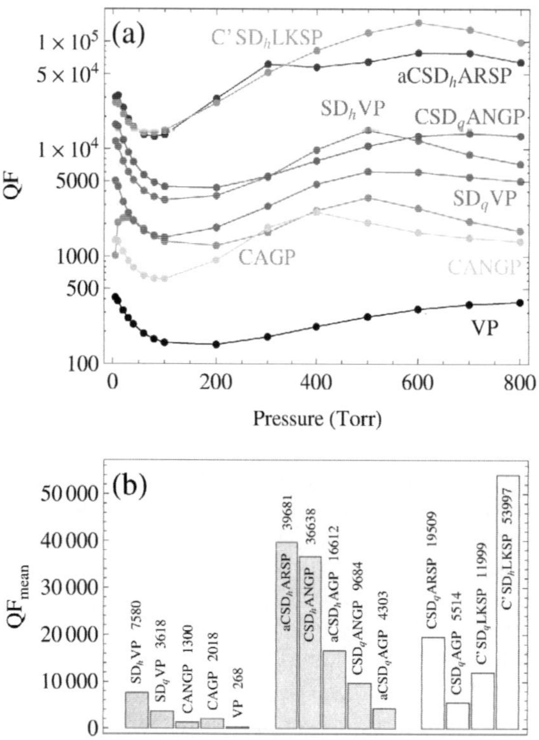

Figure 1. The quality of fit parameters QF obtained from the multi-spectrum fitting of speed-dependent Blackmore profile describing self-brodened O_2 spectra with the phenomenological line-shape models. Parameters QF dependent on the oxygen pressure (a) and their mean values (b).

3. Results

Simultaneous fitting [28, 29] of the phenomenological models to the Blackmore profile simulated at 15 pressures from 5 to 800 Torr provides model parameters, linear in pressure. This approach reduces the correlation between fitted parameters comparing to single-spectrum fits. None of these fitting models are able to reproduce the reference spectra to within a S/N greater than 6×10^4. The quality of fit (QF) value, based on the peak intensity divided by the residuals [13], indicate which models could fit spectra, well represented by the extrapolated reference profile, measured with $S/N \leq QF$. The results are presented in Fig. 1. The common parameters, such as line position, intensity, broadening and shift, exhibit a model dependence greater than their statistical uncertainties; so they would lead to systematic errors if not used with the fitted profile. Their deviations from the reference model parameters usually are inversely related to the quality of fit. The quadratic speed-dependent models, SD_q, for which fast algorithms have been developed, fit the hypergeometric reference Blackmore profile 3 times worse than their analogous SD_h versions.

These conclusions are important for improved databases [30], atmospheric remote sensing [31], trace gas metrology [32], and Doppler thermometry [33].

Acknowledgments

The research is part of the program of the National Laboratory FAMO in Toruń, Poland and is supported by the National Science Centre, Poland, Project Nos. DEC-2013/ 11/D/ST2/02663, DEC-2012/05/D/ST2/01914, 2015/18/E/ST2/00585, 2015/19/D/ST2/02195, and by the Foundation for Polish Science TEAM Project co-financed by the EU European Regional Development Fund and by MOLIM COST action. P.W. is supported by the Foundation for Polish Science START program.

References

[1] Galatry L 1961 *Phys. Rev.* **122** 1218
[2] Nelkin M and Ghatak A 1964 *Phys. Rev.* 135 p A4
[3] Rautian S G and Sobelman I I 1966 *Usp. Fiz. Nauk.* **90** 209 [Rautian S G and Sobelman I I 1967 *Sov. Phys. Usp.* **9** 701]
[4] Berman P R 1972 *J. Quant. Spectrosc. Radiat. Transfer* **12** 1331
[5] Pine A S 1999 *J. Quant. Spectrosc. Radiat. Transf.* **62** 397
[6] Robert D, Joubert P, Lance B 2000 *J. Mol. Struct.* **517-518** 393
[7] Priem D, Rohart F, Colmont J M, Wlodarczak G, Bouanich J P 2000 *J. Mol. Struct.* **517-518** 435
[8] Ciuryło R, Pine A S, Szudy J 2001 *J. Quant. Spectrosc. Radiat. Transf.* **68** 257
[9] Ngo N H, Lisak D, Tran H, Hartmann J-M 2013 *J. Quant. Spectrosc. Radiat. Transf.* **129** 89
[10] Blackmore R 1987 *J. Chem. Phys.* **87** 791
[11] Ciuryło R, Shapiro D A, Drummond J R, May A D 2002 *Phys. Rev. A* **65** 012502
[12] Wcisło P, Ciuryło R 2013 *J. Quant. Spectrosc. Radiat. Transfer* **120** 36
[13] Cygan A, Lisak D, Wójtewicz S, Domysławska J, Hodges J T, Trawiński R S, Ciuryło R 2012 *Phys. Rev. A* **85** 022508
[14] Wcisło P, Cygan A, Lisak D, Ciuryło R 2013 *Phys. Rev. A* **88** 012517
[15] Armstrong B H 1967 *J. Quant. Spectrosc. Radiat. Transf.* **7** 61
[16] Humlicek J 1979 *J. Quant. Spectrosc. Radiat. Transf.* **21** 309
[17] Schreier F 1992 *J. Quant. Spectrosc. Radiat. Transf.* **48** 743
[18] Rohart F, Mader H, Nicolaisen H W 1994 *J. Chem. Phys.* **101** 6475
[19] Boone C D, Walker K A, Bernath P F 2007 *J. Quant. Spectrosc. Radiat. Transf.* **105** 525
[20] Pine A S 1997 *J. Quant. Spectrosc. Radiat. Transf.* **57** 145
[21] Herbert F 1974 *J. Quant. Spectrosc. Radiat. Transf.* **14** 943
[22] Ciuryło R and Szudy J 1997 *J. Quant. Spectrosc. Radiat. Transf.* **57** 411
[23] Lance B, Blanquet G, Walrand J, Bouanich J P 1997 *J. Mol. Spectrosc.* **185** 262
[24] Ciuryło R, Jaworski R, Jurkowski J, Pine A S, Szudy J 2001 *Phys. Rev. A* **63** 032507
[25] Ciuryło R 1998 *Phys. Rev. A* **58** 1029
[26] Lance B and Robert D 1998 J. Chem. Phys. **109** 8283; 1999 *J. Chem. Phys.* **111** 789
[27] Shapiro D A, Ciuryło R, Jaworski R, May A D 2001 *Can. J. Phys.* **79** 1209
[28] Benner D C, Rinsland C P, Devi V M, Smith M A H, Atkins D 1995 *J. Quant. Spectrosc. Radiat. Transf.* **53** 705
[29] Pine A S, Ciuryło R 2001 *J. Mol. Spectrosc.* **208** 180
[30] Hill C, Gordon I E, Rothman L S, Tennyson J 2013 *J. Quant. Spectrosc. Radiat. Transf.* **130** 51
[31] Miller C E, Brown L R, Toth R A, Benner D C, Devi V M 2005 *C.R. Phys* **6** 876
[32] Hashiguchia K, Lisak D, Cygan A, Ciuryło R, Abe H 2013 *Sens. Actuators A* **241** 152
[33] Gianfrani L 2016 *Phil. Trans. R. Soc.* A **374** 20150047

IOP Conf. Series: Journal of Physics: Conf. Series **810** (2017) 012062 doi:10.1088/1742-6596/810/1/012062

Dispersion and relativistic corrections to the spectral line-shape models

S Wójtewicz[1], P Wcisło[1], P Amodio[2], L Gianfrani[2], D Lisak[1] and R Ciuryło[1]

[1]Institute of Physics, Faculty of Physics, Astronomy and Informatics, Nicolaus Copernicus University in Toruń, Grudziadzka 5, 87-100 Torun, Poland

[2]Department of Mathematics and Physics, Second University of Naples, Viale Lincoln 5, 81100 Caserta, Italy

E-mail: szymon@fizyka.umk.pl

Abstract. We investigate the influence of dispersion and relativistic effects on the spectral line-shape models on an example of simple Gaussian and Voigt profiles. We show that dispersion and relativistic corrections can affect the line shape even at the relative level of 10^{-5}.

1. Introduction

Recent developments in ultra-high signal-to-noise-ratio spectroscopic measurements show that the commonly used theoretical line-shape profiles should be supplemented by the do far neglected very subtle physical effects. Here we focus on the dispersion and relativistic deformations of the line-shape models. We report the dispersionally corrected Gaussian profile describing Doppler broadening and resulting from the frequency dependence of the Doppler shifting caused by dispersion [1, 2] as well as the light frequency variation over the whole spectral line shape [2]. We also derive the relativistic Voigt profile taking into account both the homogeneous and Doppler broadening in the case of spontaneous emission [3] and we discuss several properties of both profiles. These effects can affect the spectral line shapes even at the level of 10^{-5} and influence the determination of the line position at kHz level. The presented results are crucial in Doppler-width thermometry [1, 4], the determination of the interaction potentials [5], precise molecular spectroscopy for fundamental studies [6], and the development of astrophysical models.

2. Dispersionally corrected Gaussian profile

The thermal motion of the absorbing or emitting molecules leads to the Doppler broadening of the spectral line shape described by the Gaussian profile (GP). In derivation of the GP the module of the wave vector $\vec{k}$ is assumed to be constant and equal to $k_0 = \omega_0/c$ where ω_0 is the unperturbed line frequency and c is the speed of light in vacuum. This assumption results in the full Doppler width at half maximum equal to $\gamma_D = 2\sqrt{\ln 2}\ \omega_D$, where $\omega_D = k_0 v_m$, $v_m = \sqrt{2k_BT/m}$ is the most probable absorber speed, m is the absorber mass, T is the gas temperature, and k_B is the Boltzmann constant. However, the value of k within the line profile is not constant but depends on the frequency ω. In order to properly describe this frequency dependence two effects have to be taken into account: the unperturbed line frequency ω_0 should be replaced by the light frequency ω and the speed of light in vacuum c should be replaced by the speed of light in the sample $c/n_a(\omega)$ leading to the relation $k(\omega) = \omega n_a(\omega)/c$. The complex refractive index $n_a(\omega)$ consists of both nonresonant and resonant components. The latter are related to the complex line-shape function which means that the calculation of the dispersionally corrected GP requires an iteration procedure.

The simulations were made for the very strong $^{12}C^{16}O_2$ R16e transition from the $^{12}C^{16}O_2$ (00011) ← (00001) band at the temperature $T = 296$ K and pressure $p = 4$ Pa. Its wavenumber and line intensity are $\tilde{\nu}_0 = 2361.46591$ cm^{-1} and $S = 3.543 \times 10^{-18}$ cm^{-1}/(molecule/cm^2), respectively [7]. The simulations showed that the change in Doppler width is mainly caused by the nonresonant part of the refractive index while its resonant component, together with the light frequency variation over the whole profile, cause line shift. The differences between the dispersionally corrected GP and the ordinary GP for the investigated CO_2 line are presented in Fig. 1. It can be seen that in this case the dispersion corrections affect the line shape at a level of 10^{-5} and shifts the line position by about -1.1 kHz ($-1.4 \times 10^{-5}\omega_D$) while the relative change of the value of the Doppler width is only 4×10^{-8}.

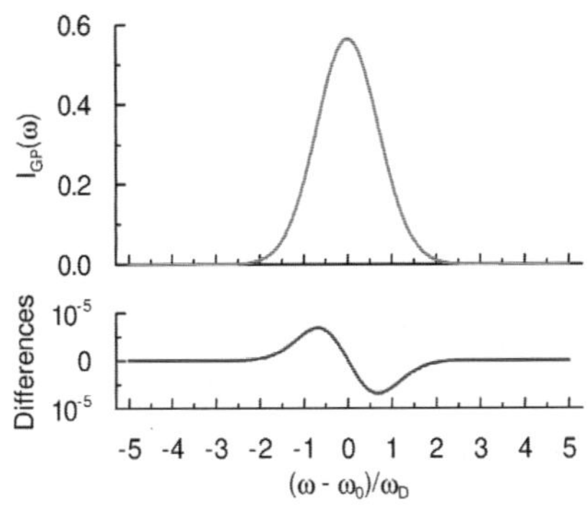

Figure 1. Top panel: the Gaussian profile. Bottom panel: differences between dispersionally corrected GP and ordinary GP for the investigated CO_2 transition.

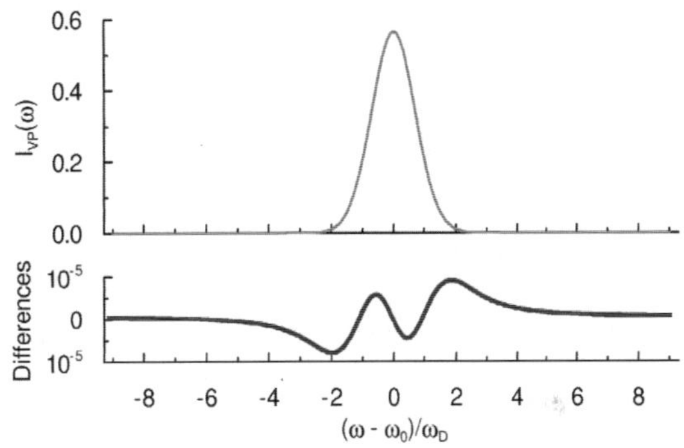

Figure 2. Top panel: the Voigt profile. Bottom panel: differences between relativistic VP and ordinary VP for $\beta_m = 10^{-4}$ and $\Gamma/\omega_D = 1$.

3. Relativistic Voigt profile

The inclusion of the relativistic effects to the spectral line-shape models was limited to the emissive [8] as well as absorptive [9] Gaussian profile describing the relativistic Doppler effect. Here we investigate the more sophisticated case of the emissive relativistic Voigt profile (VP). A detailed derivation of the relativistic VP and the assumptions made during it as well as the final expression for the profile can be found in Ref. [3]. Here we briefly describe several properties of this profile. The relativistic VP is described by three parameters: ω_D, Γ (the half collisional width at half maximum) which characterize also the classical Voigt profile, and $\beta_m = v_m/c$ which controls the magnitude of the relativistic effects. It can be shown that this formulation of the relativistic VP has proper asymptotic behavior in the cases of the classical VP (corresponding to $\beta_m \to 0$) and the relativistic GP (corresponding to $\Gamma/\omega_D \to 0$) [8]. It can be proved analytically that this profile is normalized. An alternative expression, which simplifies the numerical evaluation for the most physically meaningful weak-relativistic regime, was also derived.

Molecular hydrogen is the most obvious target for searching of the relativistic line-shape deformations due to its small mass. The differences between the relativistic VP and the classical VP for $\beta_m = 10^{-4}$ and $\Gamma/\omega_D = 1$ are presented in Fig. 2. It should be noted that β_m value of 10^{-4} is close to the dissociation limit for the molecular hydrogen and corresponds to the conditions that can be met in hot stars. On the other hand the estimated relativistic correction at room temperature is at the level of 10^{-6}.

4. Conclusions

We showed that the dispersion and relativistic corrections to the spectral line shapes are not negligible and could be observed using present spectrometers [10, 11]. These effects can affect the line shapes at the level of 10^{-5}. The resulting profiles are mainly asymmetric and can influence the line position determination at the level of a few kHz which is comparable to the accuracy of the best Doppler-limited measurements of the line position [12].

Acknowledgments

The research is part of the program of the National Laboratory FAMO in Toruń, Poland. The research was supported by the COST Action CM1405 MOLIM, Polish National Science Centre Projects No. DEC-2013/11/D/ST2/02663, No. 2015/17/B/ST2/02115, and No. 2015/19/D/ST2/02195, the Foundation for Polish Science TEAM Project co-financed by the EU European Regional Development Fund, and the Foundation for Polish Science START program.

References

[1] Gianfrani L 2016 *Phil. Trans. R. Soc. A* **374** 20150047
[2] Wójtewicz S, Wcisło P, Lisak D and Ciuryło R 2016 *Phys. Rev. A* **93** 042512
[3] Wcisło P, Amodio P, Ciuryło R, Gianfrani L 2015 *Phys. Rev. A* **91** 022508
[4] Moretti L, Castrillo A, Fasci E, De Vizia M D, Casa G, Galzerano G, Merlone A, Laporta P and Gianfrani L 2013 *Phys. Rev. Lett.* **111** 060803
[5] Kassi S, Campargue A, Pachucki K and Komasa J 2012 *J. Chem. Phys.* **136** 184309
[6] Dickenson G D, Niu M L, Salumbides E J, Komasa J, Eikema K S E, Pachucki K and Ubachs W 2013 *Phys. Rev. Lett.* **110** 193601
[7] Rothman LS *et al* 2013 *J. Quant. Spectrosc. Radiat. Transfer* **130** 4
[8] Huang Y-S, Chiue J-H, Huang Y-C and Hsiung T-C 2010 *Phys. Rev. A* **82** 010102(R)
[9] Huang Y-S 2012 *Europhys. Lett.* **97** 23001
[10] Cygan A, Lisak D, Wójtewicz S, Domysławska J, Hodges J T, Trawiński R S and Ciuryło R 2012 *Phys. Rev. A* **85** 022508
[11] Lin H, Reed Z D, Sironneau V T and Hodges J T 2015 *J. Quant. Spectrosc. Radiat. Transfer* **161** 11
[12] Truong G-W, Long D A, Cygan A, Lisak D, van Zee R D and Hodges J T 2013 *J. Chem. Phys.* **138** 094201

IOP Conf. Series: Journal of Physics: Conf. Series **810** (2017) 012063 doi:10.1088/1742-6596/810/1/012063

Comparison of the polarisation of line and continuum emission in a laser produced plasma

G A Wubetu[1, 2], T J Kelly[3], P van Kampen[1], H Fiedorowicz[2] A Bartnik[2], P Wachulak[2], W Skrzeczanowski[2], and J T Costello[1]

[1]NCPST and School of Physical Science, Dublin City University, Glasnevin, Dublin 9, Ireland

[2]Institute of Optoelectronics, Military University of Technology (MUT), Kaliskiego 2 Str. 00-908, Warsaw, Poland

[3]School of Mathematics and Physical Sciences, University of Hull, Hull HU67RX, UK

E-mail: getasew.wubetu2@mail.dcu.ie

Abstract. Anisotropic emission from a laser-produced aluminium plasma emission has been studied using time and polarisation resolved spectroscopy. The time-resolved degree of polarisation of emission from a bound – bound transition in Al^+ (466.3 nm) was compared to the time-resolved degree of polarisation from nearby continuum emission. The results are interpreted in the framework of radiative recombination and statistical imbalances in the populations of magnetic sub-states. It is found that the continuum emission is more strongly polarised than the line emission. Thus, this technique is a valid alternative to gated LIBS.

1. Introduction

Plasma polarisation spectroscopy (PPS) has been shown to be a powerful diagnostic technique which reveals information about the electron distribution function (EDF) or a population imbalance within the magnetic sub-shells of the atoms and ions within the plasma [1]. The technique of PPS has been applied to laser produced plasmas where partial polarisation has been observed in the x-ray and visible spectral ranges [2,3]. Importantly, a particular application of PPS known as polarisation-resolved laser induced breakdown spectroscopy (PRLIBS) exploits this partial polarisation in order to enhance the signal to background ratio over traditional LIBS measurements [4]. In various wide-ranging studies [4–6], it was shown that the continuum radiation tends to have a greater degree of polarisation than the line emission. For this reason, placing a polariser in front of a spectrometer results in lower background continuum, thus increasing the signal-background ratio which is a figure of merit for LIBS. The origin of the polarisation in this study was postulated to be an anisotropic electron distribution function caused by steep density gradients which result in polarized recombination radiation [6]. Similarly, studies on the visible emission from an Al plasma showed that the line emission can also be partially polarized [3]. In this study, the degree of polarisation in the Al^{2+} doublet at 569.6 and 572.3 nm was found to be on the order of 3 %. The origin of the polarisation in this study is attributed to an inherent population imbalance in the magnetic sub-shells of the ionic system arising from anisotropic recombination of higher charged ions. This study was extended in [7] where the degree of polarisation of the Al^{2+} doublet was measured with time resolution and in a variety of background gas pressures. Here, the degree of polarisation was found to oscillate with time and its origin was attributed to anisotropic electron distribution functions caused at the plasma – gas interface. In this article, we extend the studies of [6,7] and make a time-resolved comparison of the degree of polarisation of a spectral emission line in an Al plasma (the Al^+ line at 466.3 nm) and continuum emission (at 456 nm). The study thus elucidates on the nature of the polarized emission from a laser produced plasma.

2. Experimental

Figure 1 shows the experimental setup. An Nd:YAG laser (Spectron SL800, 14 ns pulse width, 1064nm wavelength) was focused onto a pure Al target at normal incidence to create a laser plasma. The on target laser fluence was measured to be ~450 J/cm^2). The target was mounted onto a precision x-y-z translation stage to ensure a fresh target after each laser shot. The plasma was imaged onto the entrance slit of a Czerny-Turner spectrometer (Chromex 500is) which was coupled to an ICCD camera (Andor i-Star) with a time resolution of ~10 ns and a mean spectral resolution of 0.17 nm. A Wollaston prism was placed in the beam path of the plasma imaging system. This created to orthogonally polarized images of the plasma on the slit. The optic axis of the experiment (defined here to mean the axis of quantization) was chosen to be normal to the target. Thus, the Wollaston prism was aligned to produce images polarized parallel and perpendicular to this (and parallel to the entrance slit of the spectrometer). The system was calibrated to account for a polarisation dependent efficiency in the spectrometer. This was achieved by imaging a small pinhole illuminated with unpolarized light through the system. This allowed for correction of the polarisation sensitivity. The degree of polarisation was then calculated from the resultant spectra from:

$$P = \frac{I^{para} - I^{Perp}}{I^{para} + I^{Perp}}$$

Where I^{para} is the intensity of the light polarized normal to the target and I^{perp} is light polarized parallel to the slit.

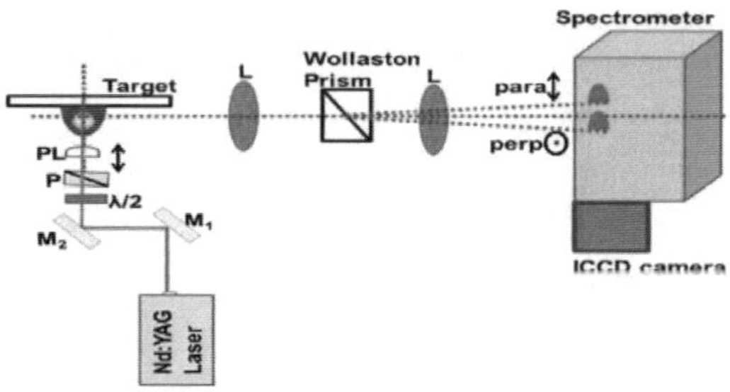

Figure 1. Schematic of experimental setup (M, mirror; λ/2, half wave plate; P, polariser; PL, planoconvex lens; and L, lens).

3. Time and polarisation resolved spectroscopy of the 466 nm line of Al^+

Figure 2a shows time and polarisation resolved spectroscopy of the Al^+ line at 466.3 nm (corresponding to the transition 4p $^1P^0_1 \rightarrow 3p^2\ ^1D_2$). It is clear from Figure 2 that, at early times, the emission is dominated by broadband continuum. At later times, line emission starts to appear. Figure 2b shows the calculated degree of polarisation at a wavelength corresponding to continuum emission (456 nm) and a wavelength corresponding to line emission (466.3 nm). The following is clear from the data: The continuum emission is more strongly polarised than the line emission, the emission is polarised along the direction of the quantisation axis (plasma expansion direction), and the degree of polarisation varies weakly over time.

This result appears to support the hypothesis of both [3] and [5] and thus we propose the following framework: Steep density gradients between the plasma and the background gas create an anisotropic electron distribution function. The continuum emission, which is comprised of either free-bound or free-free transitions is polarised due to this anisotropy. The free-free transitions can occur due to Bremsstrahlung emission which is where an electron is accelerated / decelerated in the vicinity of an ion. This qualitatively should have no preferred polarisation because the electrons are not decelerated in any preferred direction. However, free-free emission can also occur by rapid deceleration of electrons due to the background gas. These electrons should be preferentially decelerated in the direction of plasma expansion and hence be polarised along this direction. The weak polarisation of the line emission is explained using the framework of [3]. The emission of the line at 466.3 nm can arise due to excited singly charged aluminium being produced in the formation of the plasma and, again, there is no reason to assume that this has a preferred polarisation direction. Bound-bound transition emission can be polarised if there is an imbalance in the magnetic sub-levels of the upper state in the transition. The second way the line emission at 466.3 nm can occur is due to recombination of electrons with higher charged ions to create excited lower charged ions. Thus, a free electron is captured by doubly ionized aluminium ion to produce a singly charged ion in an excited state. The free electrons that participate in this process are presumably aligned due to their deceleration in a preferred direction by the background gas and thus populate a preferred magnetic sub-level upon their capture. This leads to a non-zero degree of polarisation of the line emission.

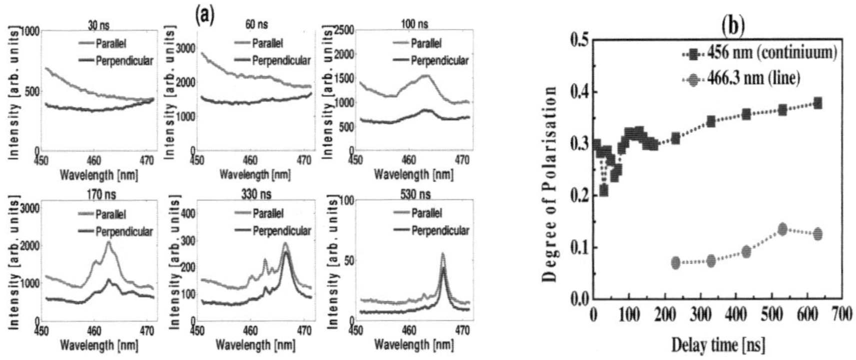

Figure 2. (a) time and polarisation resolved spectra of an Al^+ (at 466.3 nm) at a laser fluence of 452 J/cm^2 and (b) degree of polarisation as a function of gate delay.

There is one last point to be made which is that the degree of polarisation seems to vary weakly over time. One potential application of this is using plasma polarisation spectroscopy as an alternative to gated spectroscopy in LIBS. This has been alluded to in the past and expanded upon here. It seems that the continuum can be suppressed with the use of a polariser in front of the spectrometer with little loss to the intensity of the line emission.

4. Conclusion

We have compared the continuum and line emission from a laser produced plasma in terms of their respective degrees of polarisation using time resolved polarisation spectroscopy. It was found that the continuum emission is more strongly polarized than the line emission. We interpret the results in the framework of recombination of electrons with an anisotropic velocity distribution.

Acknowledgment

The work is supported by the Education, Audio-visual and Culture Executive Agency (EACEA) Erasmus Mundus Joint Doctorate Programme 'EXTATIC', Project No. 2013 – 0033 and Science Foundation Ireland Grant Nos.12/IA/1742 and 14/TIDA/2452. The work is associated with COST Action MP1208 and U.S. National Science Foundation PIRE Grant No. 1243490.

References

[1] Fujimoto T and Kazantsev S A 1997 Plasma polarization spectroscopy Review Article *Plasma Phys. Control. Fusion* **39** 1267–1294

[2] Kieffer J C, Matte J P, Chaker M, Beaudoin Y, Chien C Y, Coe S, Mourou G, Dubau J and Inal M K 1993 X-ray-line polarization spectroscopy in laser-produced plasmas *Phys. Rev. E* **48** 4648–58

[3] Kim J and Kim D E 2002 Measurement of the degree of polarization of the spectra from laser produced recombining Al plasmas *Phys. Rev. E - Stat. Nonlinear, Soft Matter Phys.* **66** 1–4

[4] Penczak J S, Liu Y and Gordon R J 2009 Polarization-resolved laser-induced breakdown spectroscopy *Opt. Lett.* **34** 494

[5] Zhao Y, Singha S, Liu Y and Gordon R J 2009 Polarization resolved laser-induced breakdown spectroscopy *Opt. Lett.* **34** 494–6

[6] Eslami Majd A, Arabanian A S and Massudi R 2010 Polarization resolved laser induced breakdown spectroscopy by single shot nanosecond pulsed Nd:YAG laser *Opt. Lasers Eng.* **48** 750–3

[7] Sharma A K and Thareja R K 2007 Anisotropic emission in laser-produced aluminum plasma in ambient nitrogen *Appl. Surf. Sci.* **253** 3113–21